Lecture Notes in Nanoscale Science and Technology

Volume 37

Lecture Notes in Nanoscale Science and Technology (LNNST) aims to report latest developments in nanoscale science and technology research and teaching – quickly, informally and at a high level. Through publication, LNNST commits to serve the open communication of scientific and technological advances in the creation and use of objects at the nanometer scale, crossing the boundaries of physics, materials science, biology, chemistry, and engineering. Certainly, while historically the mysteries in each of the sciences have been very different, they have all required a relentless step-by-step pursuit to uncover the answer to a challenging scientific question, but recently many of the answers have brought questions that lie at the boundaries between the life sciences and the physical sciences and between what is fundamental and what is application. This is no accident since recent research in the physical and life sciences have each independently cut a path to the edge of their disciplines. As both paths intersect one may ask if transport of material in a cell is biology or is it physics? This intersection of curiosity makes us realize that nanoscience and technology crosses many if not all disciplines. It is this market that the proposed series of lecture notes targets.

Suneev Anil Bansal • Virat Khanna
Nilanthy Balakrishnan • Pallav Gupta
Editors

Emerging Applications of Novel Nanoparticles

Springer

Editors
Suneev Anil Bansal
Department of R&D and New product Development
ELFROU Inc.
Gurgaon, Haryana, India

Nilanthy Balakrishnan
School of Chemical and Physical Sciences
Keele University
Newcastle-under-Lyme, Staffordshire, UK

Virat Khanna
Centre for Research Impact and Outcome
Chitkara University
Rajpura, Punjab, India

Pallav Gupta
Department of Mechanical Engineering
Amity University
Noida, Uttar Pradesh, India

ISSN 2195-2159 ISSN 2195-2167 (electronic)
Lecture Notes in Nanoscale Science and Technology
ISBN 978-3-031-57842-7 ISBN 978-3-031-57843-4 (eBook)
https://doi.org/10.1007/978-3-031-57843-4

This Springer imprint is published by the registered company Springer Nature Switzerland AG
The registered company address is: Gewerbestrasse 11, 6330 Cham, Switzerland

Paper in this product is recyclable.

Preface

This book entitled *Emerging Applications of Novel Nanoparticles* focuses on the different types of nanomaterials, their synthesis routes, and potential application in various fields. Earlier there was a trend of using materials in micron or submicron form as they were best suited for the applications of past time. However, with the advancement of time there has been a remarkable shift in the demand of society due to which, essentially, we have to alter the type of material which was being used. Nanomaterials are the materials in which the particle size goes in a range of 1-100 nm. This improves the properties by manifold. This book lays special emphasis on 2D and Mxene nanomaterials along with their application in different fields like pollutant removal, forensics, medical and biomedical applications, etc.

This book also addresses the concern over the processing of different nanomaterials and the various challenges associated with it. Other books are also available in the related area; however, this volume specifically discusses more focused approach towards the societal applications, hence the word "Novel Nanoparticles" has been specifically added to the title of the book.

The contents of this book will be highly beneficial for the students and scholars who are currently studying different UG and PG programs in mechanical engineering, civil engineering, material science, chemistry, physics as well as researchers working in industry and academia.

Seeing the requirements of day-to-day life, this book is divided into 13 chapters:

Chapter "2D Nanomaterials for Adsorption of Wastewater Pollutants" discusses in detail about the synthesis and physicochemical properties of different 2D materials. Special emphasis has been placed on the usage of 2D-based materials and related mechanisms in the adsorption of different wastewater pollutants.

Chapter "Unveiling the Power of Nanomaterials in the Area of Forensics" is dedicated towards the usage of nanomaterials in the field of nano-sensors and nano-imaging with their application in criminal investigation. Discussion has also been held on the implementation of nanomaterials in the field of evidence analysis under forensic studies.

Chapter "Buckypapers: Applications in Composite Materials" discusses the development of Buckypapers (BPs) by making use of densely packed carbon nanotube (CNT) thin films. CNTs show a good combination of electrical, thermal, and mechanical properties. Special emphasis has been laid down on the preparation of BPs and their physical and morphological properties along with the potential applications.

Chapter "Nanoparticles for Diagnosis and Treatment of Infectious Diseases" highlights the application of nanomaterials in diagnosing and treating infectious diseases. Current challenges and future research possibilities for using these nanomaterials in disease management have also been discussed in the chapter.

Chapter "$Ti_3C_2T_x$ Mxene Based Nanostructured Materials For Emerging Applications" describes the potential of using $Ti_3C_2T_x$ MXene as an important 2D nanomaterial. Apart from that, a special emphasis has been laid down in the chapter towards the usage of these Maxene in optoelectronic devices, such as photovoltaics, photodetectors, and photoelectrochemical devices, due to their exceptional electronic, optical, mechanical, and thermal properties.

Chapter "Molybdenum Disulfide: A 2D Material" discusses about molybdenum disulfide, which is an important 2D material that shows unusual properties. The chapter also showcases a comprehensive investigation of the synthesis methods in addition to its structural as well as electrical properties, along with several applications.

Chapter "Surface Functionalization of 2D MOs for Enhanced Biocompatibility and Biomedical Applications" explains about the various surface functionalization methods for developing 2D metal oxides and their applications in drug delivery, imaging, and biosensors.

Chapter "Application of a Novel Nanotherapeutic Strategy in Ayurvedic Treatment" discusses the usage of nanomaterials in ayurvedic treatment. The chapter also illustrates the nanotherapeutic approach which improves the therapeutic strategy for treatment of acute diseases like cancer, genetic diseases, neurological diseases, etc.

Chapter "Biosynthesis of Iron Oxide Nanoparticles (IONPs): Toxicity Evaluation and Applications for Magnetic Resonance Imaging and Magnetic Hyperthermia" discusses the biological synthesis and biomedical applications of iron oxide nanoparticles (IONPs). Apart from that, in the chapter, discussion has also been carried out on biosynthesized metallic nanoparticles.

Chapter "Effect of Annealing Temperature on Structural, Morphological and Optical Properties of CdZnTe Thin Films" discusses the effect of annealing temperature on structural, morphological, and optical properties of CdZnTe thin films.

Chapter "Two-Dimensional Molybdenum Disulfide Nanosheets Based Optoelectronic Devices" reports a detailed discussion on development and characterization of two-dimensional molybdenum disulfide nanosheets for the application in optoelectronic devices. Apart from that, few recent developments with respect to optoelectronic devices fabricated using MoS_2 nanosheets along with various methods of synthesizing MoS_2 nanosheets and fabricating optoelectronic devices have been discussed in the chapter.

Chapter "Photocatalytic Hydrogen Production of Perovskite Based Nanocomposites by Green Laser Irradiation Techniques" illustrates the development of photocatalyst-based on a binary direct Z-scheme STCS perovskite. The recyclability effect of the developed material has also been discussed in the chapter.

Chapter "Tuning the Physical Properties of Perovskite Multiferroic Nanoparticles for Green Energy Applications" reports the detailed review of alternative methods for tuning the properties of perovskite materials. A special case study on the behavior of BiFeO3 and transition-metal-doped BiFeO3 is also discussed as a perovskite material in the chapter.

All the 13 chapters submitted by the different authors are written in a systematic manner thereby covering all the recent references related to the various fields of nanoparticles.

All the four editors are thankful to almighty God. Apart from this, Dr. Suneev Anil Bansal, Dr. Virat Khanna, and Dr. Nilanthy Balakrishnany are also thankful to their family members for the support extended during the period of editing of this book. Dr. Pallav Gupta is also thankful to his mother (Beena Gupta), wife (Dr. Ritu Agrahari), and son (Saahas Gupta) for the encouragement as well as support extended during the entire duration of editing this book.

We editors also thankfully acknowledge the heartfelt support received from all our contributors who submitted their chapters in this volume of the book.

We are also thankful to the entire team of Springer for publishing this book in fastest possible time and in the most efficient manner.

Gurgaon, Haryana, India — Suneev Anil Bansal
Rajpura, Punjab, India — Virat Khanna
Newcastle-under-Lyme, UK — Nilanthy Balakrishnan
Noida, Uttar Pradesh, India — Pallav Gupta

Contents

2D Nanomaterials for Adsorption of Wastewater Pollutants 1
Enas Amdeha, Aesha Abd El Pasir, and Diana S. Raie

Unveiling the Power of Nanomaterials in the Area of Forensics 29
Iqra Zareef, Ahsan Riaz, Salahuddin, Muqaddas Fatima, Amna Anwar, and Anam Munawar

Buckypapers: Applications in Composite Materials 59
Fatih Turan and Mehmet Guclu

Nanoparticles for Diagnosis and Treatment of Infectious Diseases 93
Sujit Mukherjee, Debmalya Pal, Arunava Bhattacharyya, Pritam Mondal, and Subhasis Roy

$Ti_3C_2T_x$ MXene Based Nanostructured Materials for Emerging Applications 115
Lankipalli Krishna Sai, Tadisetti Taneesha, Jeotikanta Mohapatra, and Sunil Kumar Pradhan

Molybdenum Disulfide: A 2D Material 145
Neha Sharma, Prashant Kumar, and Manoj Kumar Khanna

Surface Functionalization of 2D MOs for Enhanced Biocompatibility and Biomedical Applications 175
Mohamed Abdelmonem, Emmellie Laura Albert, Ashreen Norman, Emma Ziezie Tarmizie, Samer H. Zyoud, and Che Azurahanim Che Abdullah

Application of a Novel Nanotherapeutic Strategy in Ayurvedic Treatment 199
Sumanta Bhattacharya

Biosynthesis of Iron Oxide Nanoparticles (IONPs): Toxicity Evaluation and Applications for Magnetic Resonance Imaging and Magnetic Hyperthermia 229
Mohamed Abdelmonem, Emmellie Laura Albert, Nur Khalida Rahayu Zainon, Emma Ziezie Tarmizie, Samer H. Zyoud, and Che Azurahanim Che Abdullah

Effect of Annealing Temperature on Structural, Morphological and Optical Properties of CdZnTe Thin Films 251
Sudeshna Surabhi, Kumar Anurag, and S. R. Kumar

Two-Dimensional Molybdenum Disulfide Nanosheets Based Optoelectronic Devices 267
Arun Singh Patel, Praveen Mishra, Anirban Chakraborti, and Prianka Sharma

Photocatalytic Hydrogen Production of Perovskite Based Nanocomposites by Green Laser Irradiation Techniques 301
Mohamed Jaffer Sadiq Mohamed and Mohammed Ashraf Gondal

Tuning the Physical Properties of Perovskite Multiferroic Nanoparticles for Green Energy Applications 321
Manish Kumar, Sandeep Kumar Pundir, Subhash Sharma, O. Raymond Herrera, J. M. Siqueiros, and Manish Kumar

Index 347

2D Nanomaterials for Adsorption of Wastewater Pollutants

Enas Amdeha, Aesha Abd El Pasir, and Diana S. Raie

Abbreviations

Symbol	Definition
2D	Two dimensional
Co	Initial Concentration
CNDs	Carbon nanodots
CVD	Chemical vapour deposition
DFT	Density Functional Theory
GO	Graphene oxide
HMI	Heavy metal ions
HTNs	Hierarchical titanite nanostructures
MB	Methylene blue
MG	Malachite Green
MO	Methyl orange
MOFs	Metal-Organic Frameworks
MoS2	Molybdenum disulfide
NMR	Nuclear Magnetic Resonance
NMs	Nanomaterials
NPs	Nanoparticles
PFAS	Perfluoroalkyl and polyfluoroalkyl substances
PFOA	Perfluorooctanoic Acid
PFOS	Perfluorooctane sulfonic acid

E. Amdeha (✉) · D. S. Raie
Processes Design and Development Department, Egyptian Petroleum Research Institute (EPRI), Cairo, Egypt

A. A. El Pasir
Nuclear Materials Authority, Cairo, Egypt

S. A. Bansal et al. (eds.), *Emerging Applications of Novel Nanoparticles*, Lecture Notes in Nanoscale Science and Technology 37, https://doi.org/10.1007/978-3-031-57843-4_1

pKa	Acid dissociation constants
Q	Adsorption capacity
rGO	Reduced graphene oxide
SSA	Specific Surface Area
T	Temperature
TGA	Thermal Gravimetric Analysis
TMDs	Transition Metal Dichalcogenides
UO22+	Uranyl ion
USFDA	US Food and Administration
XPS	X-ray photoelectron spectroscopy
XRD	X-ray diffraction

1 Introduction

To live a healthy life, one must have access to clean drinking water. The need for clean water supplies for home and agricultural usage is rising due to the quick development of industry and population. Water shortage is becoming a major issue that practically all countries experience to some degree, but things are getting worse as industrial sectors continue to dump toxic, harmful, and persistent contaminants into water sources (Wang et al., 2021). According to UN estimates from 2006, water deficiency will affect 70% of the world's population by 2025 as a result of ongoing increases in water pollution. The main sources of water pollution, both organic and inorganic, include human activity such as overusing fertilizers and pesticides, as well as industrial effluents (Amdeha, 2023a, 2024). The large volumes of industrial effluents that are specifically thrown into water bodies contain organic pollutants such as pesticides, herbicides, antibiotics, hydrocarbons, dyes, proteins, phenols, and detergents. These pollutants diminish the dissolved oxygen levels in the water, harming aquatic systems and the environment. (Rahmani et al., 2020).

To protect clean water resources and reduce the amount of greatly polluted wastewater that is released into water bodies, current research articles spotlight environmentally friendly and reasonably priced treatment methods for the large amount of industrial wastewater that is produced before release. Numerous treatment technologies have been implemented, such as biological treatment, flotation-coagulation, electrocoagulation, and membrane processes (Ewis et al., 2021). Even though these technologies have excellent treatment efficiency, there are often issues with them, including high energy usage, capital expenses, and difficulties scaling up. Adsorption technology has gained a lot of concern because of its ease of use, low cost and energy consumption, environmental friendliness, and high treatment efficiency (Amdeha, 2021, 2023b). Adsorption is used in the polishing stage of wastewater treatment to get rid of extremely minute particles.

Two-dimensional (2D) materials and their composites have drawn a lot of attention among the different types of adsorbents because of their remarkable qualities, which include their abundance of pore structure, chemical stability, large SSA, tunability for particular applications, ease of physical and chemical modifications, and capacity to remove a variety of contaminants (Xiong & Jivkov, 2018). Materials that are classified as 2D have atomic or molecular thicknesses; the other dimensions of these materials reach the micrometre or nanometer range. These materials' decreased dimensionality causes them to display distinctive qualities and behaviours.

Significant interest in 2D materials was aroused by the discovery of graphene in 2004, a single layer of C atoms organized in a hexagonal lattice (Geim, 2009). Many elemental or compound systems, such as graphene, graphitic carbon nitride, MOFs, TMDs, and others, can be used to create two-dimensional materials. These materials are desirable for a variety of applications due to their diverse and adjustable optical, chemical, and electrical properties (Cai et al., 2018). By adding doping, introducing flaws, or combining other materials to create heterostructures, 2D materials' characteristics can be further altered (Li et al., 2018). Chemical vapour deposition (CVD), mechanical exfoliation, and template synthesis are among the synthesis techniques for 2D materials that allow for the controlled creation of materials with the correct dimensions and properties (Kumar et al., 2022; Li et al., 2018).

Several disciplines are investigating 2D materials due to their unique features (Han et al., 2022a). They are excellent prospects for next-generation technologies and devices due to their huge surface-to-volume ratio, high carrier mobility, and variable bandgap (Zhou et al., 2021). The environmental applications of 2D-based compounds have piqued the interest of researchers worldwide, especially about their ability to adsorb contaminants in aqueous solutions. The numerous characteristics of 2D materials that make them appropriate for adsorbing a variety of different contaminants are the source of this interest.

Taking into account these facts, this chapter presents the significant studies that used 2D materials as adsorbents for the removal of different pollutants; dyes, phenols, heavy metals, radioactive materials, PFAS, and unwanted microbes. Meanwhile, the general synthesis and properties of 2D materials, and adsorption performance are discussed. The desorption studies and the suggested mechanisms are also addressed. This chapter also addresses the challenges and prospects of 2D materials as adsorbents in wastewater treatment.

2 Classification of 2D Materials

2D materials can be categorized into different groups:

Graphene Family

The graphene consists of 2D layers of sp^2 hybridized carbon atoms, arranged in a hexagonal lattice structure with a thickness of one atom. Graphene possesses unique properties e.g. a high elastic modulus, a large SSA, and high strength. Graphite is the primary precursor for graphene production and is utilized in various ways (Yadav et al., 2023). Graphene holds the distinction of being the most advanced material, finding applications in nearly every major field due to its 1-atom thick, 2D hexagonal honeycomb structure, which grants it not only exceptional thinness but also remarkable strength. Pristine graphene synthesis is challenging due to the extreme instability of single atomic layers during the synthesis process. Graphite can be oxidized to make graphene hydrophilic and to increase the interlayer gap by using a strong oxidizing chemical during the synthesis process. This feature makes the process of exfoliating graphite into graphene layers more stable and convenient. Because graphene oxide (GO) contains oxidizing molecules, its characteristics are marginally different from those of pristine graphene. One of the GO's main benefits is the addition of terminal functional groups to its surface, which, depending on the kinds of functional groups added, might increase the material's possible applications. GO functions as an electrical insulator, nevertheless, because the sp^2 links in the layers are broken. To recover graphene oxide's original characteristics and hexagonal structure, it becomes crucial to reduce graphene oxide getting (rGO) (Garg et al., 2021).

Graphitic-Carbon Nitride (g-C_3N_4)

g-C_3N_4 has a layered planar structure like graphene, with interlayer connections made possible by H-bonding or van der Waals interactions. A triazine ring (C_3N_3) or tri-s-triazine ring (C_6N_7), is formed when the carbon and nitrogen atoms in the plane make covalent connections. Furthermore, the stacking of interlayer van der Waals contacts gives g-C_3N_4 exceptional chemical stability, resistance to acids and alkalis, and insolubility in the majority of solvents. sp^2 hybridization is seen by the C and N atoms in g-C_3N_4, resulting in a highly delocalized π-conjugated system. The typical process for creating g-C_3N_4 entails heating precursors that are high in nitrogen (e.g. urea, thiourea, melamine, cyanamide, etc.). Calcination is a type of heat treatment that has several benefits, including easy control, mild reaction conditions, and low cost. The types of precursors and reaction conditions have a close relationship with the properties of the resultant g-C_3N_4 (Xing et al., 2022).

Transition metal-based dichalcogenides (TMDs)

A transition metal (M) covalently bonds to two chalcogens (X) to generate an MX_2 structure, which is the precursor of 2D TMDs. These materials have weak van der Waals forces holding their X-M-X tri-layer covalent bonding together in their crystal structure. Tungsten disulfide, tungsten ditelluride, and molybdenum disulfide are a few TMDs examples (Zhou et al., 2018). These materials' crystal structures have a significant impact on their characteristics. By creating 2D TMDs specifically suited to certain industrial needs, desired chemical, physical, electrical, optical, and mechanical properties can be attained. Thin film synthesis techniques allow for the modification of TMD properties by changing the crystal structure's layer sequence, which can include swapping out either the metal or the chalcogenides (Yadav et al., 2023).

Mxenes

Mxenes, a unique group of 2D metal-based carbides or nitrides; $M_n^{+1}AX_n$ (n = 1–3); which is produced by selectively etching the raw M (transition metal), A (metal from group III or IV), and X (N or C). The layered structure of the MAX phase, in which X fills the octahedral gaps and metal layers are stacked hexagonally, is the source of mxenes. Strong cleaning agents can be used to selectively remove the A component. Mxenes are appropriate for the elimination of different pollutants due to their high sorption selectivity (Janjhi et al., 2023).

Metal-Organic Frameworks (MOFs)
MOFs are essential for several applications. They have tunable surface area, unsaturated reactive sites, and modifiable surface characteristics. MOFs have a wide range of uses because they are unique structures created by mixing organic (ligand) and metal (inorganic) components (Peng et al., 2017). Several MOFs can be produced by altering the ligand framework. Transition metals can build 2D MOFs with desired characteristics by coordinating with various ligands.

3 2D Materials Preparation Routes

The synthesis methods for 2D nanomaterials are categorized into two routes: top-down and bottom-up approaches. The top-down route involves various techniques such as exfoliation methods (micromechanical, chemical, ultrasonic, etc.) (Demirel et al., 2018). On the other hand, the bottom-up approach includes techniques like CVD, wet chemical techniques, and template synthesis. Figure 1 illustrates different types of 2D nanomaterials preparation routes.

Top-down approach
The process of converting bulk materials into nanostructures or nanoparticles (NPs) uses a top-down methodology. Comparatively speaking, it is simpler than the bottom-up strategy since it uses bulk production techniques or reduction/separation of big materials to create the desired structure with acceptable properties. However,

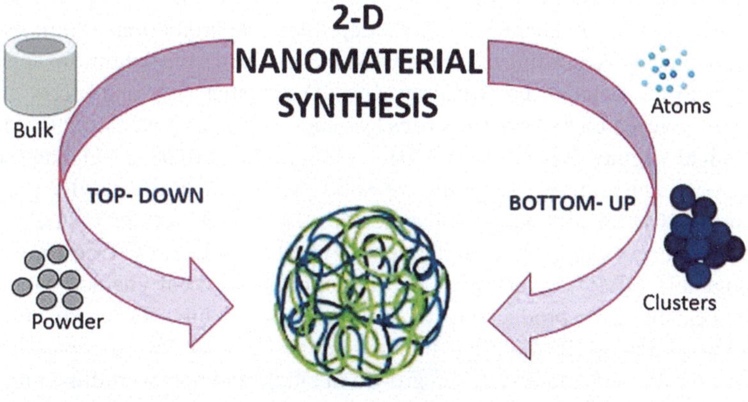

Fig. 1 Different types of 2D nanomaterials preparation routes. (Adapted with permission from (Chen et al., 2022a), © 2022 Elsevier B.V)

the primary flaw with the top-down method is the surface structure's imperfection. The top-down method produces ultrathin nanosheets through physical and chemical techniques. While the chemical top-down technique depends on processes involving heat, ion exchange, or other methods, the physical top-down method shapes van der Waals solids into single or few layers of 2D nanomaterials through the use of ultrasonic waves or mechanical force (Demirel et al., 2018).

Micromechanical exfoliation: is an easy process that preserves the crystal structure and characteristics while creating nanosheets with one or more layers. Because of its simplicity, versatility, and applicability to fundamental research, it is commonly utilized in the synthesis of 2D materials. For example, the exfoliation of 2D TMD materials was made popular when graphene and graphite were successfully separated using scotch tape. Micromechanical exfoliation is helpful for research and developing all-layered material devices, but it is not practical for large-scale production because of difficulties in managing the number and size of layers (Alam et al., 2021).

Within **ultrasonic exfoliation**, van der Waals solids are separated into single or multilayer nanosheets via lamination. Two essential elements of a successful exfoliation process are the length of the sonication and the choice of solvent. This method works better and produces more results than mechanical exfoliation does. Mass-production approachable, film-forming liquid-phase ultrasonic exfoliation offers a wide range of uses. However, it can be difficult to regulate the level of exfoliation (Liu & Li, 2020).

Bottom-up approach

Building materials atom by atom or molecule by molecule is how the bottom-up method creates nanostructures. Although this method can generate less waste, it is somewhat more costly. Bottom-up methods are widely employed in the manufacturing of industrial nanopowder. By using atomic or molecule elements that can react, expand in size, or self-assemble into more complicated frameworks, the bottom-up technique makes it possible to create nanoscale components that aid in the creation of 2D materials. There are numerous known bottom-up methods for producing 2D nanomaterials, such as chemical vapour deposition, hydrothermal synthesis, and template synthesis. According to (Ortiz & Skrabalak, 2014), the bottom-up approach is thought to be easier to use for preparing 2D nanomaterials and offers greater advantages to researchers because it uses less energy and has fewer structural flaws.

Chemical vapour deposition (CVD) is a process that produces 2D materials by reacting or disintegrating gas, liquid, or solid precursors in a carefully prepared environment. Chemically active substrates are the best choice for CVD, and the production process can be done at ambient to ultrahigh pressures. Transition metal dichalcogenide (TMD) layers with the appropriate electrical characteristics and diameters can be easily produced by the CVD process. Metals, semiconductors, and separators are among the materials that can be used to coat. Two comparable processes are used in this method: CVD growth of single and polycrystalline graphene (Alam et al., 2021).

TMDs with a variety of thicknesses, from single layers to many layers, can be prepared using **Wet chemical techniques**. Comparing the produced materials to CVD, they are much more uniform and have less deformation thickness. By introducing different chemicals throughout development, materials can be doped using wet chemical methods. Additionally, the surface of the material can be altered or smoothed by coating it with ligands. Since the TMDs are frequently employed in larger-scale manufacturing processes, wet chemical procedures might also be advantageous for them. Green chemistry and manufacturing requirements can be met by adapting solution-based processes to include ecologically acceptable precursor compounds (Miró et al., 2014). Chalcogenides are traditionally made by two wet chemical processes: hydrothermal development and solvothermal development.

Template synthesis: the key component of a template-assisted synthesis is the "template," which can be any material with nanostructured features, whose size, shape, and charge distribution substantially affect the structure guiding properties. In the template-assisted synthesis method, the template preparation step is the first. The next step is to use the template to create the needed material and, if necessary, remove it. To create the necessary material, one can employ chemical or physical methods such as addition, removal, substitution, or isomerization reactions or surface coating. Using physical techniques like dissolving or chemical methods like calcination, the template can be eliminated once the reaction is finished. The main advantage of this process is that it allows for good control over the final product's morphology, dimension, and structure. This approach has been used to create a variety of TMDs with different morphologies. Template-assisted synthesis can be further divided into three categories: colloidal, soft, and hard templates, depending on the kind of template that is accessible (Alam et al., 2021).

4 Properties of 2D Materials

The physicochemical properties of the used adsorbent determine how well 2D-based materials can remove various contaminants. Common physicochemical characteristics that have a significant impact on pollutants' adsorption include SSA, porosity, hydrophilicity, toxicity, stability, and the presence of various functional groups on adsorbent surfaces (Bilal et al., 2023). The physicochemical properties of 2D materials' components, which in turn influence their adsorption potentials, are frequently influenced by the synthesis pathways used to create them. On the surface of 2D adsorbents, several functional groups; e.g.–OH,–O, and–F; frequently form during etching processes (Zhang et al., 2018).

The adsorption of pollutants often increases in tandem with an adsorbent's expanding SSA because adsorption is mostly a surface phenomenon (Bilal et al., 2022). Gaining more surface area may result in many more pores (more active locations) that target pollutant molecules and may be able to adsorb. High SSAs typically result in greater adsorption capacity because they offer more active sites for the adsorption of contaminants.

Surface functional groups: A significant role in removing pollutants from the aqueous environment can also be played by the readily accessible surface functional groups of 2D adsorbents. The reason for this is that the (−) or (+) charged functional groups of the adsorbents have strong binding interactions with the oppositely charged pollutant molecules. These interactions frequently lead to increased pollutant uptake by the adsorbents that are used. For instance, because of the electrostatic attraction between the opposing species, the existence of (+) charged functional groups on the surfaces of the 2D adsorbents would facilitate the adsorption of the anionic dye. Given that they contribute to the different chemical and physical features of the 2D adsorbents, the existence of diverse surface functional groups is crucial for the application of 2D adsorbents (Sajid, 2021).

Another factor that significantly influenced an adsorbent's capacity for adsorption was its **surface wettability**. The targeted pollutant's selective adsorption can be significantly influenced by the wettability of the used 2D adsorbents (Bilal et al., 2023).

5 2D Materials as Adsorbents for Wastewater Pollutants

To sustain a healthy lifestyle, clean drinking water must be readily available (Baig et al., 2019). Continuous wastewater discharge from several sources needs to be treated to enhance water quality, lessen diseases and health issues, and provide an ecosystem that is conducive to aquatic life.

5.1 Adsorption of Organic Pollutants

Adsorption of Dyes

Organic pollutants, including dyes, are used for a variety of purposes and can be fatal if disposed of improperly in untreated natural water. Many sectors, such as textile, paper industries, printing, food processing, leather, and dye manufacturing, utilize dyes often. However, throughout the dyeing process, more than 10–15% of the dye output is lost, and as a result, a significant amount of the dye ends up in wastewater. Because of their poisonous and carcinogenic qualities, they are regarded as a significant source of water contamination (Amdeha & Salem, 2022; Deriase et al., 2021; El-Salamony et al., 2023).

According to (Mantasha et al., 2020), 2D MOFs, $[M\ (ox)\ (bpy)]_n$ $[H_2ox$ = oxalic acid, bpy = 4,4′-bipyridine and M = Cu (1) or Co (2)], were synthesised via solvothermal (Fig. 2). MB and MO, two organic dyes that are commonly regarded as water contaminants, are adsorbed onto the two MOFs. Remarkably, the cationic dye MB exhibited selectivity towards the molecular frameworks. The framework can be recycled up to five times and stays stable following the adsorption of dye. The frameworks remove the maximum amount of MB (96–98%) at equilibrium, with a

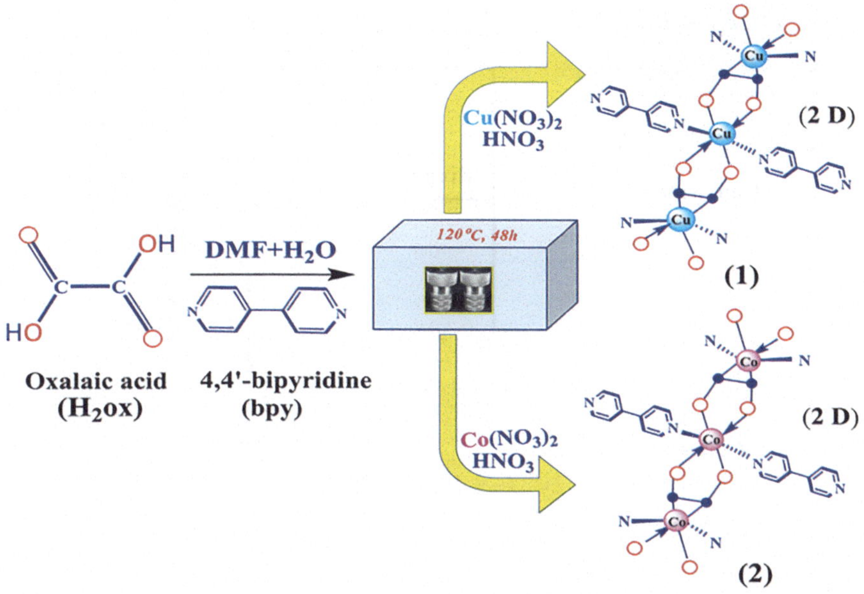

Fig. 2 Synthetic route of 1 and 2. (Adapted with permission from (Mantasha et al., 2020), © 2020 Elsevier B.V)

pH of neutral and T = 25 °C. According to the kinetic studies, the adsorption will likely proceed through a pseudo-second-order process.

$Ti_3C_2T_x$ MXene was employed as an adsorbent by (Kadhom et al., 2022) to remove Malachite Green (MG) dye, taking into account the dye's chromatic and leuco forms. The Freundlich model got the highest R^2, 0.97, and was the best-fitting isotherm model. The model with the pseudo-second order has the highest R^2 of 0.99. Results showed that removal effectiveness rises with increasing adsorbent dose; at 0.09 g in a 50 ml solution, removal efficiency reached 94.1%. It is noteworthy to observe that the removal increases with changes in pH, with the minimum efficiency seen at pH = 6. This was explained by the dye's leuco nature, which causes it to become colourless and more difficult to detect as pH rises. The results indicate that removal is higher at low pH values and decreases as pH increases, but colour loss makes these differences intangible.

Through the use of amphiphilic CNDs, (Yin et al., 2022) effectively exfoliated hydrophobic bulk MoS_2 into 2D nanosheets, Fig. 3. Higher than that of the literature, the optimal dose of the obtained 2D-MoS_2 is 0.41 mg/ml. Two dimensions of 2D-MoS_2 are 142.6 (±2.3) nm in length and 4 in terms of layers, on average. CNDs can be instantaneously adsorbed on the surface of 2D-MoS_2 to produce 2D-MoS_2/CNDs. Because of CNDs, the amount of 2D-MoS_2 remains 90% of the initial concentration even after standing for 1000 hours. Because there are many adsorption sites and high dispersion, the 2D-MoS_2/CNDs composite has a great Q (1137 mg/g) and an incredibly quick adsorption rate (one minute) for the removal of MB.

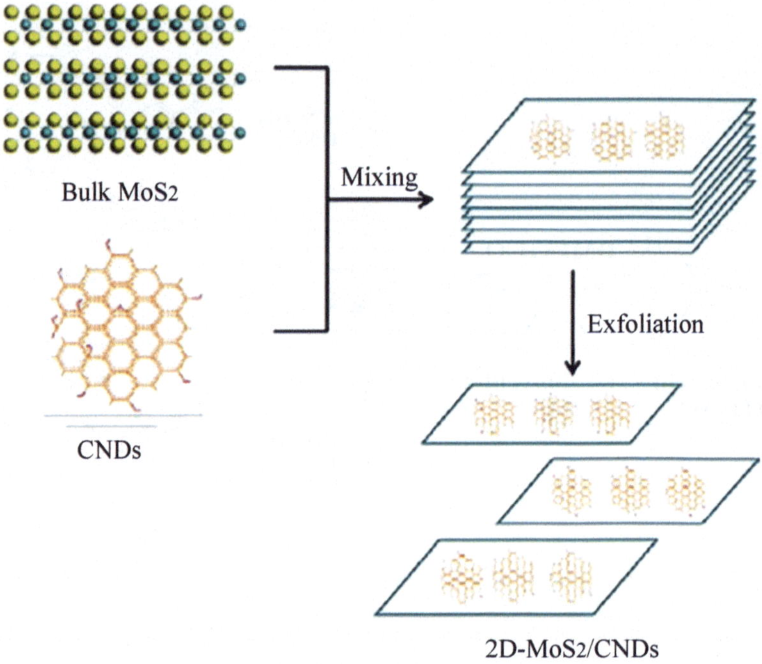

Fig. 3 CNDs as dispersing agents for exfoliating and dispersing MoS_2. (Adapted with permission from (Yin et al., 2022), © 2022 Elsevier B.V)

Adsorption of Phenols

Organic substances combined with benzene rings and OH–groups to form additional aromatic compounds are known as phenolic pollutants. Starting with phenol, several polymers can be synthesized. Phenolic pollutants have serious detrimental impacts on human health and are regarded as priority pollutants. High levels of phenols are found in the effluents produced by various industries, including fertilizer, plastic, coking, pharmaceutical, and refinery industries, as well as home trash (Rout & Jena, 2022c). In bottled drinking waters, the USFDA has set a phenol concentration of 0.001 mg/L and in groundwater, 0.01 mg/L. For this reason, before water is released into the environment, the content of phenolic contaminants in it should be decreased.

To remove phenol, rGO was used as a sorbent; at 25 °C and neutral pH (Rout & Jena, 2022a), its sorption capacity was 602 mg/g. It was shown that the reduction of GO improved the adsorption performance. The regeneration studies demonstrate that following five cycles, the phenol removal stayed at 78.3%. To aid in the sorption of phenolic chemicals e.g. 2-CP, phenol, and 2,4-DCP, (Rout & Jena, 2022b) have prepared an adsorbent composed of rGO-modified epichlorohydrin cross-linked beta-cyclodextrin (rGO-βCD-ECH). According to the authors, this composite's maximum adsorption was 674, 659, and 703 mg/g, for these phenols, respectively. By forming a host-guest inclusion complex with the phenolic compounds, the modification of βCD may improve its adsorption capacity.

Adsorption of PFAS as emerging pollutants

A group of highly fluorinated aliphatic compounds known as Perfluoroalkyl and poly-fluoroalkyl substances (PFAS) are those in which fluorine atoms partially (poly-fluoroalkyl acids) or totally (perfluoroalkyl acids) replace H–atoms affixed to carbon chains. Fluorinated carbon chains are not the only functional groups found in PFAS—terminal functional groups can be carboxylic, phosphonic, sulfonic, sulfonamide, phosphinic, or ether sulfonic. According to (Phong Vo et al., 2020), PFAS are best suited for use in many applications e.g. fire fighting foams, waterproof fabrics, stain repellents, food packaging, cleaning products, non-stick cookware, electrical wire coverings, and metal plating because of their special combination of chemical and thermal stability and hydrophilic-lipophilic characterization. 4730 registered PFAS have been synthesized over the previous 70 years as a result of these numerous applications (Pauletto & Bandosz, 2022). Research has been done on numerous human exposure pathways to PFAS, Fig. 4.

Depending on how long their carbon chains are, PFAS are divided into subclasses. Perfluoroalkyl sulfonic acids (PFSA; $C \geq 6$) and perfluoroalkyl carboxylic acids (PFCA; $C \geq 8$) are referred to here as "long-chain" acids. According to (Pauletto & Bandosz, 2022), PFCA; $C \leq 7$ and PFSA; $C \leq 5$ are referred to as "short-chain" compounds. Emerging persistent organic pollutants are those that contain perfluoroalkyl acids (PFAA).

Fig. 4 Sources of PFAS. (Adapted with permission from (Yadav et al., 2022), © 2022 Elsevier B.V)

The octanol-water partition coefficients(K_{ow}) show that PFAS becomes more hydrophobic as the C–F chain length increases. The functional groups that are linked to the hydrophilic C–F chains, despite their hydrophobic nature, make PFAS soluble in water (Gagliano et al., 2020). PFAS are found in the environment in their (−)charged as shown by the low values of pK_a. According to (Saeidi et al., 2021) these physicochemical characteristics demonstrate the bioaccumulative, persistent, and spreadable nature of PFAS under many circumstances. According to (Wu et al., 2021) exposure to PFAS can impact an individual's immune system, making them more vulnerable to cause COVID-19. In many countries, the amount of PFAS in water is higher than what is advised. Thus, to protect the health of everyone on the planet, effective methods for removing PFAS from polluted water must be implemented.

According to (Yang et al., 2020), perfluorooctanoic acid (PFOA) was found to adsorb onto MIL-100-Fe and MIL-101-Fe. When $C_0 = 1000$ mg/L, Q was 370 mg/g for MIL-101-Fe and 349 mg/g for MIL-100-Fe, according to their experimental data. The ligands were 1,3,5-benzene tricarboxylic acid and 1,4-benzene dicarboxylic acid for MIL-100-Fe and MIL-101-Fe, respectively, and the pore structure was identified as the cause of this difference. Only MIL-101-Fe had triangular pores, which may have led to the increased PFOA adsorption. The authors proposed several mechanisms for the adsorption behaviour, in which the Lewis acid-base complexation was proposed as the primary adsorption process.

As described by (Barpaga et al., 2019), Per fluoro octane sulfonic acid (PFOS) adsorption on MIL-101-Cr and MIL-101-Fe was compared. The micelles most likely formed under these circumstances, as the C_0 of PFOS was 4141 mg/L. The exposed terminal sulfonic groups were therefore more likely to interact with the adsorbent surfaces than the fluorinated carbon chain of PFOS. MIL-101-Cr was able to adsorb PFOS to a larger extent and double as rapidly as MIL-101-Fe, according to the results of fluorine NMR.

5.2 Adsorption of Inorganic Pollutants

Adsorption of Heavy Metals

Pesticides, mining, electroplating, painting, and other industrial operations release a significant amount of heavy metal ions (HMI) (Hg^{2+}, Cd^{2+}, Pb^{2+}, Cr^{3+}, Cr^{6+}, etc.) into the aquatic ecosystem. The removal of more than one hundred HMI from wastewater requires the development of effective and eco-friendly technology due to the potential health risks posed by dissolved HMI, which can arise from their toxicity, persistence, and bioaccumulation. According to (Zhang et al., 2023), adsorption presents a lot of promise because of its low cost and simplicity of use.

The $Ti_3C_2T_x$ MXene/chitosan/lignosulfonate adsorbent (MCL) was created by (Othman et al., 2023) using chitosan/lignosulfonate nanospheres as a simple way to decorate $Ti_3C_2T_x$ MXene sheets. This renewable and biodegradable addition can enhance the MXenes'aqueous stability and biocompatibility. To give MXne sheets a range of surface functions, high SSA, and antioxidant properties, chitosan/lignosulfonate nanospheres were stabilized on their surface. In multi-metal systems, competitive adsorption demonstrated that MCL had a preferential adsorption affinity toward different HMIs; in neutral pH settings, the MCL adsorption removal followed a sequence of $Pb^{2+} > Cr^{6+} \approx Cu^{2+} > Ni^{2+} \approx Co^{2+}$. On the ions Cu^{2+} and Cr^{6+}, a moderate reduction was noted. It is proposed that surface complexation, ion exchange, and electrostatic interaction work synergistically to form the adsorption mechanism.

Green synthesis and in-situ polymerization techniques were utilized by (Raza et al., 2023) to build a unique form of conducting polymer based on 2D materials. Glycidyl methacrylate (GMA) was used to functionalize the 2D molybdenum disulfide (MoS_2), resulting in the formation of MoS_2@PGMA. The MoS_2@PGMA is then mixed with polyaniline (PANI) by the grafting method, as shown in, Fig. 5, to create conducting polymer grafted thin film nanosheets with a thickness of micrometer size. The maximum Q for Cu^{2+} and Cd^{2+} was 307 mg/g and 214 mg/g, respectively, producing amazing results. The prepared nanosheets can be successfully recycled and recovered for five cycles. It is anticipated that the established process for preparing MoS_2@PGMA/PANI and manufacturing the 2D materials would work for a variety of further applications.

2D iron oxide-hydroxide (FeOOH) nanoparticles, which are inexpensive and environmentally benign composites are considered promising materials for HMIs removal. Due to the naturally non-layered structure, it is currently difficult to create 2D FeOOH adsorbents with high Q for Cr (VI) removal. In this regard, (Wu et al.,

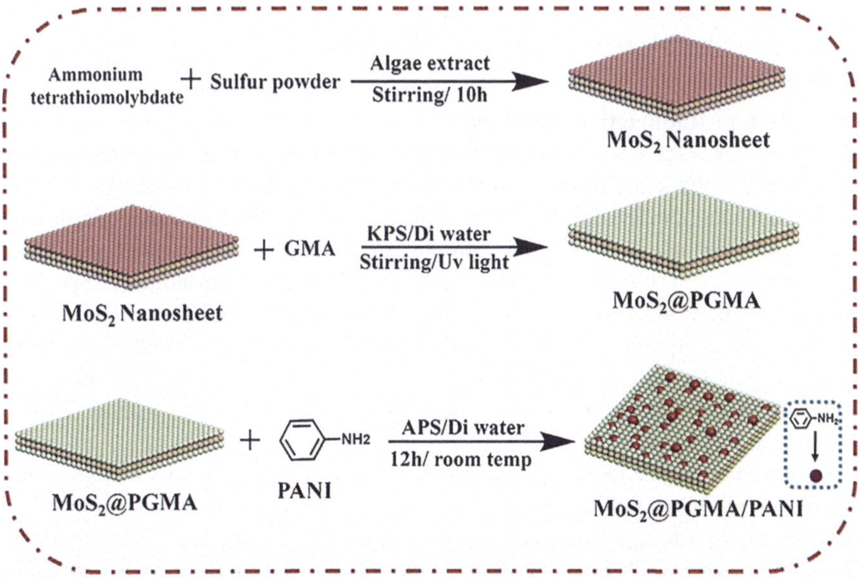

Fig. 5 Preparation of MoS_2, MoS_2@PGMA, and MoS_2@PGMA/PANI. (Adapted with permission from (Raza et al., 2023), © 2023 Elsevier B.V)

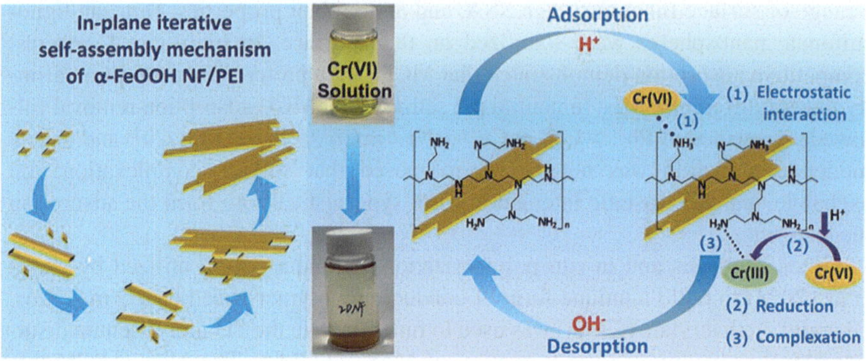

Fig. 6 Self-assembly mechanism of 2D α-FeOOH NF/PEI and the adsorption and desorption of Cr (VI) ions. (Adapted with permission from (Wu et al., 2023), © 2023 Elsevier B.V)

2023) prepared a unique 2D single-layer nano-raft-like α-FeOOH (α-FeOOH NF) functionalized with polyethyleneimine (PEI) via a simple one-pot hydrothermal method, featuring ultrathin nanowires that are oriented parallel to one another, Fig. 6. Anisotropic development of α-FeOOH served as intermediates and iterative seeds in the formation of the 2D NF nanostructure, which was discovered to be the result of an in-plane iterative self-assembly mechanism. The 2D α-FeOOH NF has a high SSA and porous structure that makes it highly effective at capturing Cr(VI) ions in water. Because of its special structure and PEI modification, it demonstrated high Q for the Cr(VI) removal, a quick adsorption kinetic rate, and good reusability. The suggested removal mechanism included both adsorption and reduction processes. The fifth cycle showed a removal effectiveness of 83.9% whereas the Q was 67 mg/g.

Adsorption of Radioactive pollutants

Nuclear power is seen as a viable substitute for addressing the growing demand for energy due to the scarcity of conventional fossil resources (Chen et al., 2022b). However, throughout the development of nuclear energy, a huge number of radionuclides with long half-lives and great mobility are released into the water, including uranium (^{235}U, ^{238}U), cesium (^{137}Cs), thorium ($^{23}2Th$), and strontium (^{90}Sr). The emission of these radionuclides causes high levels of heat, radiation, toxicity, and corrosiveness in aquatic environments. Because of their long-term radiological and chemical toxicity, these radionuclides pose serious dangers and long-term threats to the health of humans and the environment, both above and below ground. This is true even at low concentrations.

Radiation exposure in aquatic environments can result in various malignancies of different organs, neurological problems, and birth defects. There are rumors that these regrettable, long-term effects became evident following the Fukushima and Chernobyl disasters. Therefore, to remove radionuclides from the environment, it is always required to develop effective, environmentally friendly, and radiation-resistant adsorbents (Nezami et al., 2023). Uranium has always been the primary component of nuclear fuel energy. The extended half-life of uranium (4.45×10^9 years) makes it a non-essential radioactive element that poses a risk to both people and the environment. The U^{6+} is quite adaptable. Scientists have developed adsorbent materials to remove this pollutant because it quickly adsorbs in pure water, generating the far more stable UO_2^{2+} (Yu et al., 2019).

The most efficient adsorbent for UO_2^{2+} adsorption is $Ti_3C_2(OH)_2$, with Q = 595 mg/g. It was shown that UO_2^{2+} bound to deprotonated oxygen adsorptive sites were favoured over protonated molecules that may be accessed on surface hydroxylated sites (Zhang et al., 2020). A straightforward technique for creating and preserving pure and embedded MXene in under-hydrated conditions was presented by (Wang et al., 2017b). This method increases the gap between the layer of $Ti_3C_2T_x$ MXene. Using a hydrated intercalation strategy, the researchers demonstrated an unexpectedly substantial enhancement in $Ti_3C_2T_x$'s adsorption capacity for radioactive isotope rejection. This procedure, known as "radioactive imprisoning," was based on rapid low-temperature post-adsorption calcination.

By carefully controlling the interlayer spacing between the layers of MXene, they were able to confirm that U^{6+} could be captured within the multilayer $Ti_3C_2T_x$. The soaked $Ti_3C_2T_x$ NMs with DMSO significantly improved their adsorption behaviour for U^{6+}, and Q = 214 mg/g, as compared to dry $Ti_3C_2T_x$. (Sun et al., 2018) experimented on $Ti_3C_2T_x$ /g-C_3N_4 to confirm the effect of the –OH surface functional group on the MXene-based adsorbent's ability to adsorb. During their examination, they confirmed that MXene with a –OH surface group has strong bonding energy between 3.3 and 4.5 eV. This suggests that MXene with this type of surface group could be used as a good absorbents to reject radioactive materials from nuclear effluent.

Even at low concentrations, the major nuclear by-product Europium poses a health risk to the general public because it is harsh and must be released into the environment by nuclear reactors. (Zhang et al., 2019) prepared hierarchical titanite nanostructures (HTNs) using an in situ oxidation conversion process, starting with $Ti_3C_2T_x$ MXene precursor. The generated HTNs appear to be exceedingly robust and have large sorption capacities for Eu^{3+} because of their extensively interchangeable guest cations and well-maintained multilayer architecture. Effective adsorption of $Eu^{3+} > 200$ mg/g was demonstrated by the produced HTNs. Eu^{3+} was desorption by an ion-exchange mechanism, where H^+, Na^+, and K^+ were exchanged by solid H–bonds and electrostatic attraction. Verification of Eu^{3+} displacement for Na^+ or K^+ ions in HTN nanolayers has been confirmed by XPS and XRD. When it comes to environmental remediation, these HTNs are appealing substitutes for removing and cleansing actinides and trivalent lanthanides from aqueous solutions. During the experiment, several parameters are provided that impact how well HTNs work.

Thorium is still regarded as a major resource and the most promising substitute energy source for ^{233}U due to its fertile nature. From the Ti_2AlC precursor, (Li et al., 2019) synthesized $Ti_3C_2T_x$ -hydrated and $Ti_3C_2T_x$ -dry. The studies' ultimate conclusion is that $Ti_3C_2T_x$ -hydrated MXene has Q = 213 mg/g, which is significantly higher than the adsorption capacities of many other popular inorganic adsorbents. Furthermore, the $Ti_3C_2T_x$-hydrated material exhibits exceptional adsorption selectivity against Th^{4+} when the right metal ions are present. According to the present research, Th^{4+} pre-concentration and extraction could benefit from the use of $Ti_3C_2T_x$-hydrated.

6 Adsorption Mechanism

It is necessary to evaluate mechanism studies to have a thorough understanding of the adsorption of potentially hazardous contaminants. An understanding of the adsorption mechanism requires the application of some spectroscopic techniques, adsorption isotherms and kinetics, and theoretical computations like DFT. According to (Wu et al., 2019), the surface functional groups and numerous reaction sites of new nanomaterials might help with the effective removal of various contaminants.

The type of the adsorbate and the adsorbent surface properties are related to the adsorption mechanism. When additional materials are added to an adsorbent, the surface characteristics and interaction are altered, which ultimately impacts the adsorption mechanism. The most often reported adsorption mechanisms for 2D adsorbents against various water pollutants include ion exchange, H–bonding, n-π interaction, surface complexation, electrostatic interaction, van der Waals, hydrophobic interaction, and chemical adsorption (e.g., covalent bond). The structure, surface characteristics, and functional groups of the 2D adsorbents, as well as the chemical characteristics of the pollutant, determine the dominant adsorption mode. Additionally, as the adsorbent surface properties can be changed by the adsorption circumstances, such as pH and ionic strength, these factors have an impact on the adsorption mechanism (Ewis et al., 2022).

Ionic bonds are formed through **electrostatic interaction**, which consists of both electrostatic attraction and repulsion. It developed between negative anions and positive cations as a result of electron loss and gain in atoms. Analyzing the impact of solution pH on the adsorbent's ability for adsorption is important. This is so because changes in the pH of the solution can affect the charges on the surface of the adsorbent, which in turn affects the electrostatic interaction (Ewis et al., 2022).

As GO, for example, provides a benzene ring, the **n-π and π-π interactions** were in charge of the adsorption process. Unsaturated bonds on both the aromatic rings of the adsorbent and the adsorbate interact weakly through the π-π interaction (Chua et al., 2021).

Ion exchange is the process by which ions of the same charge are exchanged between the adsorbent and the adsorbate. To keep the solution electrically neutral, it happens when the adsorbent adsorbs ions from the solution and releases equivalent ions back into the solution. A form of dipole-dipole attraction known as **hydrogen bonding** occurs when an electronegative atom, such as N, O, or F, bonds to a H atom. **Surface complexation**: may be a factor in how minerals and clay interact with organic contaminants. According to (Yang et al., 2019), surface complexation is the result of the adsorbent and adsorbate interacting electrostatically while the

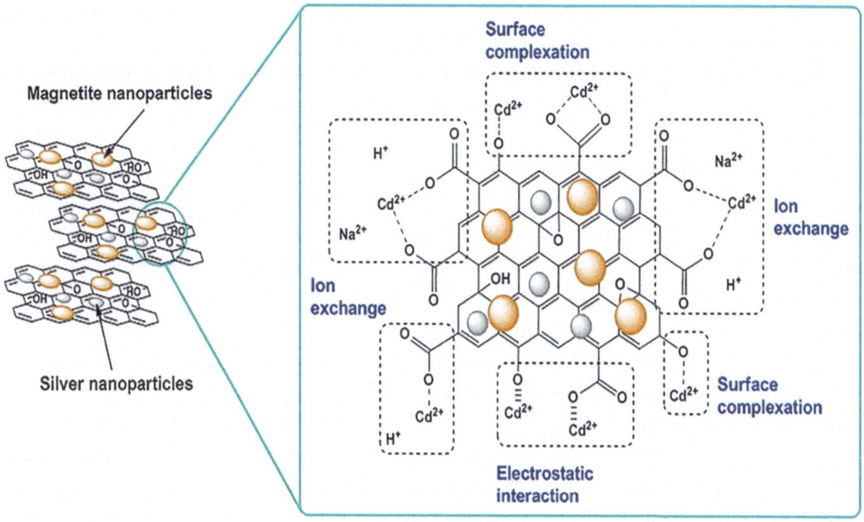

Fig. 7 Suggested adsorption mechanisms for Cd^{2+} on the rGO/magnetite/silver NH sample. (Adapted with permission from (Park et al., 2019), © 2019 Elsevier B.V)

adsorbate ions maintain their hydration sphere. The primary forces behind adsorption were surface complexation, ion exchange, and electrostatic attraction Fig. 7.

7 Desorption and Regeneration Studies

An essential step in managing the cost of the adsorbent, environmental risks, and the hazardous sludge disposal issue is ensuring that pollutants can be recovered from the loaded 2D adsorbents and that they can be reused. A variety of eluents, including HCl, NH_4OH, HNO_3, H_2SO_4, NaOH, Na_2CO_3, KOH, DI-water, and Na_2EDTA, have been applied in the literature to aid in the desorption of pollutants that have been adsorbed by 2D adsorbents (Bilal & Ihsanullah, 2022). When choosing an appropriate adsorbent, one of the most crucial factors is its renewability. Adsorbents with high contaminant adsorption on their surface at equilibrium, more than the quantity of contaminant left in the water solution, typically exhibit sluggish or ineffective desorption/regeneration, which restricts their use in wastewater treatment (Vakili et al., 2019).

If adsorbents are not recycled, the adsorption process becomes more expensive and there is a chance that the contaminant will be discharged into the environment during disposal or storage of the worn adsorbent. By recovering pollutants from the used adsorbent, the cost of the adsorption process can be decreased because the regenerated adsorbent can be used continuously. Consequently, there are significant financial and environmental benefits to recycling the wasted adsorbent through the implementation of a suitable regeneration technique.

Chemical regeneration involves applying a solvent at a specific pH and temperature, which breaks the adsorption equilibrium between the solvent, adsorbent, and adsorbate. Chemical solvents can be either organic, like methanol or inorganic, like HCl and NaOH. For organic pollutants with a low boiling point and high concentration, chemical regeneration is an appropriate regeneration technique. Furthermore, the primary factor influencing the effectiveness of the regeneration process is the adsorbate's solubility in the solvent. Solvent regeneration poses a primary concern as it has the potential to modify the structure of the adsorbent, hence impacting its efficiency in subsequent adsorption cycles (Momina et al., 2020). One crucial thing to keep in mind is that the chemical regeneration process is hampered and/or ineffective in regenerating the composite when the pollutant is firmly adsorbed on the surface of the composite and within its micropores. This is an important problem for composites that get highly bound to organic pollutants and have a micropore-like structure. Consequently, it is advisable to take into account alternate regeneration methods. As part of the regeneration process, the concentration of the solvent is a crucial consideration. An increased concentration of solvent improves regeneration. Regenerating adsorbents is not aided by exceptionally high solvent concentrations, nevertheless. Consequently, appropriate research should be done on the influence of solvent concentration. For example, research has shown that MB desorption is improved when the concentration of HCl is increased from 0.001 M to 0.1 M. In addition, temperature has a significant role in the regeneration of 2D composites. By changing the adsorbate's molecular activity and decreasing the adsorbate–adsorbent interaction, temperature can accelerate the rate of desorption (Mansour et al., 2018). Furthermore, heating increases the organic contaminants' solubility in the solvent. Another crucial element is the timing of desorption. Since more interaction between the adsorbate and the solvent is possible, longer desorption times generally result in faster desorption rates. The predicted duration for 2D composites ranged from 3 to 24 hours. Nonetheless, the majority of these investigations purposefully prolonged the desorption period to guarantee that the system attains the desorption equilibrium. Therefore, to achieve the highest desorption efficiency during the regeneration process, it is necessary to analyze temperature, solvent concentration, the kind of interaction between the adsorbate and the adsorbent, and the contact time.

The **ultrasonication** procedure comprised soaking the used adsorbent for an hour in 10 millilitres of pure ethanol and then ultrasonicating it for 30 minutes.

During **thermal regeneration**, the adsorbent was subjected to a high temperature, which could break the connection that holds the adsorbate and adsorbent together. The adsorbent's weight could be decreased and its form changed by the thermal heat produced. Additionally, because thermal regeneration involves extremely high temperatures, adsorbents with limited mechanical strength and low thermal stability cannot be thermally regenerated.

8 Antimicrobial Activity of 2D Materials

Waterborne diseases result in the deaths of about five million people annually (Fahimirad et al., 2021). This dramatic impact makes finding antimicrobial agents a hot topic for research. Antimicrobials are inhibitory substances for the growth of microorganisms (Di Martino, 2022). The development of new antimicrobial substances is crucial for overcoming such diseases.

2D materials as novel, effective antimicrobial substances may provide new solutions to the microbial infection problem. Such 2D materials have some advantages for antimicrobial applications compared to other structures. The ultra-large surface area of 2D materials makes them suitable carriers for antimicrobial drugs. 2D materials interact with microbial cells via physical and chemical modes, which dramatically reduces the ability of the microorganisms to develop antimicrobial resistance. The photothermal and photocatalytic properties of some 2D materials enable them to eradicate microorganisms regardless of drug resistance (Sun & Wu, 2018).

Graphene, as a 2D carbon material, has gained attention for its antimicrobial properties. Graphene materials are easy to prepare and are renewable. They have unique catalytic properties, a large specific surface area and mechanical strength (Kumar et al., 2019). The antibacterial activity of graphene-based materials towards *Escherichia coli* (*E. coli*), including graphite and its oxide, graphene oxide, and reduced graphene oxide, was explored (Liu et al., 2011). The density of the graphene edges was the main parameter that contributed to the antibacterial property of the graphene films (Pham et al., 2015). Graphene nanosheets form pores in the bacterial cell wall, causing a subsequent osmotic imbalance and cell death (Pham et al., 2015).

The antibacterial properties of MXenes have been explored in various studies, highlighting their effectiveness against different types of bacteria and their applications in biomedical engineering, separation membranes, and wastewater treatment. MXenes are a new class of 2D antibacterial nanomaterials introduced by (Rasool et al., 2016). $Ti_3C_2T_x$ showed higher antimicrobial activity than graphene oxide toward *E. coli* and *Bacillus subtilis*. The direct contact between $Ti_3C_2T_x$ and bacterial cells can disrupt cellular membranes, leading to cell damage and death (Rasool et al., 2016). AgNPs-decorated $Ti_3C_2T_x$ membranes have anti-biofouling ability (Pandey et al., 2018). The bactericidal mechanism of MXenes is still largely unexplored.

MOFs have gained attention in antibacterial applications due to their unique properties and potential for controlled release of antibacterial agents. MOF-based bimetallic nanocomposites have also shown superior antibacterial activities against *E. coli* and *B. subtilis* compared to chemical antibiotics (Kim et al., 2023). The antibacterial activity of MOFs can release active biocidal agents in a sustained manner, contributing to their excellent antibacterial activity (Chattopadhyay et al., 2021). MOFs can be designed to contain organic antibacterial ligands that can be released in a controlled manner, leading to antibacterial activity (Li et al., 2021). The self-degrading MOFs with organic antibacterial ligands have also demonstrated antibacterial activities through the controlled release of these ligands (Liu et al., 2019). MOFs can involve metal ions, such as copper ions, which have antibacterial behaviour (Fu et al., 2020). The slow release of copper ions from MOFs contributes to their improved antibacterial activity against bacteria like *Staphylococcus aureus* and *E. coli* (Fu et al., 2020). MOFs can disrupt the integrity of bacterial cell membranes, leading to drug internalisation and antibacterial effects. MOFs perform as catalytically active sites, mimicking enzymes and acting as photosensitisers, all contributing to their antibacterial activity (Han et al., 2022b). MOFs can act as photocatalysts, generating reactive oxygen species (ROS) upon light irradiation, which exhibit antimicrobial effects (Li et al., 2021). The broad-spectrum antibacterial action of MOFs against various oral infections is attributable to the synergistic effects of released ions from MOFs and ROS (Li et al., 2021).

The antimicrobial activity of g-C_3N_4 depends on its photocatalytic properties (Sun & Wu, 2018). g-C_3N_4 has a high surface area, which allows for increased contact between the material and microorganisms, enhancing its antimicrobial efficacy (Mallikarjuna et al., 2021). g-C_3N_4 has been found to generate ROS under light irradiation, such as hydroxyl radicals (•OH), which have strong antimicrobial properties (Niu et al., 2012). These ROS can damage the microbial cell membranes and DNA, leading to their inactivation (Mallikarjuna et al., 2021).

Doping can enhance the antimicrobial activity of g-C_3N_4. Loading silver (Ag) nanoparticles onto g-C_3N_4 improved its antimicrobial activity against multidrug-resistant pathogens (Mallikarjuna et al., 2021). Similarly, modifying g-C_3N_4 with copper (Cu) ions enhances its visible light photocatalytic properties, which can contribute to its antimicrobial activity (Li et al., 2020). In addition, g-C_3N_4 nanosheets have peroxidase-like activity, which can be applied for disinfection (Tian et al., 2013). Incorporating H_2O_2 with *g*-C_3N_4@AuNPs inhibits biofilm formation (Wang et al., 2017a).

9 Challenges and Future Perspectives

There have been encouraging possibilities for treating wastewater using 2D materials as adsorbents. Even still, there are several significant obstacles connected to 2D adsorbents that need more future perspectives:

- Certain synthetic 2D materials lack functional groups and are therefore only suitable as single-layer adsorbents, which limits their ability to be enhanced and improved for better adsorption capacity.
- In the laboratory, 2D adsorbents are prepared, but before they can be sold or commercialized, they are encapsulated, which may cause them to lose some of their surface characteristics and large SSA.
- A theoretical calculation is necessary to determine the relationship between pollutants and 2D adsorbents and how to enhance the sorption performance.

- Analyzing the competing and synergistic interactions between adsorbents and pollutants in real wastewater is crucial because most research has concentrated on synthetic solutions, whereas real wastewater is complicated and comprises a variety of pollutants.
- For batch adsorption procedures, the majority of the research is performed. However, before applying 2D adsorbents to real treatment processes, it is necessary to learn how well these materials work in continuous adsorption processes.
- To assess the effectiveness and practical performance of each adsorbent, a comprehensive investigation of regeneration and reusability must be conducted.
- The intrinsic relationship between the adsorbent microstructure and the adsorption enhancement effect has to be further explored.

10 Conclusion

In summary, 2D materials are a class of materials with atomic or molecular thickness and unique properties arising from their reduced dimensionality. They have a main interest because of their diverse characteristics and various potential applications. The synthesis and manipulation of 2D materials continue to be active areas of research, to harness their unique properties for technological advancements.

This chapter displays how 2D nanomaterials as adsorbents can be used to remove hazardous pollutants from wastewater. 2D materials include graphene, g-C_3N_4, TMDs, and MOFs. Different approaches, mostly top-down or bottom-up, can be used to synthesise these 2D materials. It was discovered that the presence of various surface functional groups, surface wettability, stability of the material, and SSA, among other physicochemical features of 2D adsorbents, had a significant impact on the pollutants' ability to be adsorbed.

For wastewater treatment and pollution remediation, adsorption is a very promising and successful technique. Many contaminants, such as dyes, phenols, PFAS, heavy metals, and radioactive pollutants, have been studied in this chapter. 2D adsorbents have demonstrated exceptional adsorption capabilities, and choosing the right adsorbents is essential to optimizing pollutant removal. It is necessary to tune the initial pollutant concentration, adsorbent dose, pH, contact time, and temperature to get the best removal results from adsorption methods. Designing and optimizing adsorption systems is also aided by knowledge of appropriate regeneration procedures and adsorption mechanisms. 2D materials showed antimicrobial activity which may provide a promising solution to the microbial infection problem.

Finally, further research is needed to design and regulate 2D adsorbents, reuse them effectively, lower environmental risks, increase their adsorption capacity, optimize their design, investigate the mechanism of interaction between pollutants and adsorbents, create adsorption processes for real wastewater, and commercialize 2D adsorbents.

References

Alam, S., Asaduzzaman Chowdhury, M., Shahid, A., Alam, R., & Rahim, A. (2021). Synthesis of emerging two-dimensional (2D) materials—Advances, challenges and prospects. *FlatChem, 30*, 100305. https://doi.org/10.1016/j.flatc.2021.100305

Amdeha, E. (2021). Recovery of nanomaterials from agricultural and industrial wastes for water treatment applications. In A. S. H. Makhlouf & G. A. M. Ali (Eds.), *Topics in mining, metallurgy and materials engineering* (pp. 385–417). Springer. https://doi.org/10.1007/978-3-030-68031-2_14

Amdeha, E. (2023a). Smart nanomaterials for photo-catalytic applications. In S. A. Bansal, V. Khanna, N. Balakrishnan, & P. Gupta (Eds.), *Diversity and applications of new age nanoparticles* (pp. 112–154). IGI Global. https://doi.org/10.4018/978-1-6684-7358-0.ch005

Amdeha, E. (2023b). Biochar-based nanocomposites for industrial wastewater treatment via adsorption and photocatalytic degradation and the parameters affecting these processes. *Biomass Conversion and Biorefinery, 1*, 1–26. https://doi.org/10.1007/s13399-023-04512-2

Amdeha, E. (2024). Sensors for heavy metals and dyes detection for water analysis. In *Handbook of nanosensors* (pp. 1–35). Springer. https://doi.org/10.1007/978-3-031-16338-8_64-1

Amdeha, E., & Salem, M. S. (2022). Facile green synthesis of ZnO supported on exfoliated graphite for photocatalytic degradation of dye under UV and visible-light irradiation. *Egyptian Journal of Chemistry, 65*(13), 557–569. https://doi.org/10.21608/EJCHEM.2022.130532.5931

Baig, N., Ihsanullah, S., & M., & Saleh, T. A. (2019). Graphene-based adsorbents for the removal of toxic organic pollutants: A review. *Journal of Environmental Management, 244*, 370–382. https://doi.org/10.1016/j.jenvman.2019.05.047

Barpaga, D., Zheng, J., Han, K. S., Soltis, J. A., Shutthanandan, V., Basuray, S., McGrail, B. P., Chatterjee, S., & Motkuri, R. K. (2019). Probing the sorption of Perfluorooctanesulfonate using mesoporous metal-organic frameworks from aqueous solutions. *Inorganic Chemistry, 58*(13), 8339–8346. https://doi.org/10.1021/acs.inorgchem.9b00380

Bilal, M., & Ihsanullah, I. (2022). What makes MXenes emergent materials for the adsorption of heavy metals from water? A critical review. *Journal of Water Process Engineering, 49*, 103010. https://doi.org/10.1016/j.jwpe.2022.103010

Bilal, M., Ihsanullah, I., Younas, M., & Shah, M., U. H. (2022). Recent advances in applications of low-cost adsorbents for the removal of heavy metals from water: A critical review. *Separation and Purification Technology, 278*, 119510. https://doi.org/10.1016/j.seppur.2021.119510

Bilal, M., Khan, U., & Ihsanullah, I. (2023). MXenes: The emerging adsorbents for the removal of dyes from water. *Journal of Molecular Liquids, 385*, 122377. https://doi.org/10.1016/j.molliq.2023.122377

Cai, Z., Liu, B., Zou, X., & Cheng, H. M. (2018). Chemical vapor deposition growth and applications of two-dimensional materials and their heterostructures. *Chemical Reviews, 118*(13), 6091–6133. https://doi.org/10.1021/acs.chemrev.7b00536

Chattopadhyay, K., Mandal, M., & Maiti, D. K. (2021). Smart metal-organic frameworks for biotechnological applications: A mini-review. *ACS Applied Bio Materials, 4*(12), 8159–8171. https://doi.org/10.1021/acsabm.1c00982

Chen, J. R., Chen, J. R., & Su, C. K. (2022a). Solution Foaming–Treated 3D-Printed monolithic packing for enhanced solid phase extraction of trace metals. *Talanta, 241*, 123237. https://doi.org/10.1016/j.talanta.2022.123237

Chen, X., Huang, L., Liu, J., Song, D., & Yang, S. (2022b). Peak shaving benefit assessment considering the joint operation of nuclear and battery energy storage power stations: Hainan case study. *Energy, 239*, 121897. https://doi.org/10.1016/j.energy.2021.121897

Chua, S. F., Nouri, A., Ang, W. L., Mahmoudi, E., Mohammad, A. W., Benamor, A., & Ba-Abbad, M. (2021). The emergence of multifunctional adsorbents and their role in environmental remediation. *Journal of Environmental Chemical Engineering, 9*(1), 104793. https://doi.org/10.1016/j.jece.2020.104793

Demirel, M. C., Vural, M., & Terrones, M. (2018). Composites of proteins and 2D nanomaterials. *Advanced Functional Materials, 28*(27), 1704990. https://doi.org/10.1002/adfm.201704990

Deriase, S. F., El-Salamony, R. A., Amdeha, E., & Al-Sabagh, A. M. (2021). Statistical optimization of photocatalytic degradation process of methylene blue dye by SnO–TiO2–AC composite using response surface methodology. *Environmental Progress and Sustainable Energy, 40*(5), e13639. https://doi.org/10.1002/ep.13639

Di Martino, P. (2022). Antimicrobial agents and microbial ecology. *AIMS Microbiology, 8*(1), 1–4. https://doi.org/10.3934/microbiol.2022001

El-Salamony, R. A., Amdeha, E., El Shafey, A. M., & Al Sabagh, A. M. (2023). Preparation and characterisation of Ce-doped SiO_2 nano-materials as effective photo-catalyst under visible light. *International Journal of Environmental Analytical Chemistry, 103*(4), 868–883. https://doi.org/10.1080/03067319.2020.1865328

Ewis, D., Ismail, N. A., Hafiz, M. A., Benamor, A., & Hawari, A. H. (2021). Nanoparticles functionalized ceramic membranes: Fabrication, surface modification, and performance. *Environmental Science and Pollution Research, 28*(10), 12256–12281. https://doi.org/10.1007/s11356-020-11847-0

Ewis, D., Ba-Abbad, M. M., Benamor, A., & El-Naas, M. H. (2022). Adsorption of organic water pollutants by clays and clay minerals composites: A comprehensive review. *Applied Clay Science, 229*, 106686. https://doi.org/10.1016/j.clay.2022.106686

Fahimirad, S., Fahimirad, Z., & Sillanpää, M. (2021). Efficient removal of water bacteria and viruses using electrospun nanofibers. *Science of the Total Environment, 751*, 141673. https://doi.org/10.1016/j.scitotenv.2020.141673

Fu, L. Q., Chen, X. Y., Cai, M. H., Tao, X. H., Fan, Y. B., & Mou, X. Z. (2020). Surface engineered metal-organic frameworks (MOFs) based novel hybrid Systems for effective wound healing: A review of recent developments. *Frontiers in Bioengineering and Biotechnology, 8*, 576348. https://doi.org/10.3389/fbioe.2020.576348

Gagliano, E., Sgroi, M., Falciglia, P. P., Vagliasindi, F. G. A., & Roccaro, P. (2020). Removal of poly- and perfluoroalkyl substances (PFAS) from water by adsorption: Role of PFAS chain length, effect of organic matter and challenges in adsorbent regeneration. *Water Research, 171*, 115381. https://doi.org/10.1016/j.watres.2019.115381

Garg, M., Vishwakarma, N., Sharma, A. L., & Singh, S. (2021). Different types and intense classification of 2D materials. In *Materials horizons: From nature to nanomaterials* (pp. 11–28). Springer Nature. https://doi.org/10.1007/978-981-16-3322-5_2

Geim, A. K. (2009). Graphene: Status and prospects. *Science, 324*(5934), 1530–1534.

Han, Y., Gao, L., Zhou, J., Hou, Y., Jia, Y., Cao, K., Duan, K., & Lu, Y. (2022a). Deep elastic strain engineering of 2D materials and their twisted bilayers. *ACS Applied Materials and Interfaces, 14*(7), 8655–8663. https://doi.org/10.1021/acsami.1c23431

Han, D., Liu, X., & Wu, S. (2022b). Metal organic framework-based antibacterial agents and their underlying mechanisms. *Chemical Society Reviews, 51*(16), 7138–7169. https://doi.org/10.1039/d2cs00460g

Janjhi, F. A., Ihsanullah, I., Bilal, M., Castro-Muñoz, R., Boczkaj, G., & Gallucci, F. (2023). MXene-based materials for removal of antibiotics and heavy metals from wastewater- a review. *Water Resources and Industry, 29*, 100202. https://doi.org/10.1016/j.wri.2023.100202

Kadhom, M., Kalash, K., & Al-Furaiji, M. (2022). Performance of 2D MXene as an adsorbent for malachite green removal. *Chemosphere, 290*, 133256. https://doi.org/10.1016/j.chemosphere.2021.133256

Kim, D., Park, K. W., Park, J. T., & Choi, I. (2023). Photoactive MOF-derived bimetallic silver and cobalt nanocomposite with enhanced antibacterial activity. *ACS Applied Materials and Interfaces, 15*(19), 22903–22914. https://doi.org/10.1021/acsami.3c01529

Kumar, P., Huo, P., Zhang, R., & Liu, B. (2019). Antibacterial properties of graphene-based nanomaterials. *Nanomaterials, 9*(5), 737.

Kumar, P., Dey, A., Roques, J., Assaud, L., Franger, S., Parida, P., & Biju, V. (2022). Photoexfoliation synthesis of 2D materials. *ACS Materials Letters, 4*(2), 263–270. https://doi.org/10.1021/acsmaterialslett.1c00651

Li, H., Li, Y., Aljarb, A., Shi, Y., & Li, L. J. (2018). Epitaxial growth of two-dimensional layered transition-metal dichalcogenides: Growth mechanism, controllability, and scalability. *Chemical Reviews, 118*(13), 6134–6150. https://doi.org/10.1021/acs.chemrev.7b00212

Li, S., Wang, L., Peng, J., Zhai, M., & Shi, W. (2019). Efficient thorium(IV) removal by two-dimensional Ti_2CT_x MXene from aqueous solution. *Chemical Engineering Journal, 366*, 192–199. https://doi.org/10.1016/j.cej.2019.02.056

Li, X., Wang, B., Yin, W., Di, J., Xia, J., Zhu, W., & Li, H. (2020). Cu2+ modified g-C3N4 photocatalysts for visible light photocatalytic properties. *Wuli Huaxue Xuebao/Acta Physico – Chimica Sinica, 36*(3), 1902001. https://doi.org/10.3866/PKU.WHXB201902001

Li, R., Chen, T., & Pan, X. (2021). Metal-organic-framework-based materials for antimicrobial applications. *ACS Nano, 15*(3), 3808–3848. https://doi.org/10.1021/acsnano.0c09617

Liu, Y., & Li, R. (2020). Study on ultrasound-assisted liquid-phase exfoliation for preparing graphene-like molybdenum disulfide nanosheets. *Ultrasonics Sonochemistry, 63*, 104923. https://doi.org/10.1016/j.ultsonch.2019.104923

Liu, S., Zeng, T. H., Hofmann, M., Burcombe, E., Wei, J., Jiang, R., Kong, J., & Chen, Y. (2011). Antibacterial activity of graphite, graphite oxide, graphene oxide, and reduced graphene oxide: Membrane and oxidative stress. *ACS Nano, 5*(9), 6971–6980. https://doi.org/10.1021/nn202451x

Liu, X., Yan, Z., Zhang, Y., Liu, Z., Sun, Y., Ren, J., & Qu, X. (2019). Two-dimensional metal-organic framework/enzyme hybrid Nanocatalyst as a benign and self-activated cascade reagent for in vivo wound healing. *ACS Nano, 13*(5), 5222–5230. https://doi.org/10.1021/acsnano.8b09501

Mallikarjuna, K., Vattikuti, S. V. P., Manne, R., Manjula, G., Munirathnam, K., Mallapur, S., Marraiki, N., Mohammed, A., Reddy, L. V., Rajesh, M., & Razab, M. K. A. A. (2021). Sonochemical synthesis of silver quantum dots immobilized on exfoliated graphitic carbon nitride nanostructures using ginseng extract for photocatalytic hydrogen evolution, dye degradation, and antimicrobial studies. *Nanomaterials, 11*(11), 2918. https://doi.org/10.3390/nano11112918

Mansour, F., Al-Hindi, M., Yahfoufi, R., Ayoub, G. M., & Ahmad, M. N. (2018). The use of activated carbon for the removal of pharmaceuticals from aqueous solutions: A review. *Reviews in Environmental Science and Biotechnology, 17*(1), 109–145. https://doi.org/10.1007/s11157-017-9456-8

Mantasha, I., Hussain, S., Ahmad, M., & Shahid, M. (2020). Two dimensional (2D) molecular frameworks for rapid and selective adsorption of hazardous aromatic dyes from aqueous phase. *Separation and Purification Technology, 238*, 116413. https://doi.org/10.1016/j.seppur.2019.116413

Miró, P., Han, J. H., Cheon, J., & Heine, T. (2014). Hexagonal transition-metal chalcogenide nanoflakes with pronounced lateral quantum confinement. *Angewandte Chemie – International Edition, 53*(46), 12624–12628. https://doi.org/10.1002/anie.201404704

Momina, Mohammad, S., & Suzylawati, I. (2020). Study of the adsorption/desorption of MB dye solution using bentonite adsorbent coating. *Journal of Water Process Engineering, 34*, 101155. https://doi.org/10.1016/j.jwpe.2020.101155

Nezami, S., Ghaemi, A., & Yousefi, T. (2023). Application of titanium carbide/nitride (MXene)-based NPs in adsorption of radionuclides and heavy metal ions for wastewater remediation: A review. *Case Studies in Chemical and Environmental Engineering, 7*, 100326. https://doi.org/10.1016/j.cscee.2023.100326

Niu, P., Liu, G., & Cheng, H. M. (2012). Nitrogen vacancy-promoted photocatalytic activity of graphitic carbon nitride. *Journal of Physical Chemistry C, 116*(20), 11013–11018. https://doi.org/10.1021/jp301026y

Ortiz, N., & Skrabalak, S. E. (2014). On the dual roles of ligands in the synthesis of colloidal metal nanostructures. *Langmuir, 30*(23), 6649–6659. https://doi.org/10.1021/la404539p

Othman, Z., Mackey, H. R., & Mahmoud, K. A. (2023). MXene/chitosan/lignosulfonate (MCL) nanocomposite for simultaneous removal of Co(II), Cr(VI), Cu(II), Ni(II) and Pb(II) heavy metals from wastewater. *2D Materials, 10*(2), 024004. https://doi.org/10.1088/2053-1583/acbd62

Pandey, R. P., Rasool, K., Madhavan, V. E., Aïssa, B., Gogotsi, Y., & Mahmoud, K. A. (2018). Ultrahigh-flux and fouling-resistant membranes based on layered silver/MXene (Ti 3 C 2 T x) nanosheets. *Journal of Materials Chemistry A, 6*(8), 3522–3533.

Park, C. M., Wang, D., Han, J., Heo, J., & Su, C. (2019). Evaluation of the colloidal stability and adsorption performance of reduced graphene oxide–elemental silver/magnetite nanohybrids for selected toxic heavy metals in aqueous solutions. *Applied Surface Science, 471*, 8–17. https://doi.org/10.1016/j.apsusc.2018.11.240

Pauletto, P. S., & Bandosz, T. J. (2022). Activated carbon versus metal-organic frameworks: A review of their PFAS adsorption performance. *Journal of Hazardous Materials, 425*, 127810. https://doi.org/10.1016/j.jhazmat.2021.127810

Peng, Y., Huang, Y., Zhu, Y., Chen, B., Wang, L., Lai, Z., Zhang, Z., Zhao, M., Tan, C., Yang, N., Shao, F., Han, Y., & Zhang, H. (2017). Ultrathin two-dimensional covalent organic framework nanosheets: Preparation and application in highly sensitive and selective DNA detection. *Journal of the American Chemical Society, 139*(25), 8698–8704. https://doi.org/10.1021/jacs.7b04096

Pham, V. T. H., Truong, V. K., Quinn, M. D. J., Notley, S. M., Guo, Y., Baulin, V. A., Al Kobaisi, M., Crawford, R. J., & Ivanova, E. P. (2015). Graphene induces formation of pores that kill spherical and rod-shaped bacteria. *ACS Nano, 9*(8), 8458–8467. https://doi.org/10.1021/acsnano.5b03368

Phong Vo, H. N., Ngo, H. H., Guo, W., Hong Nguyen, T. M., Li, J., Liang, H., Deng, L., Chen, Z., & Hang Nguyen, T. A. (2020). Poly-and perfluoroalkyl substances in water and wastewater: A comprehensive review from sources to remediation. *Journal of Water Process Engineering, 36*, 101393. https://doi.org/10.1016/j.jwpe.2020.101393

Rahmani, H., Lakzian, A., Karimi, A., & Halajnia, A. (2020). Efficient removal of 2,4-dinitrophenol from synthetic wastewater and contaminated soil samples using free and immobilized laccases. *Journal of Environmental Management, 256*, 109740. https://doi.org/10.1016/j.jenvman.2019.109740

Rasool, K., Helal, M., Ali, A., Ren, C. E., Gogotsi, Y., & Mahmoud, K. A. (2016). Antibacterial activity of $Ti_3C_2T_x$ MXene. *ACS Nano, 10*(3), 3674–3684. https://doi.org/10.1021/acsnano.6b00181

Raza, S., Hameed, M. U., Ghasali, E., Hayat, A., Orooji, Y., Lin, H., Karaman, C., Karimi, F., & Erk, N. (2023). Algae extract delamination of molybdenum disulfide and surface modification with glycidyl methacrylate and polyaniline for the elimination of metal ions from wastewater. *Environmental Research, 221*, 115213. https://doi.org/10.1016/j.envres.2023.115213

Rout, D. R., & Jena, H. M. (2022a). Removal of phenol from aqueous solution using reduced graphene oxide as adsorbent: Isotherm, kinetic, and thermodynamic studies. *Environmental Science and Pollution Research, 29*(21), 32105–32119. https://doi.org/10.1007/s11356-021-17944-y

Rout, D. R., & Jena, H. M. (2022b). Synthesis of novel epichlorohydrin cross-linked β-cyclodextrin functionalized with reduced graphene oxide composite adsorbent for treatment of phenolic wastewater. *Environmental Science and Pollution Research, 29*(48), 73444–73460. https://doi.org/10.1007/s11356-022-21018-y

Rout, D. R., & Jena, H. M. (2022c). Polyethylene glycol functionalized reduced graphene oxide coupled with zinc oxide composite adsorbent for removal of phenolic wastewater. *Environmental Research, 214*, 114044. https://doi.org/10.1016/j.envres.2022.114044

Saeidi, N., Kopinke, F. D., & Georgi, A. (2021). What is specific in adsorption of perfluoroalkyl acids on carbon materials? *Chemosphere, 273*, 128520. https://doi.org/10.1016/j.chemosphere.2020.128520

Sajid, M. (2021). MXenes: Are they emerging materials for analytical chemistry applications?—A review. *Analytica Chimica Acta, 1143*, 267–280. https://doi.org/10.1016/j.aca.2020.08.063

Sun, W., & Wu, F. G. (2018). Two-dimensional materials for antimicrobial applications: Graphene materials and beyond. *Chemistry—An Asian Journal, 13*(22), 3378–3410. https://doi.org/10.1002/asia.201800851

Sun, Y., Jin, D., Sun, Y., Meng, X., Gao, Y., & Dall'Agnese, Y., Chen, G., & Wang, X. F. (2018). G-C3N4/Ti3C2TX (MXenes) composite with oxidized surface groups for efficient photocatalytic hydrogen evolution. *Journal of Materials Chemistry A, 6*(19), 9124–9131. https://doi.org/10.1039/c8ta02706d

Tian, J., Liu, Q., Asiri, A. M., Qusti, A. H., Al-Youbi, A. O., & Sun, X. (2013). Ultrathin graphitic carbon nitride nanosheets: A novel peroxidase mimetic, Fe doping-mediated catalytic performance enhancement and application to rapid, highly sensitive optical detection of glucose. *Nanoscale, 5*(23), 11604–11609. https://doi.org/10.1039/c3nr03693f

Vakili, M., Deng, S., Cagnetta, G., Wang, W., Meng, P., Liu, D., & Yu, G. (2019). Regeneration of chitosan-based adsorbents used in heavy metal adsorption: A review. *Separation and Purification Technology, 224*, 373–387. https://doi.org/10.1016/j.seppur.2019.05.040

Wang, Z., Dong, K., Liu, Z., Zhang, Y., Chen, Z., Sun, H., Ren, J., & Qu, X. (2017a). Activation of biologically relevant levels of reactive oxygen species by Au/g-C_3N_4 hybrid nanozyme for bacteria killing and wound disinfection. *Biomaterials, 113*, 145–157. https://doi.org/10.1016/j.biomaterials.2016.10.041

Wang, L., Tao, W., Yuan, L., Liu, Z., Huang, Q., Chai, Z., Gibson, J. K., & Shi, W. (2017b). Rational control of the interlayer space inside two-dimensional titanium carbides for highly efficient uranium removal and imprisonment. *Chemical Communications, 53*(89), 12084–12087. https://doi.org/10.1039/c7cc06740b

Wang, J., Wang, W., Ai, Z., Li, M., Li, H., Peng, W., Zhao, Y., & Song, S. (2021). Adsorption toward Pb(II) occurring on three-dimensional reticular-structured montmorillonite hydrogel surface. *Applied Clay Science, 210*, 106153. https://doi.org/10.1016/j.clay.2021.106153

Wu, Y., Pang, H., Liu, Y., Wang, X., Yu, S., Fu, D., Chen, J., & Wang, X. (2019). Environmental remediation of heavy metal ions by novel-nanomaterials: A review. *Environmental Pollution, 246*, 608–620. https://doi.org/10.1016/j.envpol.2018.12.076

Wu, Q., Coumoul, X., Grandjean, P., Barouki, R., & Audouze, K. (2021). Endocrine disrupting chemicals and COVID-19 relationships: A computational systems biology approach. *Environment International, 157*, 106232. https://doi.org/10.1016/j.envint.2020.106232

Wu, H., Tufa, L. T., Kwon, J., Choi, Y., & Lee, J. (2023). Facile one-pot synthesis of polyethyleneimine functionalized α-FeOOH nanoraft consisted of single-layer parallel-aligned ultrathin nanowires for efficient removal of Cr (VI): Synergy of reduction and adsorption. *Science of the Total Environment, 857*, 159446. https://doi.org/10.1016/j.scitotenv.2022.159446

Xing, J., Wang, N., Li, X., Wang, J., Taiwaikuli, M., Huang, X., Wang, T., Zhou, L., & Hao, H. (2022). Synthesis and modifications of g-C_3N_4-based materials and their applications in wastewater pollutants removal. *Journal of Environmental Chemical Engineering, 10*(6), 108782. https://doi.org/10.1016/j.jece.2022.108782

Xiong, Q., & Jivkov, A. P. (2018). Anion diffusion in clay-rich sedimentary rocks – A pore network modelling. *Applied Clay Science, 161*, 374–384. https://doi.org/10.1016/j.clay.2018.05.010

Yadav, S., Ibrar, I., Al-Juboori, R. A., Singh, L., Ganbat, N., Kazwini, T., Karbassiyazdi, E., Samal, A. K., Subbiah, S., & Altaee, A. (2022). Updated review on emerging technologies for PFAS contaminated water treatment. *Chemical Engineering Research and Design, 182*, 667–700. https://doi.org/10.1016/j.cherd.2022.04.009

Yadav, A., Kumar, H., Shrama, R., & Kumari, R. (2023). Synthesis, processing, and applications of 2D (nano)materials: A sustainable approach. *Surfaces and Interfaces, 39*, 102925. https://doi.org/10.1016/j.surfin.2023.102925

Yang, W., Shi, X., Wang, J., Chen, W., Zhang, L., Zhang, W., Zhang, X., & Lu, J. (2019). Fabrication of a novel bifunctional nanocomposite with improved selectivity for simultaneous nitrate and phosphate removal from water. *ACS Applied Materials and Interfaces, 11*(38), 35277–35285. https://doi.org/10.1021/acsami.9b08826

Yang, Y., Zheng, Z., Ji, W., Xu, J., & Zhang, X. (2020). Insights to perfluorooctanoic acid adsorption micro-mechanism over Fe-based metal organic frameworks: Combining computational calculation with response surface methodology. *Journal of Hazardous Materials, 395*, 122686. https://doi.org/10.1016/j.jhazmat.2020.122686

Yin, X., Teng, A., Zeng, Z., Meng, H., & Wu, W. (2022). Facile and novel preparation of 2D-MoS2/carbon nanodots composite with high adsorption efficiency and capacity for dye removal. *Chemical Engineering Journal, 450*, 138319. https://doi.org/10.1016/j.cej.2022.138319

Yu, S., Wei, D., Shi, L., Ai, Y., Zhang, P., & Wang, X. (2019). Three-dimensional graphene/titanium dioxide composite for enhanced U(VI) capture: Insights from batch experiments, XPS spectroscopy and DFT calculation. *Environmental Pollution, 251*, 975–983. https://doi.org/10.1016/j.envpol.2019.04.127

Zhang, Y., Wang, L., Zhang, N., & Zhou, Z. (2018). Adsorptive environmental applications of MXene nanomaterials: A review. *RSC Advances, 8*(36), 19895–19905. https://doi.org/10.1039/c8ra03077d

Zhang, P., Wang, L., Yuan, L. Y., Lan, J. H., Chai, Z. F., & Shi, W. Q. (2019). Sorption of Eu(III) on MXene-derived titanate structures: The effect of nano-confined space. *Chemical Engineering Journal, 370*, 1200–1209. https://doi.org/10.1016/j.cej.2019.03.286

Zhang, S., Liao, S., Qi, F., Liu, R., Xiao, T., Hu, J., Li, K., Wang, R., & Min, Y. (2020). Direct deposition of two-dimensional MXene nanosheets on commercially available filter for fast and efficient dye removal. *Journal of Hazardous Materials, 384*, 121367. https://doi.org/10.1016/j.jhazmat.2019.121367

Zhang, P., He, M., Teng, W., Li, F., Qiu, X., Li, K., & Wang, H. (2023). Ordered mesoporous materials for water pollution treatment: Adsorption and catalysis. *Green Energy & Environment*. https://doi.org/10.1016/j.gee.2023.11.001

Zhou, J., Lin, J., Huang, X., Zhou, Y., Chen, Y., Xia, J., Wang, H., Xie, Y., Yu, H., Lei, J., Wu, D., Liu, F., Fu, Q., Zeng, Q., Hsu, C. H., Yang, C., Lu, L., Yu, T., Shen, Z., et al. (2018). A library of atomically thin metal chalcogenides. *Nature, 556*(7701), 355–359. https://doi.org/10.1038/s41586-018-0008-3

Zhou, H., Tan, J., Yang, L., Wang, J., Ding, B., Pan, Y., Yu, X., Liu, M., Yang, C., Qiu, L., Cheng, H. M., & Liu, B. (2021). Independent thickness and lateral size sorting of two-dimensional materials. *Science China Materials, 64*(11), 2739–2746. https://doi.org/10.1007/s40843-021-1690-2

Unveiling the Power of Nanomaterials in the Area of Forensics

Iqra Zareef, Ahsan Riaz, Salahuddin, Muqaddas Fatima, Amna Anwar, and Anam Munawar

1 Introduction

Concept of nanotechnology and the nanoparticles was first discussed by the Feynman in 1959, in his famous lecture 'there is a plenty of room at the bottom' (Goddard III et al., 2002). The concept of nanotechnology and its importance was known for a long period of time but the further studies were hindered due to lack of technology needed by scientists to study structures of these particles. Later on the invention of electron microscopy (TEM and SEM) has led to further developments in the nanoscience (Subedi, 2014). Nanomaterials basically represent a distinct class of particles having at least one dimension lesser than 100 nm (Verhagen & Kelarakis, 2020).

1.1 Properties Imparted by Nanomaterials

The physiological and chemical properties of material at nanoscale differ from that of bulk state due to high surface to area ratio which makes them excellent catalytic agents. Nanoparticles can possibly possess quantum effects. This quantum confinement engenders an alteration in energy band structures and emission spectra, therefore nanoparticles find application in sectors such as LED displays and advanced

I. Zareef · A. Riaz · Salahuddin · A. Anwar · A. Munawar (✉)
Univeresity of Health Sciences, Lahore, Punjab, Pakistan
e-mail: anam.munawar@uhs.edu.pk

M. Fatima
Government College University Faisalabad, Faisalabad, Punjab, Pakistan

S. A. Bansal et al. (eds.), *Emerging Applications of Novel Nanoparticles*, Lecture Notes in Nanoscale Science and Technology 37, https://doi.org/10.1007/978-3-031-57843-4_2

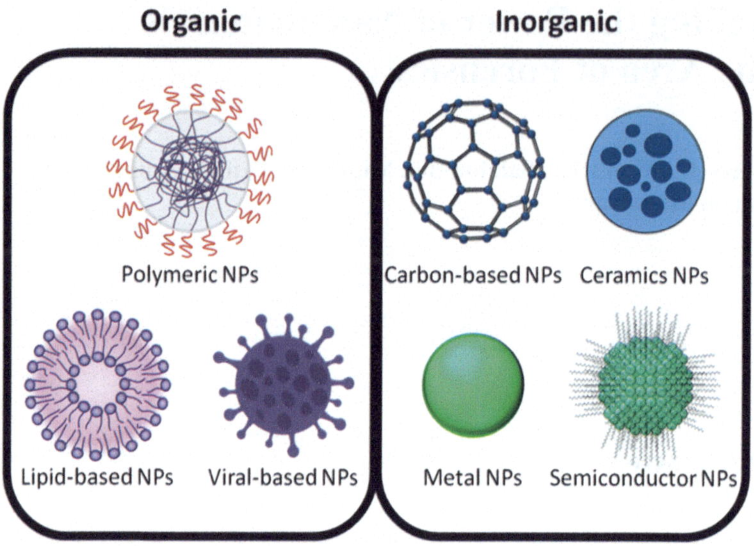

Fig. 1 Different types of nanoparticles (Cheng et al., 2022)

bio-imaging. Nanoparticles also possess a distinctive feature of self-assembly and can make unique and intricate structures. This property is the basis for development of nano-devices (Ibrahim Khan & Khan, 2019; Guisbiers et al., 2012). Mechanical properties of nanomaterials are often characterized by unparalleled strength and flexibility. Carbon nanotubes and grapheme are the perfect examples for the fabrication of lightweight but robust materials. Similarly the optical properties of nanoparticles are harnessed efficaciously in applications ranging from vivid LED displays to intricate biological imaging modalities. Biocompatibility is also another yet fascinating property of nanoparticles so they hold potential for applications in diverse medical contexts, encompassing drug delivery with heightened precision and imaging agents for enhanced diagnostic modalities. All these properties and many other make nanomaterials exert their potential in various industries, including food science, agriculture, marine science, environmental chemistry, veterinary medicine, medicine detection, and forensics (Singh et al., 2022). Figure 1 is depicting the different types of nanoparticles.

1.2 Challenges of Forensic Sciences

Forensic science is a multidisciplinary field that applies scientific principles to investigate crimes, analyze evidence, and provide expert opinions in legal proceedings. It encompasses various specialties, such as forensic chemistry, forensic biology, forensic toxicology, and fingerprint analysis, among others. Fingerprint analysis has traditionally relied on methods such as powdering, chemical treatments, and latent print development. However, these traditional methods have limitations in terms of accuracy, sensitivity, and efficiency. Nanotechnology has emerged as a promising solution to overcome these limitations and revolutionize the field of fingerprint analysis in forensic science (Paikrao et al., 2022). Forensics also face constrains in other areas for example telling the exact time of death, detection of explosive residue, detection of drugs to avoid Drug facilitated crimes (DFC). Advancements in nanotechnology has the power to fundamentally revolutionize the investigative processes of forensic science. Incorporation of nanotechnology in conventional forensic tools and technologies has led to accurate and efficient analysis of crime scenes.

2 Nanomaterials in Forensic Analysis: Basics

Nanomaterials are the substances that have one or more dimension less than 100 micrometer. Their size ranges from 1–100 micrometer. Discovery of nanoparticles has revolutionized various fields including medicine, material sciences and electronics. Nanoparticles have also had a profound impact in forensics. With their high sensitivity and precision, nanoparticles have analyzed complex materials, identified trace evidence and uncovered hidden traces. Incorporation of nanotechnology in forensics has made significant advancements in the three main areas of detection, analysis and imaging.

The greatest contribution of nanotechnology in the field of forensics is its ability to selectively detect the trace evidence. Functionalized nanoparticles have been engineered to selectively target specific substances such as blood, semen, saliva or drugs as well as visualize the faintest traces of evidence under alternate light source. These techniques have provided a breakthrough in crime scene investigation, making it possible to identify critical evidence that may have been overlooked before (Hallikeri et al., 2012).

Moreover, development of nanoscale sensors have facilitated real-time monitoring of various volatile compounds in crime scene investigations. These sensors can detect substances like explosives or narcotics with unparalleled accuracy, providing law enforcement with the tools needed to respond swiftly to potential threats.

2.1 *Ethical Considerations*

As with any technological advancement, the integration of nanotechnology into forensics raises ethical and legal questions. Privacy concerns emerge when discussing the use of nanoscale sensors for surveillance or tracking individuals. Moreover, issues related to evidence admissibility and potential challenges to the accuracy of nanoscale analysis must be carefully addressed to maintain the integrity of the criminal justice system. Table 1 is providing the detail about application of nanotechnology in trace chemistry.

Table 1 Nanotechnology applications in trace evidence analysis

Nanotechnology application	Target evidence	Benefits	Challenges
Nanoparticles for Blood Detection	Blood traces	Enhanced visualization of invisible stains	Selectivity for specific blood types
Quantum Dots for Latent Prints	Latent fingerprints	Improved fingerprint detection	Stability and potential toxicity
Nanoscale Imaging of Tool Marks	Tool marks on surfaces	High-resolution visualization	Sample preparation, surface complexities
Nanoparticle-Enhanced Hair Analysis	Hair fibers	Increased contrast for identifying features	Standardization, differentiation
Detection of Biological Pathogens	Biohazard substances	Rapid identification in crime scenes	Specificity, false positives
Nanosensors for Explosives Detection	Explosive residues	Real-time detection and identification	Environmental conditions, calibration
Nanomaterials for Drug Traces	Drug residues	Sensitivity in detecting trace amounts	Cross-contamination, sample handling

3 Nanotechnology-Enhanced Fingerprint Analysis

3.1 Fingerprint Analysis

Since the beginning of time, criminologists have relied on fingerprints as a distinctive kind of evidence. Latent fingerprints are technically defined as deposits produced by a combination of sweating and the actions of a number of sebaceous glands in the skin. These deposits are essentially invisible and must be revealed using certain optical techniques, development, and powders. It has long been proven that fingerprints can be used to identify people. Since they respond under certain situations, and may be created using a variety of colored materials, such as carbon soot on a light backdrop or aluminum flake on a dark background. However, these products have significant flaws including fuzzy fingerprint pictures. Materials based on nanotechnology are used to tackle this issue.

3.2 Nanomaterials for Fingerprint Analysis

The use of nanomaterials for fingerprint analysis was first used in 1970. In 1980, latent fingerprints on the surface of porous paper were frequently seen using gold and silver nanoparticles. However, over the past 20 years, a number of other nanoparticles have gained popularity for the detection of latent fingerprints on porous and non-porous surfaces, including silicon dioxide, fluorescent starch-based carbon, aluminum oxide, Eu + 3-doped Al2O3, and CdS (a photoluminescent nanocrystal capped with dioctysulfo-succinate) (Prasad et al., 2020).

Research findings have demonstrated the benefits of zinc oxide nanoparticles in the domain of fingerprint analysis. These nanoparticles possess inherent fluorescence, contributing to the augmentation of fingerprint visibility and clarity. Beyond their luminescent qualities, ZnO nanoparticles can function well under wet conditions, a trait that conventional nanomaterials find challenging (Kaushik et al., 2017).

Further innovation is observed through the integration of calcium selenide nanoparticles, whose minute size, often as diminutive as 10 nm, is matched by their inherent fluorescence upon exposure to ultraviolet (UV) light. The synergy between these nanoparticles and UV light lends itself to enhanced fingerprint visualization, translating to heightened accuracy and detail. The integration of calcium selenide nanoparticles effectively elevates the quality of fingerprint images, contributing to advancements in the precision of forensic identification.

3.3 Techniques for Fingerprint Detection

The identification and analysis of sweat and oil residues in fingerprints are of paramount significance in the realm of forensic science, providing critical insights into an individual's interactions with surfaces and objects. Nanotechnology has provided a transformative platform in this area by offering sensitive and selective techniques which are advantages over conventional techniques.

3.3.1 Nanoscale Imaging Techniques

Nanoscale imaging methods, for example Scanning Electron Microscopy (SEM) and Atomic Force Microscopy (AFM), have surpassed the limitations inherent to optical microscopy. SEM provides intricate high-resolution images, revealing the subtle details of fingerprint patterns and residue distribution. Conversely, AFM furnishes a tactile examination, enabling the mapping of surface topography influenced by sweat and oil deposits. These techniques provide a heightened understanding of residue spatial distribution and morphology, facilitating enhanced identification and differentiation of individuals.

3.3.2 Fluorescent Nanoparticles

The design of nanoparticles with inherent fluorescence and selective affinity for sweat and oil molecules has been advancement in forensics. These nanoparticles exhibit precise molecular bonding with target substances, resulting in distinctive fluorescence emissions upon interaction. The utility of fluorescent nanoparticles allows for the precise localization and visualization of sweat and oil residues, even within intricate fingerprint patterns. The pivotal advantage of fluorescence lies in its capacity to discriminate between distinct compounds, thus facilitating accurate

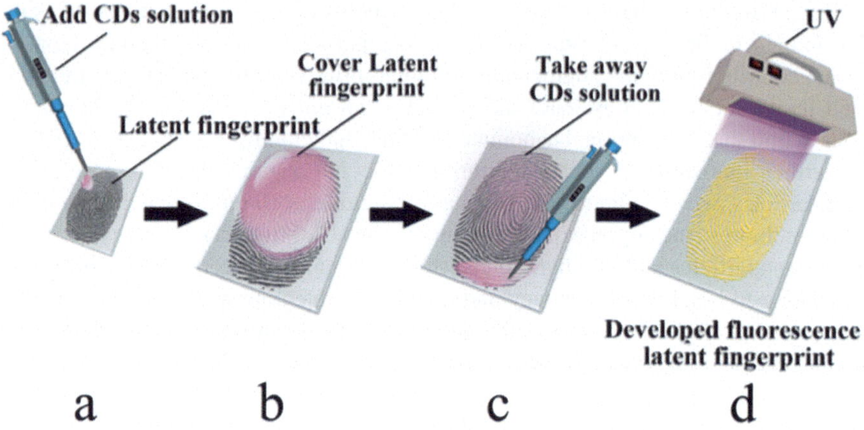

Fig. 2 The fingerprint development process (**a**) C dot solution is added to the surface. (**b**) Solution removed by pipette (**c**) visualize the surface under UV light (**d**) develop the reference image. (Verhagen & Kelarakis, 2020)

identification (Lodha et al., 2016). Figure 2 is providing the application of fluorescent nanoparticles for the development of finger prints.

3.3.3 Nanoscale Chemical Analysis

The integration of nanoscale chemical an analysis technique, including Raman spectroscopy and Mass Spectrometry (MS), has engendered a paradigm shift in sweat and oil residue identification. Raman spectroscopy elucidates molecular information by measuring vibrational energy levels in the fingerprint. It offers a non-invasive analysis approach, unveiling the chemical composition of residues. On the other hand, MS provides high-resolution mass analysis, thereby enabling the identification of specific molecules present in trace amounts. These methodologies collectively contribute to a holistic characterization of residue compounds, augmenting the precision of forensic analysis.

3.3.4 Quantum Dots

Quantum dots, nanoscale semiconductor particles with exceptional optical properties, have emerged as a revolutionary tool in the realm of fingerprint analysis. These tiny yet powerful structures exhibit tunable fluorescence emissions, making them uniquely suited for enhancing the detection of sweat and oil residues within fingerprints. When coupled with fingerprint residues, quantum dots can emit distinct wavelengths of light upon interaction with target molecules, significantly elevating the sensitivity and precision of fingerprint detection. Their application in fingerprinting underscores their potential to redefine the landscape of forensic science by offering unparalleled capabilities in residue visualization and identification.

4 Nanoparticles in Trace Evidence Analysis

Trace evidence analysis a critical analysis in forensic investigations and involves the analysis of small and unnoticeable particles that are transferred during the commission of a crime. Due to this, nanoparticles have gained significant attention in their application in the field of forensics because of their unique physical and chemical properties. In this section, we will discuss the application of nanoparticles in trace evidence analysis and highlight their contributions to forensic science.

4.1 Nanomaterials for Trace Evidence Detection Including Hair, Fiber and Paint

Among several nanoparticles, zinc oxide (ZnO) nanoparticles have shown promising results in forensic investigations in locating anonymous evidence because of their high sensitivity and precision (Malik & Hassan, 2022). Recent studies about analysis and identification of hair trace evidence have been focused on identifying their chemical modifications. Several studies have utilized nanoparticles to preconcentrate the extracted hair dye and detect with the help of a composite electrode (Boll et al., 2017; Correa et al., 2016). In recent years, majority of research has explored new ways to improve the existing techniques for hair analysis by application of nanoparticles in these techniques. Much of these techniques include the characterization of ZnO nanoparticles using scanning electron microscopy (SEM), transmission electron microscopy (TEM), X-ray diffraction (XRD), energy-dispersive X-ray spectroscopy (EDX), Fourier-transform infrared spectroscopy (FTIR), and Raman-IR radiation (Correa et al., 2016). Analysis of fibers and paints and the identifications of the dye used in them are traditionally done by comparison microscopy. However, use of nanoparticles to supplement this comparison analysis by utilizing chemometrics and databases has greatly helped with the examination. The newest technique that is being employed for this purpose is the spatially offset Raman spectroscopy (SORS) (Mistek et al., 2019). Utilization of nanoparticles to aid in the characterization of these traditional techniques allows for the discovery and gathering of crucial data that was previously difficult to obtain (Malik & Hassan, 2022).

4.2 Nanomaterials for Trace Illicit Drugs

Magnetic nanoparticles functionalized with KH570 and Toluene with their specific density of 23 mg/ml are used to detect morphine in the human hair samples. These surface-modified magnetic nanoparticles are characterized for the drug detection on the basis of rapid and sensitive method of magnetic solid-phase extraction with diode array detection (Boojaria et al., 2015). Commercially available magnetic nanoparticles functionalized with the help of certain amino acids (hapten-BSA conjugate) by strobers method are used for the detection of illicit drugs in a rapid lateral flow immunoassay without sample pretreatment. It provides both single and multiplex assays at a point-of-need analysis of sample (Guteneva et al., 2019). Quantum dots of carbon, graphene, and silicon are being utilized for screening and sensing illicit drugs to aid in forensic investigations (Ganesan & Nagaraaj, 2020). Such that, graphene-based quantum dots have been reported to detect cyanide ions (Hou et al., 2013) and Mn-doped ZnS quantum dots have been reported to successfully detect cocaine in saliva (Chantada–Vázquez et al., 2018). The schematic representation of the detection of illicit drugs (morphine and fentanyl drug) by the functionalized magnetic nanoparticles on assay strip is presented in Fig. 3.

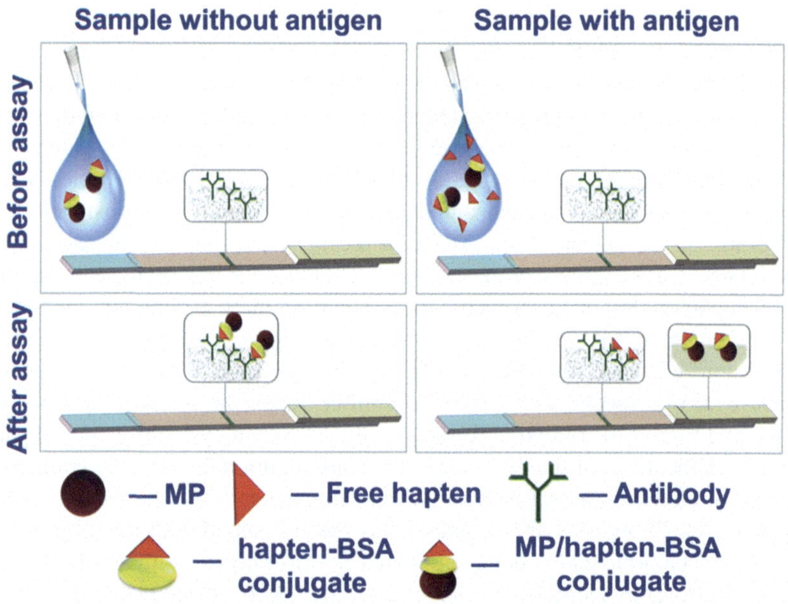

Fig. 3 Lateral flow (LF) assay. MP with covalently bound conjugate of antigen and BSA added to samples containing known concentrations of a drug of abuse. (Guteneva et al., 2019) (Copyright © 2019, Springer-Verlag GmbH Austria, part of Springer Nature. All rights reserved)

4.3 *Nanomaterials for Trace Explosives*

Magnetic nanoparticles (MNPs) functionalized into CD/Fe3O4@MIPs via a reverse micro-emulsion method and structured into spherical core-shell structure are used as a composite material to detect trinitophenol (M. Wang et al., 2019). Another technique includes modification and functionalization of magnetic nanoparticles with methacrylic acid and TNT-imprinted polymeric shell for the detection of explosives (Alizadeh, 2014). Single-crystalline semiconductor CdS nano slabs have also been used for explosives detection and functionalized TiO2 nanorods for the detection of organo-phosphorus chemical warfare agents (Tambo & Ablateye, 2020). Right now, highly sensitive, rapid and low cost methods for the trace explosive detection are required to face the increasingly severe security situation and the increased use of explosive materials. Several nanoparticles with their limit of detection (LOD) for their targeted explosive material as well as specific material used for their functionalization to increase their sensitivity are given in Table 2.

4.4 *Nanomaterials for GSR Analysis*

Gold nanoparticles have been used to detect barium (Ba2+) in Gunshot Residue (GSR) analysis. A study published in 2021 developed a hand-held device for such detection by using gold nanoparticles capped sodium malonate for rapid detection of barium found in the gunshot residue. The results demonstrated high sensitivity and selectivity for detection of barium and were shown to be unaffected by interfering agents such as As3+, Sb3+, Cu2+, Pb2+, Hg2+, Ca2+, Cr3+, Mg2+ (Shrivastava et al., 2021).

5 Nanomaterials in DNA Profiling

Different types of nanomaterials including nanochips, nano-biosensors and nanoprobes are being utilized for the detection of DNA. Traditional techniques like PCR are being incorporated with silver nanomaterials (AuNPs) to enhance their levels of sensitivity and selectivity, resulting in nanomaterial-assisted PCR (nano-PCR) (Pan et al., 2012). Magnetic nanoparticles, silica-based magnetic nanoparticles, and copper nanoparticles have been used for the extraction of DNA from biological evidence as well as gold nanoparticle protein nanopores have been used for the detection of single-stranded DNA (Tambo & Ablateye, 2020). SiNPs have been used for the DNA detection and analysis similar to silver nanoparticles because of their unique geometric properties and higher stability (Wang & Liu, 2007). Dendrimers, nanowires, and quantum dots have also been developed and utilized for DNA detection (Rawtani et al., 2019).

Table 2 Summary of several functionalized nanoparticles for Trace Explosives Detection

Nanoparticles	Functionalization	Targeted explosive material	Limit of Detection (LOD)	References
AgNPs/AuNRs	4-ATP (4-aminothiophenol)	RDX (cyclotrimethylenetrinitramine) HMX (octogen)	0.39 mg/L 0.61 mg/L	Guven et al. (2017)
AuTNPs colloids	mono-6-thio-β-cyclodextrin	DNT (2, 4-dinitro-toluen)	10–11 M	Xu et al. (2011)
AgNSs colloids	DDTC	PA (2, 4, 6-trinitrophenol)	8.96 ng/cm2	Chen et al. (2016)
AuNPs colloids	Cysteine	TNT (2, 4, 6-trinitrotoluene)	2 × 10–12 M	Dasary et al. (2009)
Ag-ZnO NFs	4-ATP	TNT (2, 4, 6-trinitrotoluene)	10–9 M	He et al. (2014)
AuNPs colloids	Cysteamine	TNT (2, 4, 6-trinitrotoluene)	22.7 μg/L	Jamil et al. (2015)
AuNPs/TiO2 NR	Not documented	TNT (2, 4, 6-trinitrotoluene)	10–7 M	Samransuksamer et al. (2018)
PC/AgNSs hybrid nanofiber membrane	4-ATP	TNT (2, 4, 6-trinitrotoluene)	2.05 × 10–13 M	Li et al. (2017)
polydopamine@Au nanowaxberry	polydopamin	DNT (2, 4-dinitro-toluen)	1 pM	Chen et al. (2018)
AgNSs@silicon	PABT	TNT (2, 4, 6-trinitrotoluene)	1pM	Chen et al. (2017)
ZnO-Ag NRs	4-ATP	TNT (2, 4, 6-trinitrotoluene)	5 × 10–9 M	He et al. (2015)

5.1 Carbon Black Nanomaterials

Carbon Black has recently piqued the interest of researchers, which differs from carbon nanotubes and nanopores because of its exploitative functionalization (Arduini et al., 2020). Carbon black nanomaterials are used in biosensors to detect various biological materials, including DNA, enzymes, and antibodies (Aydın et al., 2018).

5.2 Optimization of Analytical Techniques

Traditional techniques that are used for the detection and analysis of DNA including the colorimetric techniques are now being optimized and incorporated with nanoparticles to increase their sensitivity and selectivity. For such purpose, AuNPs, and AgNPs have been used to functionalize Halloysite Nanotubes (HNT) for the detection of DNA (Vikrant et al., 2019).

6 Nano-scale Drug Analysis in Toxicology

Nanotechnology has had a significant impact on various fields, including drug identification and toxicology analysis. Its unique properties at the nanoscale have enabled the development of innovative techniques and tools that enhance the sensitivity, specificity, and efficiency of drug identification and toxicology analysis (Kumar et al., 2019). Figure 4 is representing the applications of nanosensors in forensics.

Here's how nanotechnology is playing a role in these areas:

6.1 Improved Sensitivity and Selectivity

Due to their size and structure, nanomaterials have unique characteristics, such as a high surface area-to-volume ratio, which can be used to improve sensitivity and selectivity in toxicity investigation and drug identification. To increase the precision of detection, functionalized nanoparticles can be created to attach selectively to target molecules (Kesarwani et al., 2020).

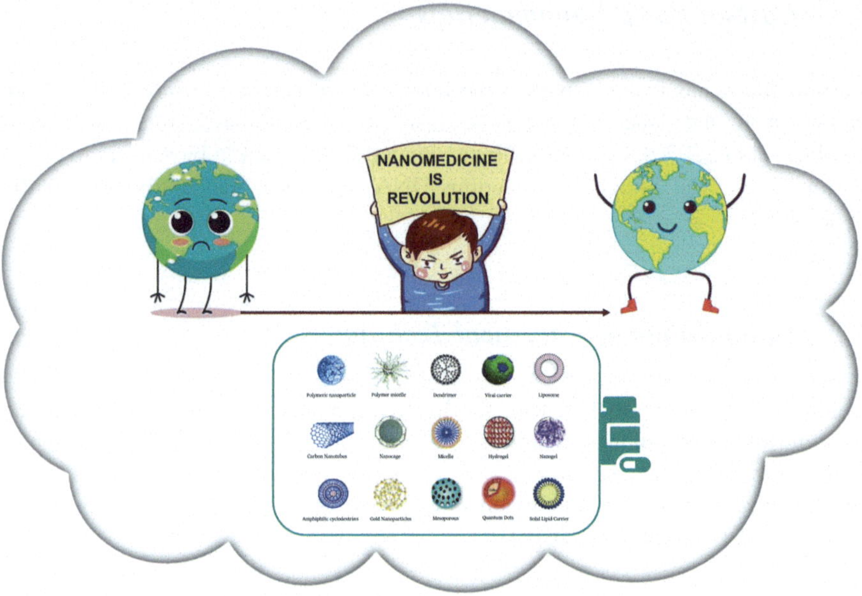

Fig. 4 Schematic representation demonstrating the importance of nanomaterials in drug identification and toxicological analysis, copied with permission from Ref. (Joseph et al., 2023) (Copyright 2023, Nanomaterials)

6.2 *Nanoparticle-Based Sensors*

Highly sensitive sensors for the detection of drugs and dangerous chemicals have been made possible thanks to nanotechnology. With the use of particular ligands that attach to target molecules, nanoparticles can be functionalized, such as gold or quantum dot nanoparticles. When these nanoparticles engage with the target, they can then emit detectable signals like fluorescence or colour changes, enabling quick and accurate identification (D'Aurelio et al., 2020).

6.3 *Lab-on-a-Chip Devices*

The creation of lab-on-a-chip devices that consolidate numerous analytical procedures onto a single chip at a smaller scale has been made possible by nanotechnology. These tools can carry out numerous processes of toxicological and drug identification analysis, such as sample preparation, separation, and detection, on a compact, portable platform (Kapoor et al., 2020).

6.4 *Drug Delivery Systems*

Nanotechnology has transformed drug delivery by creating nanoparticles that can encapsulate pharmaceuticals and deliver them to specific locations in the body. These nanoparticles have the potential to improve drug efficacy, reduce side effects, and promote targeted therapy. They also contribute to toxicological analysis by helping researchers analyze the distribution and effects of toxins in biological systems (Kheraldine et al., 2021).

6.5 *Biosensors and Biomarkers*

Nanotechnology has made it possible to create biosensors that can recognize particular biomarkers connected to dangerous outcomes or pharmaceutical responses. These biosensors allow for the immediate monitoring of cellular responses as well as the early detection of harmful effects (Pant et al., 2021).

7 Nanomaterials in Ballistics and Firearms Analysis

Nanomaterials have the potential to significantly impact various fields, including ballistics and firearms analysis. Figure 5 is providing the details on the application of nanomaterials for the bullet resistance and to absorb the energy.

Here's how nanomaterials could be applied in these areas:

7.1 *Improved Bullet Design*

By enhancing the aerodynamics of bullets, raising their velocities, and maintaining constant trajectories, nanomaterials can improve their performance. Bullet surfaces with nano-coatings might have less friction, which would improve accuracy and extend effective ranges (Obradović et al., 2014).

7.2 *Reinforced Firearms Components*

Nanocomposites may be utilized to reinforce weapon parts, enhancing their toughness and wear resistance. This might make firearms last longer and require less frequent maintenance (Baggili & Behzadan, 2019).

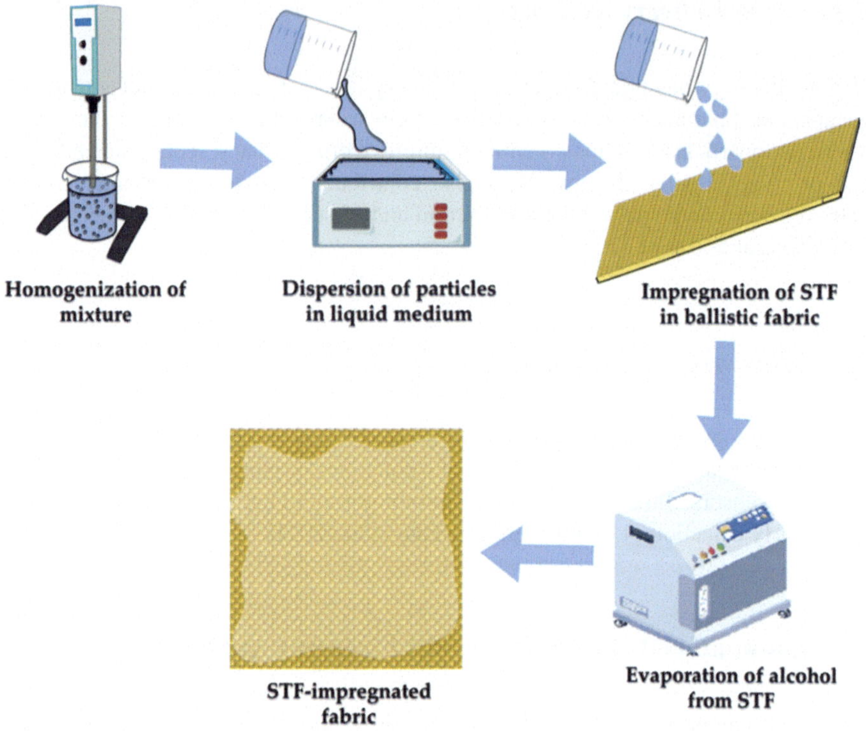

Fig. 5 Scheme of preparation of shear thickening fluid. (Reproduced with permission from Ref. (Ribeiro et al., 2022) Copyright 2022, polymers)

7.3 Enhanced Ammunition

To enhance qualities like energy transfer upon impact, nanomaterials could be used in ammunition. Depending on the intended purpose, this can lead to either more controlled penetration or more potent stopping force (Rubio-Monterde et al., 2022).

7.4 Forensic Analysis

Ballistic evidence may be subjected to more sensitive and accurate forensic processing thanks to nanotechnology. Nanoparticles enhance the visibility of gunshot residue, making it easier to recognize and analyze. Investigations into crimes might benefit from this (Paikrao et al., 2022).

7.5 Traceable Markers

Bullets and cartridges could be better traced back to their source by using markers made of nanoparticles with special chemical or physical properties. This might help with crime scene cleanup. Figure 6 is providing the schematic of ballistic test apparatus.

7.6 Smart Gun Technologies

Nanosensors that are built inside firearms may be able to offer real-time information on many factors, including performance, temperature, and pressure. This information might aid in enhancing the care and upkeep of firearms (Kumar et al., 2019).

7.7 Reduced Recoil and Noise

Nanomaterials might enable the development of suppressors that are more efficient at reducing both recoil and noise. This improves user comfort and safety (Wu et al., 2022).

7.8 Detection and Identification

Advanced explosives and illegal guns detection techniques may use nanomaterials in the future. Highly sensitive nanosensors may be able to identify minute quantities of particular compounds linked to explosives or guns (Bhatt et al., 2020).

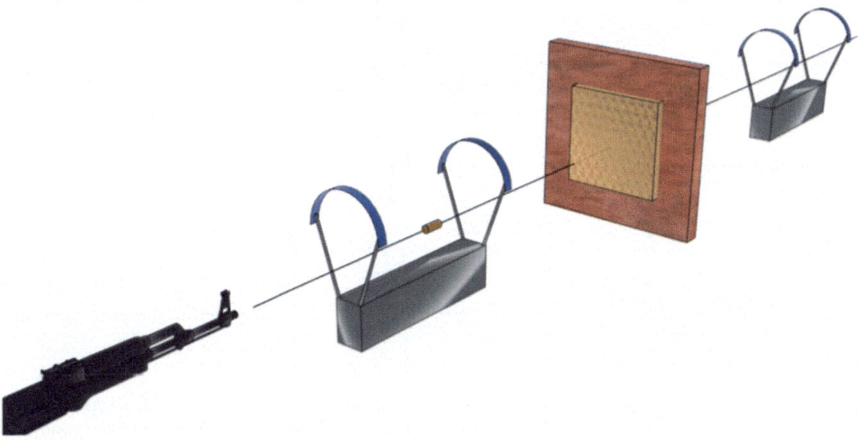

Fig. 6 Schematic figure of ballistic test apparatus. (Taken with permission from Ref. (Ribeiro et al., 2022) Copyright 2022, polymers)

7.9 Material Analysis

Nanotechnology offers the analyses of components of firearms at the nanoscale level. This might entail looking at material deterioration, wear patterns, and spotting signs of tampering (Jandt & Watts, 2020).

It's important to note that while the potential applications of nanomaterials in ballistics and firearms analysis are promising, there are also ethical, legal, and safety considerations that need to be addressed. Additionally, the integration of nanotechnology into these areas may require extensive research, testing, and regulatory oversight to ensure the responsible and safe deployment of these technologies (Chauhan et al., 2021).

8 Nanomaterials in Digital Forensics

Nanomaterials have the potential to revolutionize various fields, including digital forensics. Digital forensics involves the recovery and analysis of digital evidence for legal purposes, such as in criminal investigations or civil litigation. Nanomaterials can be integrated into this process in several ways, offering enhanced capabilities and novel techniques for data extraction, preservation, and analysis. The typical Illustration of compressed sensing based on memristor crossbar arrays for typical edge computing application is given in the Fig. 7.

Here's how nanomaterials could impact digital forensics:

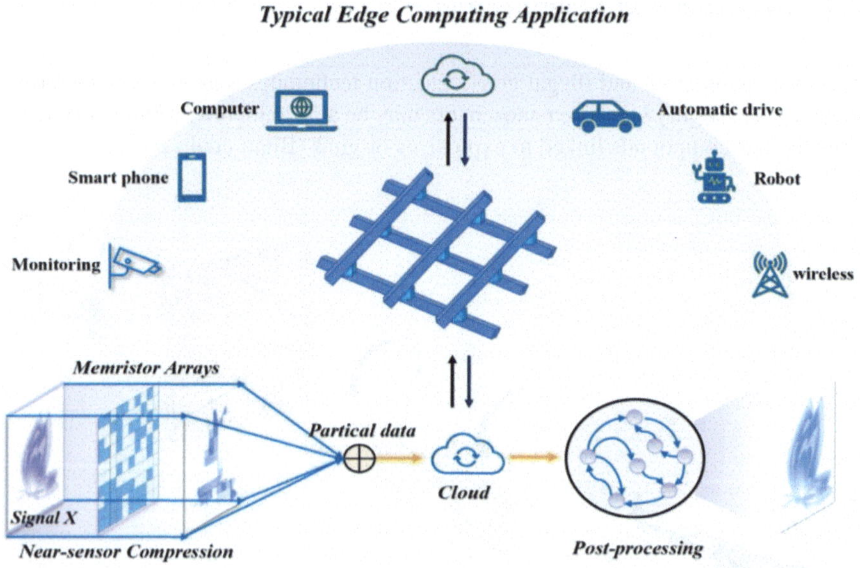

Fig. 7 Illustration of compressed sensing based on memristor crossbar arrays for typical edge computing application. (Reproduced with permission from Ref. (Wang et al., 2023) Copyright 2022, Wiley-VCH)

8.1 Data Storage and Preservation

Digital evidence can be stored on ultra-durable, high-density storage media thanks to the usage of nanomaterials. Storage devices can be shielded against environmental elements including heat, moisture, and electromagnetic radiation, which can deteriorate data over time, by using nanoparticles and nano-coatings (Atlam et al., 2020).

8.2 Nano-sized Sensors

To detect physical tampering, unexpected environmental conditions, or unauthorized access to digital devices, nanoscale sensors may be used. Devices could incorporate these sensors to track their status and spot any unauthorized efforts to compromise them (Saglam et al., 2021).

8.3 Nanoparticle-Based Imaging

High-resolution imaging techniques can be created using nanoparticles to retrieve deleted or hidden data from digital storage devices. This would make it possible to see residual magnetic or electronic signals that conventional approaches might not pick up on (Paikrao et al., 2022).

8.4 Nanotechnology-Enhanced Data Recovery

Data from broken or malfunctioning storage devices could be physically recovered using nanoscale probes or instruments. These devices may be able to extract information that would otherwise be unreachable by manipulating individual bits at the nanoscale level (Easttom, 2021).

8.5 Forensic Analysis of Nanomaterials

It might be required to add nanomaterial analysis into digital forensics as nanotechnology spreads. Understanding the origin, handling, and potential modification of evidence may be possible by detecting and analyzing nanoparticles present on equipment, such as nanoscale pollutants or coatings (Bhatt et al., 2020).

8.6 *Anti-Tampering Measures*

Anti-tampering mechanisms can be developed with the use of nanomaterials. Digital devices could become more resistant to unauthorized access or modifications by including nanoparticles that change characteristics when tampered with (Easttom, 2021).

8.7 *Authentication and Encryption*

Nanomaterials having distinctive optical, magnetic, or electrical characteristics may be the basis for future authentication and encryption technologies. These materials might allow for tamper-proof and secure communication between users' gadgets (Atlam et al., 2020).

8.8 *Forensic Nanotechnology Laboratories*

The analysis of digital evidence at the nanoscale may need specialized labs with cutting-edge nanotechnology technologies. To find hidden or manipulated data, these labs may employ procedures such as scanning probe microscopy, nanoanalytical methods, and sophisticated spectroscopy (Subramani et al., 2022).

8.9 *Challenges and Ethical Considerations*

Although nanoparticles have many potential advantages, their use in digital forensics also presents difficulties with standardization, dependability, cost, and possibly ethical issues. Furthermore, it will be essential to make sure that nanomaterial-based techniques follow all applicable laws and regulations (Yang et al., 2015).

9 Nanotechnology for Crime Scene Investigation

Crime scene have key important in forensic sciences. Crime scene is a place where a crime has been occurred. Nano forensic approaches make it very easy for law enforcement agencies to collect trace evidence from crime scene. Due to Nano forensic advancement it is easy to maintain chain of custody. Chain of custody have prime importance in forensic sciences (Houck & Siegel, 2009). Application of nanotechnology in interrogation of crime scene are discussed below.

9.1 *Integration of Nanomaterials into Crime Scene Investigation Procedures*

Nano-tags and nanowires (in open places) can be used to secure the place of crime scene. To keep the record who visited place of crime and to prevent entry of outsiders in place of crime scene. Nano-tags also can be used to preserve evidence. This help to prevent tampering or mishandling of evidence and ensure its integrity throughout the investigation process. Nano-tags can also be used to track the chain of custody of evidence (Easttom, 2021).

Nanocoatings could be used to preserve evidence from degradation through different environmental factors. Nanocoatings aids to maintain the integrity of the evidence for longer periods, aiding in ongoing investigations or future reexaminations. Nano fluorescent sprays are used to enhance the surface of evidence for example analysis of blood stain, shoe prints and other body fluids which are not visible to naked eyes (Dube et al., 2020). These nanoparticles could fluoresce under specific wavelengths of light, making it easier for investigators to locate and document evidence. Nanoparticles in form of sprays and powders applied for fingerprints collection Nanotechnology enable the creation of highly detailed and unique identifiers that go beyond traditional fingerprint analysis (Prabakaran & Pillay, 2021). Nanoscale patterns on surfaces could be used as evidence, providing an even more accurate way to link individuals to crime scenes.

Nanoparticles can be remold according to require unique properties. According to properties of nanoparticles it can be used to collect different chemicals from crime scene which are present in trace amount. There are special nanosensors for collection of trace amount of chemicals. For example nanoparticle of different properties are developed to collect explosive, toxins, drugs and poisons from crime scene (Rawtani et al., 2019). These sensors integrated into portable devices, allowing investigators to quickly identify potentially hazardous substances on-site. In data storage and retrieval nanotechnology offers the potential for extremely compact and durable data storage. Crime scene data, including images, videos, and analysis reports, could be stored in nanoscale storage devices, ensuring long-term preservation and easy retrieval (Easttom, 2021).

9.2 *Potential of Nanosensors in Forensic Sciences for Evidence Detection, Collection and Analysis*

Using nanomaterials to develop sensors for detecting hidden evidence at crime scenes could significantly enhance forensic investigations. Nanotechnology-enabled sensors could provide sensitive and selective detection of trace substances, aiding law enforcement agencies in uncovering concealed evidence. In following paragraph some possible application of nanosensors are discussed (Chauhan et al., 2017).

9.2.1 Bodily Fluids and DNA

Nanoscale immunosensor can identify and locate bodily fluids like blood, saliva, or sweat, even in cases where these fluids are not immediately visible. Additionally, Nano biosensors could detect DNA fragments, providing a tool to identify potential suspects (Dube et al., 2020).

9.2.2 Biological Threat Detection

Nanosensors could detect biological agents like bacteria, viruses, or toxins. By functionalizing nanomaterials with biomolecules that interact specifically with these agents, the sensors could provide early warning of potential biohazards (Mulhall, 2010).

9.2.3 Chemical Detection

Metallic and electrochemical sensors are design to detect specific chemicals associated with crime scenes, such as explosives, or accelerants. Nanomaterials with specific properties, such as enhanced reactivity or selectivity, could enhance the sensor's sensitivity (Khanna, 2008).

9.2.4 Gunshot Residue

Nanomaterial-based sensors can identify gunshot residue particles that are not easily visible to the naked eye. These sensors could differentiate between gunshot residue and other particles commonly found in urban environments (Promsuwan et al., 2020).

9.2.5 Toxic Chemicals

Sensors equipped with nanomaterials can detect toxic substances like poisons or chemical agents. This capability could assist in cases involving intentional poisoning or hazardous materials (Buzea et al., 2007).

9.2.6 Drugs and Narcotics

Functionalized nanoparticles engineered to target illicit drugs and narcotics are incorporated in nanosensors to detect trace amounts of these substances even if they are in very small quantity (Kumar et al., 2018).

9.2.7 Covert Surveillance

Nanoparticles incorporated nanosensors can also be utilized in covert surveillance equipment to monitor specific locations which can greatly help the law enforcement gather information greatly.

9.2.8 Remote Sensing and Mapping

Nanosensors can also be used in various locations for monitoring of specific substances in real-time. The collected data can be used to create a "chemical map" of that area which can aid the investigators in understanding the movement of specific substances (Caygill et al., 2012).

9.2.9 Integration with Drones and Robotics

Nanosensors can also be integrated in drones and robotic systems for crime scene investigations for remote or autonomous crime scene analysis. These sensors can be used to scan large-scale areas for hidden evidence without having any direct human intervention.

9.3 *Nanoscale Imaging Techniques for Mapping Crime Scenes*

Nanoscale imaging techniques incorporated with functionalized nanoparticles have the potential to accurately provide a detailed map of crime scenes which can help crime scene investigators in gathering evidence. Some of the nanoscale imaging techniques have been described below:

9.3.1 Scanning Electron Microscopy (SEM)

Scanning Electron Microscope (SEM) is an imaging technique that utilizes focused beams of electrons to scan a surface. It reveals the fine details of surfaces, their textures and structure in the form of a high-resolution image. SEM technique can be used to analyze various types of evidence including fibers, hairs, gunshot residue particles, paint fragments and tool marks (Rawtani & Hussain, 2023).

9.3.2 Atomic Force Microscopy (AFM)

The working principle of Atomic Force Microcopy is that it works by scanning a sharp probe over the given sample surface and measures the forces between the probe and the sample. It is able to provide a three-dimensional image with the resolution pixels of nanometers. Because of this, it can be used to analyze intricate surface of fabric weaves, textured fingerprints or various tool marks (Yadavalli & Ehrhardt, 2021).

9.3.3 Field Scanning Optical Microscopy (NSOM or SNOM)

NSOM/SNOM is an optical imaging technique and is used in forensic investigations to analyze the detailed features of samples and evidences that are collected in trace amounts. It provides a high resolution image down to a nanoscale. This technique can be used to analyze surface textures of chemical compounds or fluorescent markers (Ganesh, 2016).

9.3.4 Super-Resolution Optical Microscopy

This technique combines stimulated emission depletion (STED) microscopy and structured illumination microscopy (SIM) which provides high resolution images that are not possible using the conventional optical microscopy. These methods can be used to analyze the detailed patterns of trace evidence and fluorescent dyes at the nanoscale (Schmidt et al., 2012).

9.3.5 Confocal Microscopy

Confocal microscopy provides a detailed 3D maps of various surfaces. It is useful in forensic investigations by analyzing various complex layers and surfaces of multilayer costing of paints and fabrics.

9.3.6 Raman Microscopy

This technique provides detailed information about the chemical compositions of different compounds and materials. Therefore, raman microscopy is useful in analyzing the composition of different compounds that are collected as evidence from crime scenes (Doty & Lednev, 2018).

9.3.7 Electron Backscatter Diffraction (EBSD)

Electron backscatter diffraction is used to analyze the crystalline structures of different compounds. This technique can be used to inspect the crystalline nature of different materials like metals, ceramics or other trace evidence collected from the crime scenes (Fitzpatrick et al., 2009).

9.3.8 Infrared Microscopy

Infrared microscopy is used to analyze the molecular composition of various organic materials such as drugs or explosives. Such materials are collected from crime scenes in trace amounts and require nanoscale analysis of their chemical makeup. When these nanoscale techniques are used to analyze the trace evidences obtained from crime scenes, it is important to maintain the integrity of these evidences so that correct results can be obtained. Additionally, the integration of these techniques into the standard investigative process requires specialized training and equipment, as well as careful consideration of the ethical and legal implications of the data collected (Jales et al., 2021).

10 Challenges and Ethical Considerations

Use of nanotechnology have great potential and application in Nano forensic. However there are certain challenges, limitation and ethical consideration which are discussed here. Nanomaterial are highly sensitive and require certain environmental conditions for sample processing so vigorous environmental condition are required for results. Special instrument are required such as atomic microscopy, scanning electron microscopy etc. which are not present in all forensic labs sample only could be run in high tech labs (van Asten, 2014). One concern is the potential for privacy violations if nanosensors are used to collect information about individuals without their knowledge or consent. Another concern is the possibility of false positives or false negatives if the technology is not properly calibrated or if the evidence is mishandled. In addition, there may be concerns about the long-term effects of exposure to nanoparticles on investigators and other personnel who work with them. Some nanomaterial are toxic to human health and could enter into human body through skin this should be addressed with special protocols to handle. It is important to ensure that appropriate safety measures are in place and that any potential risks are carefully evaluated before using nanotechnology in crime scene investigation (Constantino et al., 2014).

It is important to keep in mind the potential impact of nanotechnology on the right to privacy and innocent till proven guilty presumption of an individual. Another important aspect that should be addressed is the potential or unintended risks (if any) of these engineered nanomaterials on the environment when they are used for the analysis of any crime scene (Uddin, 2010). Moreover, there should be proper public awareness and education campaigns to make sure that the individuals understand the implications of these technologies. Ethical considerations should come first to ensure that utilization of nanotechnology for forensic analysis does not come at the cost of any individual rights.

11 Future Discretions and Innovations

Use of nanomaterials in forensic sciences has revolutionized the analytical platform and made the conventional techniques obsolete and time-taking in this fast-paced era. An overview of the future impact that nanomaterials will likely have on the field of forensics is given here:

Enhanced Evidence Collection Nanomaterials will be able to aid in more specific and targeted evidence collection, especially in the case of trace evidence, from areas that are hard to analyze by using conventional techniques.

Increased Sensitivity and Precision Functionalized nanoparticles with high sensitivity and selectivity will aid in detecting and analyzing trace amounts of evidence in cases that are difficult to investigate using conventional techniques.

Improved Analytical Techniques Nanomaterials have already paved the way for novel analytical techniques for characterizing evidence. They provide enhanced accuracy and sensitivity.

Rapid On-Site Analysis Nanomaterials integrated portable nanosensors will enable the crime scene investigators to perform on-site analysis of evidence samples which will ultimately accelerate the investigation process and reduce evidence contamination.

Data-Rich Investigations Combination of nanotechnology with data analytics and AI will aid the forensic investigators to get valuable insights from the data collected from previously documented investigations.

Improved Accuracy in DNA Analysis Nanotechnology integrated DNA analysis methods such as nanopore sequencing have already enabled faster and accurate identification analysis of individuals and species.

Real-Time Monitoring and Decision Support Integration of nanotechnology with AI for the purpose of forensic analysis could enable us with real-time monitoring of substances and compounds during forensic investigations.

Cross-Disciplinary Collaboration Collaboration of forensic scientists with different disciplines such as nanotechnology, legal professionals, data analytics and AI experts will certainly lead to innovative solutions.

Advancements in Evidence Packaging and Preservation Materials integrated with functionalized nanomaterials could enhance evidence packaging, ensuring sample integrity, traceability, and protection against contamination.

Global Standardization and Best Practices There is need for collaborative efforts to establish standardized protocols for nanotechnology based forensic analysis techniques.

12 Conclusion

Use of nanomaterials in forensic sciences and forensic analysis will offer fast and accurate analysis, reduction in contamination risks and evidence loss. However, the development process of nanotechnology integrated nanosensors would require careful development, testing and validation procedures to make sure that they have accurate specificity and sensitivity. The next step in employing these nano-based sensors for forensic use would be to address the ethical considerations and legal concerns in the real-world scenarios. Aside from this, there is no doubt that nanotechnology has provided a new platform to forensic analysis techniques and revolutionized the conventional techniques that are used to detect and analyze evidences such as biological materials, fingerprints, trace materials and gunshot residues. Nanotechnology has enabled the identification, visualization and differentiation of forensic evidence techniques with unprecedented accuracy. Nanotechnology has also made possible various innovative toxicity testing platforms and highly sensitive drug detection and analysis methods. As new advancements are made in nanotechnology, the potential for future innovations in the areas of forensic analysis techniques will also continue to grow. The ongoing researches and developments in the forensic DNA profiling techniques equipped with nanosensors will reshape the investigative processes.

References

Alizadeh, T. (2014). Preparation of magnetic TNT-imprinted polymer nanoparticles and their accumulation onto magnetic carbon paste electrode for TNT determination. *Biosensors and Bioelectronics, 61*, 532–540.

Arduini, F., Cinti, S., Mazzaracchio, V., Scognamiglio, V., Amine, A., & Moscone, D. (2020). Carbon black as an outstanding and affordable nanomaterial for electrochemical (bio) sensor design. *Biosensors and Bioelectronics, 156*, 112033.

van Asten, A. C. (2014). On the added value of forensic science and grand innovation challenges for the forensic community. *Science & Justice, 54*(2), 170–179.

Atlam, H. F., Alenezi, A., Alassafi, M. O., Alshdadi, A. A., & Wills, G. B. (2020). Security, cybercrime and digital forensics for IoT. In *Principles of Internet of Things (IoT) ecosystem: Insight paradigm* (pp. 551–577). Springer.

Aydın, M., Aydın, E. B., & Sezgintürk, M. K. (2018). A highly selective electrochemical immunosensor based on conductive carbon black and star PGMA polymer composite material for IL-8 biomarker detection in human serum and saliva. *Biosensors and Bioelectronics, 117*, 720–728.

Baggili, I., & Behzadan, V. (2019). Founding the domain of AI forensics. ArXiv Preprint ArXiv:1912.06497.

Bhatt, P. V., Pandey, G., Tharmavaram, M., Rawtani, D., & Mustansar Hussain, C. (2020). Nanotechnology and taggant technology in forensic science. In *Technology in forensic science: Sampling, analysis, data and regulations* (pp. 279–301). Wiley.

Boll, M. S., Doty, K. C., Wickenheiser, R., & Lednev, I. K. (2017). Differentiation of hair using ATR FT-IR spectroscopy: A statistical classification of dyed and non-dyed hairs. *Forensic Chemistry, 6*, 1–9.

Boojaria, A., Masrournia, M., Ghorbani, H., Ebrahimitalab, A., & Miandarhoie, M. (2015). Silane modified magnetic nanoparticles as a novel adsorbent for determination of morphine at trace levels in human hair samples by high-performance liquid chromatography with diode array detection. *Forensic Science, Medicine, and Pathology, 11*, 497–503.

Buzea, C., Pacheco, I. I., & Robbie, K. (2007). Nanomaterials and nanoparticles: Sources and toxicity. *Biointerphases, 2*(4), MR17–MR71.

Caygill, J. S., Davis, F., & Higson, S. P. J. (2012). Current trends in explosive detection techniques. *Talanta, 88*, 14–29.

Chantada–Vázquez, M. P., de–Becerra–Sánchez, C., Fernández–del–Río, A., Sánchez–González, J., Bermejo, A. M., Bermejo–Barrera, P., & Moreda–Piñeiro, A. (2018). Development and application of molecularly imprinted polymer – Mn-doped ZnS quantum dot fluorescent optosensing for cocaine screening in oral fluid and serum. *Talanta, 181*, 232–238. https://doi.org/10.1016/j.talanta.2018.01.017

Chauhan, V., Singh, V., & Tiwari, A. (2017). Applications of nanotechnology in forensic investigation. *The International Journal of Life-Sciences Scientific Research, 3*(3), 1047–1051.

Chauhan, I., Verma, M., & Yasir, M. (2021). Nanoethics and nanotechnology. In *Nanotechnology in medicine: Toxicity and safety* (pp. 349–372). Wiley.

Chen, J., Shi, Y., Zhang, M., & Zhan, J. (2016). Diethyldithiocarbamate (DDTC) induced formation of positively charged silver nanoparticles for rapid in situ identification of inorganic explosives by surface enhanced Raman spectroscopy. *RSC Advances, 6*(57), 51823–51829.

Chen, N., Ding, P., Shi, Y., Jin, T., Su, Y., Wang, H., & He, Y. (2017). Portable and reliable surface-enhanced Raman scattering silicon chip for signal-on detection of trace trinitrotoluene explosive in real systems. *Analytical Chemistry, 89*(9), 5072–5078.

Chen, D., Zhu, X., Huang, J., Wang, G., Zhao, Y., Chen, F., Wei, J., Song, Z., & Zhao, Y. (2018). Polydopamine@ gold nanowaxberry enabling improved SERS sensing of pesticides, pollutants, and explosives in complex samples. *Analytical Chemistry, 90*(15), 9048–9054.

Cheng, T.-M., Chu, H.-Y., Huang, H.-M., Li, Z.-L., Chen, C.-Y., Shih, Y.-J., Whang-Peng, J., Cheng, R. H., Mo, J.-K., & Lin, H.-Y. (2022). Toxicologic concerns with current medical nanoparticles. *International Journal of Molecular Sciences, 23*(14), 7597.

Constantino, R., Stewart, C., Campbell, P., Moynihan, B., Kagan, S., Daugherty, P., Thimsen, K., Tiefel, L., O'Brien, C., & Johnson, J. (2014). Evidence collection for the unconscious and unconsented patient. *Open Journal of Nursing, 4*(04), 287–295.

Correa, G. T., Tanaka, A. A., Pividori, M. I., & Zanoni, M. V. B. (2016). Use of a composite electrode modified with magnetic particles for electroanalysis of azo dye removed from dyed hair strands. *Journal of Electroanalytical Chemistry, 782*, 26–31.

D'Aurelio, R., Chianella, I., Goode, J. A., & Tothill, I. E. (2020). Molecularly imprinted nanoparticles based sensor for cocaine detection. *Biosensors, 10*(3), 22.

Dasary, S. S. R., Singh, A. K., Senapati, D., Yu, H., & Ray, P. C. (2009). Gold nanoparticle based label-free SERS probe for ultrasensitive and selective detection of trinitrotoluene. *Journal of the American Chemical Society, 131*(38), 13806–13812.

Doty, K. C., & Lednev, I. K. (2018). Raman spectroscopy for forensic purposes: Recent applications for serology and gunshot residue analysis. *TrAC Trends in Analytical Chemistry, 103*, 215–222.

Dube, S., Tharmavaram, M., Pandey, G., Rawtani, D., & Mustansar Hussain, C. (2020). Sensors for the detection of biological fluids. In *Technology in forensic science: Sampling, analysis, data and regulations* (pp. 239–258). Wiley.

Easttom, C. (2021). *Digital forensics, investigation, and response*. Jones & Bartlett Learning.

Fitzpatrick, R. W., Raven, M. D., & Forrester, S. T. (2009). A systematic approach to soil forensics: Criminal case studies involving transference from crime scene to forensic evidence. In *Criminal and environmental soil forensics* (pp. 105–127). Springer.

Ganesan, M., & Nagaraaj, P. (2020). Quantum dots as nanosensors for detection of toxics: A literature review. *Analytical Methods, 12*(35), 4254–4275. https://doi.org/10.1039/D0AY01293A

Ganesh, E. N. (2016). Application of nanotechnology in forensic science. *Environment, 1*(1).

Goddard, W. A., III, Brenner, D., Lyshevski, S. E., & Iafrate, G. J. (2002). *Handbook of nanoscience, engineering, and technology*. CRC Press.

Guisbiers, G., Mejía-Rosales, S., & Deepak, F. L. (2012). Nanomaterial properties: Size and shape dependencies. *Journal of Nanomaterials, 2012*, 20.

Guteneva, N. V., Znoyko, S. L., Orlov, A. V., Nikitin, M. P., & Nikitin, P. I. (2019). Rapid lateral flow assays based on the quantification of magnetic nanoparticle labels for multiplexed immunodetection of small molecules: Application to the determination of drugs of abuse. *Microchimica Acta, 186*, 1–9.

Guven, B., Eryilmaz, M., Üzer, A., Boyaci, I. H., Tamer, U., & Apak, R. (2017). Surface-enhanced Raman spectroscopy combined with gold nanorods for the simultaneous quantification of nitramine energetic materials. *RSC Advances, 7*(59), 37039–37047.

Hallikeri, V. R., Bai, M., & Kumar, A. G. V. (2012). Nanotechnology-the future armour of forensics: A short review. *Journal of the Scientific Society, 39*(1), 10–11.

He, X., Wang, H., Li, Z., Chen, D., & Zhang, Q. (2014). ZnO–Ag hybrids for ultrasensitive detection of trinitrotoluene by surface-enhanced Raman spectroscopy. *Physical Chemistry Chemical Physics, 16*(28), 14706–14712.

He, X., Wang, H., Li, Z., Chen, D., Liu, J., & Zhang, Q. (2015). Ultrasensitive SERS detection of trinitrotoluene through capillarity-constructed reversible hot spots based on ZnO–Ag nanorod hybrids. *Nanoscale, 7*(18), 8619–8626.

Hou, X., Zeng, F., Du, F., & Wu, S. (2013). Carbon-dot-based fluorescent turn-on sensor for selectively detecting sulfide anions in totally aqueous media and imaging inside live cells. *Nanotechnology, 24*(33), 335502.

Houck, M. M., & Siegel, J. A. (2009). *Fundamentals of forensic science*. Academic.

Ibrahim Khan, K. S., & Khan, I. (2019). Nanoparticles: Properties, applications and toxicities. *Arabian Journal of Chemistry, 12*(7), 908–931.

Jales, J. T., Barbosa, T. M., de Medeiros, J. R., de Lima, L. A. S., de Lima, K. M. G., & Gama, R. A. (2021). Infrared spectroscopy and forensic entomology: Can this union work? A literature review. *Journal of Forensic Sciences, 66*(6), 2080–2091.

Jamil, A. K. M., Izake, E. L., Sivanesan, A., & Fredericks, P. M. (2015). Rapid detection of TNT in aqueous media by selective label free surface enhanced Raman spectroscopy. *Talanta, 134*, 732–738.

Jandt, K. D., & Watts, D. C. (2020). Nanotechnology in dentistry: Present and future perspectives on dental nanomaterials. *Dental Materials, 36*(11), 1365–1378.

Joseph, T. M., Kar Mahapatra, D., Esmaeili, A., Piszczyk, Ł., Hasanin, M. S., Kattali, M., Haponiuk, J., & Thomas, S. (2023). Nanoparticles: Taking a unique position in medicine. *Nanomaterials, 13*(3), 574.

Kapoor, A., Balasubramanian, S., Muthamilselvi, P., Vaishampayan, V., & Prabhakar, S. (2020). Lab-on-a-Chip devices for water quality monitoring. In *Nanosensor technologies for environmental monitoring* (pp. 455–469). Springer.

Kaushik, M., Mahendru, S., Chaudhary, S., & Kukreti, S. (2017). DNA fingerprints: Advances in their forensic analysis using nanotechnology. *Journal of Forensic Biomechanics, 8*(1), 1–4.

Kesarwani, S., Parihar, K., Sankhla, M. S., & Kumar, R. (2020). Nano-forensic: New perspective and extensive applications in solving crimes. *Latent in Applied Nanobioscience, 10*(1), 1792–1798.

Khanna, V. K. (2008). Nanoparticle-based sensors. *Defence Science Journal, 58*(5).

Kheraldine, H., Rachid, O., Habib, A. M., Al Moustafa, A.-E., Benter, I. F., & Akhtar, S. (2021). Emerging innate biological properties of nano-drug delivery systems: A focus on PAMAM dendrimers and their clinical potential. *Advanced Drug Delivery Reviews, 178*, 113908.

Kumar, V., Kumar, P., Pournara, A., Vellingiri, K., & Kim, K.-H. (2018). Nanomaterials for the sensing of narcotics: Challenges and opportunities. *TrAC Trends in Analytical Chemistry, 106*, 84–115.

Kumar, N., Dixit, A., Kumar, N., & Dixit, A. (2019). Role of nanotechnology in futuristic warfare. In *Nanotechnology for defence applications* (pp. 301–329). Springer.

Li, Y., Lu, R., Shen, J., Han, W., Sun, X., Li, J., & Wang, L. (2017). Electrospun flexible poly (bisphenol A carbonate) nanofibers decorated with Ag nanoparticles as effective 3D SERS substrates for trace TNT detection. *Analyst, 142*(24), 4756–4764.

Lodha, A. S., Pandya, A., & Shukla, R. K. (2016). Nanotechnology: An applied and robust approach for forensic investigation. *Forensic Research & Criminology International Journal, 2*(1), 44.

Malik, A. A., & Hassan, S. M. (2022). The ZnO nanoparticles: Synthesis, characterization and applications in forensics. *International Journal for Electronic Crime Investigation, 6*(1), 8.

Mistek, E., Fikiet, M. A., Khandasammy, S. R., & Lednev, I. K. (2019). Toward Locard's exchange principle: Recent developments in forensic trace evidence analysis. *Analytical Chemistry, 91*(1), 637–654. https://doi.org/10.1021/acs.analchem.8b04704

Mulhall, D. (2010). *Our molecular future: How nanotechnology, robotics, genetics and artificial intelligence will transfor M our world*. Prometheus Books.

Obradović, V., Stojanović, D. B., Jančić-Heinemann, R., Živković, I., Radojević, V., Uskoković, P. S., & Aleksić, R. (2014). Ballistic properties of hybrid thermoplastic composites with silica nanoparticles. *Journal of Engineered Fibers and Fabrics, 9*(4), 1914. https://doi.org/10.1177/155892501400900412

Paikrao, H. M., Tajane, D. S., Patil, A. S., & Dipale, A. D. (2022). Applications of nanotechnology in forensic science. In *Engineered nanomaterials for innovative therapies and biomedicine* (pp. 257–276). Springer.

Pan, D., Mi, L., Huang, Q., Hu, J., & Fan, C. (2012). Genetic analysis with nanoPCR. *Integrative Biology, 4*(10), 1155–1163. https://doi.org/10.1039/c2ib20076g

Pant, A., Mackraj, I., & Govender, T. (2021). Advances in sepsis diagnosis and management: A paradigm shift towards nanotechnology. *Journal of Biomedical Science, 28*(1), 1–30.

Prabakaran, E., & Pillay, K. (2021). Nanomaterials for latent fingerprint detection: A review. *Journal of Materials Research and Technology, 12*, 1856–1885.

Prasad, V., Lukose, S., Agarwal, P., & Prasad, L. (2020). Role of nanomaterials for forensic investigation and latent fingerprinting—A review. *Journal of Forensic Sciences, 65*(1), 26–36. https://doi.org/10.1111/1556-4029.14172

Promsuwan, K., Kanatharana, P., Thavarungkul, P., & Limbut, W. (2020). Nitrite amperometric sensor for gunshot residue screening. *Electrochimica Acta, 331*, 135309.

Rawtani, D., & Hussain, C. M. (2023). *Modern forensic tools and devices: Trends in criminal investigation*. Wiley.

Rawtani, D., Tharmavaram, M., Pandey, G., & Hussain, C. M. (2019). Functionalized nanomaterial for forensic sample analysis. *TrAC Trends in Analytical Chemistry, 120*, 115661. https://doi.org/10.1016/j.trac.2019.115661

Ribeiro, M. P., da Silveira, P. H. P. M., de Oliveira Braga, F., & Monteiro, S. N. (2022). Fabric impregnation with shear thickening fluid for ballistic Armor polymer composites: An updated overview. *Polymers, 14*(20), 4357.

Rubio-Monterde, A., Quesada-González, D., & Merkoçi, A. (2022). Toward integrated molecular lateral flow diagnostic tests using advanced micro-and nanotechnology. *Analytical Chemistry, 95*(1), 468–489.

Saglam, N., Korkusuz, F., & Prasad, R. (2021). *Nanotechnology applications in health and environmental sciences*. Springer.

Samransuksamer, B., Horprathum, M., Jutarosaga, T., Kopwitthaya, A., Limwichean, S., Nuntawong, N., Chananonnawathorn, C., Patthanasettakul, V., Muthitamongkol, P., & Treetong, A. (2018). Facile method for decorations of Au nanoparticles on TiO2 nanorod arrays toward high-performance recyclable SERS substrates. *Sensors and Actuators B: Chemical, 277*, 102–113.

Schmidt, D. A., Kopf, I., & Bruendermann, E. (2012). A matter of scale: From far-field microscopy to near-field nanoscopy. *Laser & Photonics Reviews, 6*(3), 296–332.

Shrivastava, P., Singh, B. P., Jain, S. K., Jain, V. K., & Nagpal, S. (2021). A novel approach to detect barium in gunshot residue using a handheld device: A forensic application. *Analytical Methods, 13*(38), 4379–4389. https://doi.org/10.1039/D1AY01272J

Singh, R., Sharma, A., Saji, J., Umapathi, A., Kumar, S., & Daima, H. K. (2022). Smart nanomaterials for cancer diagnosis and treatment. *Nano Convergence, 9*(1), 21.

Subedi, S. K. (2014). An introduction to nanotechnology and its implications. *Himalayan Physics, 5*, 78–81.

Subramani, I. G., Perumal, V., Gopinath, S. C. B., Fhan, K. S., & Mohamed, N. M. (2022). Organic-inorganic hybrid nanoflower production and analytical utilization: Fundamental to cutting-edge technologies. *Critical Reviews in Analytical Chemistry, 52*(7), 1488–1510.

Tambo, F., & Ablateye, D. N. O. (2020). A review on the role of emerging revolutionary nanotechnology in forensic investigations. *Journal of Applied and Natural Science, 12*(4), 582–591.

Uddin, M. (2010). Nanotechnology: Social and ethical considerations.

Verhagen, A., & Kelarakis, A. (2020). Carbon dots for forensic applications: A critical review. *Nanomaterials, 10*(8), 1535.

Vikrant, K., Bhardwaj, N., Bhardwaj, S. K., Kim, K.-H., & Deep, A. (2019). Nanomaterials as efficient platforms for sensing DNA. *Biomaterials, 214*, 119215. https://doi.org/10.1016/j.biomaterials.2019.05.026

Wang, Y., & Liu, B. (2007). Silica nanoparticle assisted DNA assays for optical signal amplification of conjugated polymer based fluorescent sensors. *Chemical Communications, 34*, 3553–3555. https://doi.org/10.1039/B705936A

Wang, M., Fu, Q., Zhang, K., Wan, Y., Wang, L., Gao, M., Xia, Z., & Gao, D. (2019). A magnetic and carbon dot based molecularly imprinted composite for fluorometric detection of 2, 4, 6-trinitrophenol. *Microchimica Acta, 186*, 1–11.

Wang, R., Zhang, W., Wang, S., Zeng, T., Ma, X., Wang, H., & Hao, Y. (2023). Memristor-based signal processing for compressed sensing. *Nanomaterials, 13*(8), 1354.

Wu, S., Xu, Z., Hu, C., Zou, X., & He, X. (2022). Numerical simulation study of ballistic performance of Al2O3/aramid-carbon hybrid FRP laminate composite structures subject to impact loading. *Ceramics International, 48*(5), 6423–6435.

Xu, J. Y., Wang, J., Kong, L. T., Zheng, G. C., Guo, Z., & Liu, J. H. (2011). SERS detection of explosive agent by macrocyclic compound functionalized triangular gold nanoprisms. *Journal of Raman Spectroscopy, 42*(9), 1728–1735.

Yadavalli, V. K., & Ehrhardt, C. J. (2021). Atomic force microscopy as a biophysical tool for nanoscale forensic investigations. *Science & Justice, 61*(1), 1–12.

Yang, N., Chen, X., Ren, T., Zhang, P., & Yang, D. (2015). Carbon nanotube based biosensors. *Sensors and Actuators B: Chemical, 207*, 690–715. https://doi.org/10.1016/j.snb.2014.10.040

Buckypapers: Applications in Composite Materials

Fatih Turan and Mehmet Guclu

1 Introduction

Carbon nanotubes (CNTs) possess high electrical (Collins & Avouris, 2000) and thermal conductivity (Balandin, 2011) as well as superior mechanical properties (Li & Kröger, 2012) so that they have attracted widespread attention to create functional composite materials. CNT modified functional composites have been widely used in many fields such as electromagnetic interference (EMI) shielding (Hu et al., 2019b), heat resistance (Zangrossi et al., 2022), and sensor applications (Smajda et al., 2007a). The CNTs should be evenly dispersed within matrix phase (Kessler et al., 2003) or evenly coated on to fabric surface (Malekimoghadam et al., 2023) to reach the desired functional properties in composites. However, the tendency to form agglomerates due to their high surface energy makes the dispersion of CNTs in composites can be quite challenging, especially at high CNT contents (Amrollahi et al., 2009).

Densely packed carbon nanotube (CNT) thin films are called buckypapers (BPs). In BPs, CNTs are held together in the form of entangled array by van der Waals interactions at the tube–tube junctions (Chu et al., 2014) . Their flexible nature and ease of handling of BPs overcome the dispersion problem of CNTs in composites. Thanks to having outstanding properties such as low density, high electrical and thermal conductivity, BPs are used in wide range of applications such as

F. Turan
Department of Mechanical Engineering, Faculty of Engineering,
Eskisehir Technical University, Eskisehir, Turkey

M. Guclu (✉)
Department of Mechanical Engineering, Faculty of Engineering,
Istanbul University-Cerrahpaşa, Istanbul, Turkey
e-mail: mehmet.guclu@iuc.edu.tr

S. A. Bansal et al. (eds.), *Emerging Applications of Novel Nanoparticles*, Lecture Notes in Nanoscale Science and Technology 37, https://doi.org/10.1007/978-3-031-57843-4_3

membranes (Sears et al., 2010; Zhang, 2008), electrodes (Barisci et al., 2004; Che et al., 2013; Jia et al., 2012), actuators (Chen et al., 2012b; Cottinet et al., 2011; Suppiger et al., 2008), sensors (Rein et al., 2011; Zhang et al., 2015b), heat conductors (Chen et al., 2012a; Zhang et al., 2012) and for structural reinforcement in polymer composites (Falzon et al., 2015; Gou, 2006; Han et al., 2014, 2015; Wan Dalina et al., 2019; Wang et al., 2004).

Existing research has clearly indicated that the characteristics of BPs are subject to the influence of numerous parameters. These factors include the type of carbon CNTs being employed, which can be either single-walled or multi-walled (Muramatsu et al., 2005); the dimensions of the CNTs, involving both their length and diameter (Chen et al., 2012a); the level of purity exhibited by the CNTs (Roy et al., 2012; Suppiger et al., 2008; Vohrer et al., 2004); chemical functionalization of the CNTs (Sreekumar & Thomas, 2008); and the degree of alignment achieved (Zhang et al., 2012).In general, CNTs with smaller diameters and higher purity results in higher tensile strength in BPs. Conversely, the utilization of CNTs with larger diameters tends to yield BPs featuring increased porosity and reduced tensile strength (Sears et al., 2010).The process of functionalizing CNTs presents the potential to enhance the interactions taking place between neighboring CNTs within the BP structure. This enhancement consequently contributes to the improved mechanical performance and conductivity of the BPs, albeit within certain limits (Smajda et al., 2007b; Zhang et al., 2013; Zhang et al., 2004).

Moreover, the choice of solvent medium for dispersing the CNTs, as well as the specific conditions employed during synthesis and subsequent post-treatment steps, have important influence on some physical characteristics such as the distribution of pore sizes and the density at which nanotubes are packed within the BP structure. Collectively, these variables significantly influence the final properties exhibited by the BPs.

The purpose of this chapter is to provide a comprehensive review of BPs including preparation methods of BPs, morphological properties of BPs, physical properties of BPs, and applications of BPs in composite materials.

2 Preparation Methods of BPs

The techniques for producing BPs can be categorized into two primary groups (H. Zhu & Wei, 2008): wet methods and dry methods. The wet method offers several key benefits, including straightforward operation, the production of high-quality products, predictability, controllability of outcomes, and suitability for laboratory-scale manufacturing. On the other hand, the dry method is commonly favored for cost-effective large-scale commercial productions. However, the dry method is notably more complicated and tends to yield lower product quality compared to the wet method.

2.1 Wet Method

There are two main steps in the wet method, including preparation and filtration of the CNT suspension (Frizzell et al., 2005; Skákalová et al., 2005). During the preparation phase, CNTs are evenly distributed within a solvent using dispersion techniques such as mechanical stirring and ultrasonication. In the subsequent filtration stage, the suspension of CNTs and solvent is passed through a filter film to completely evaporate the solvent. Subsequently, the obtained BP film is washed by deionized water, followed by drying the water to eliminate any remaining moisture, resulting in the formation of the BP film. Buckypapers exhibit characteristics like permeable membranes, featuring a free space that accounts for 60-70% of the overall paper volume. The procedure for creating the BP film through the wet method is depicted in Fig. 1. The most commonly used solvents include deionized water (Kim et al., 2006; Wang et al., 2004) and N–N dimethylformamide (DMF) (Kang et al., 2006; Smajda et al., 2007a) while the most commonly used dispersants are Triton-X100 (Coluci et al., 2008; Xing et al., 2008), sodium dodecyl sulfate (SDS) (Hong & Tai, 2008; Wang et al., 2011), polyvinylpyrrolidone (PVP) (Coleman et al., 2003).

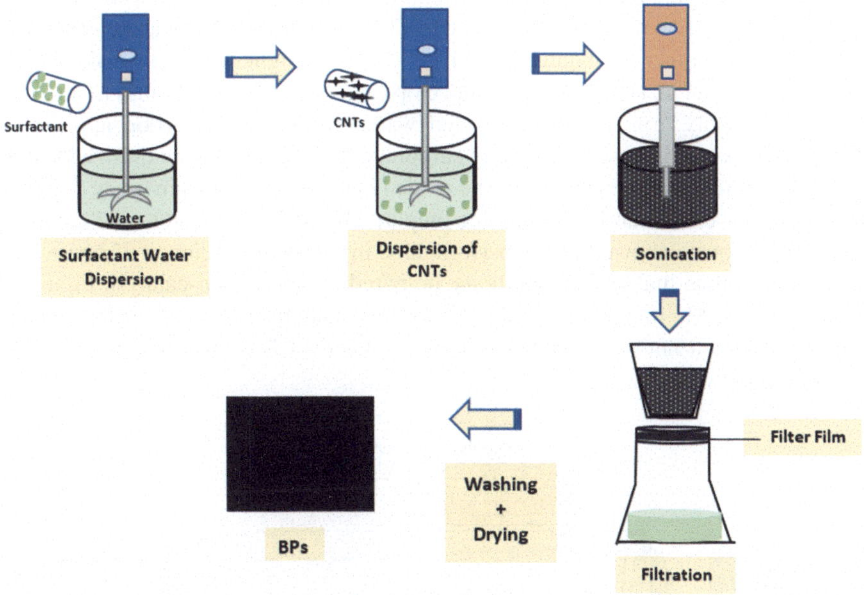

Fig. 1 A schematic illustration of the fabrication of wet method

2.2 *Dry Method*

Dry techniques encompass methods like chemical vapor deposition (CVD) (Bai et al., 2023; Li et al., 2008; Pierson, 2021; Wang et al., 2008). CVD is a procedure wherein a solid substance is deposited from a gas phase (Fig. 2). The production of CNTs, whether single-walled or multi-walled, using CVD encompasses catalytic disintegration of a carbon precursor such as hydrocarbons on a nanostructured catalyst composed of transition metals like Co, Ni, or Fe. Whereas this method is low cost and suitable for large scale fabrication, the impurities from the precursor might negatively affect desired properties of BPs such as electrical and thermal conductivity.

3 Morphological Properties of BPs

The morphological observation BPs and CNTs in BPs can be achieved by microscopic techniques such as scanning electron microscopy (SEM), atomic force microscopy (AFM), transmission electron microscopy (TEM), and optical microscopy (OM). The pore content, void distribution, CNT dispersion level of the BP can be observed by SEM while the individual CNT morphology in BPs can be observed by TEM. On the other hand optical miscroscopy give insight in a macro scale observation. Alterary et al. (2022) aimed to develop innovative BP membranes to remove the heavy metals from the contaminated water. Multi-walled carbon nanotubes (MWCNTs) were dispersed in glycerol carbonate (GC) surfactant synthesized using two different metal oxide catalysts, zinc oxide (ZnO) and magnesium oxide (MgO). The results revealed that BP membrane prepared by MWCNTs/ZnO nano-catalyst glycerol carbonate showed better performance in terms of efficiency of removing heavy metals than that of BO membrane prepared by MgO nano-catalyst. Figure 3 shows the SEM images of BP membranes manufactured by using MgO nano-catalyst and ZnO-nano-catalyst. It is seen that the CNTs are uniformly distributed in the BP surface.

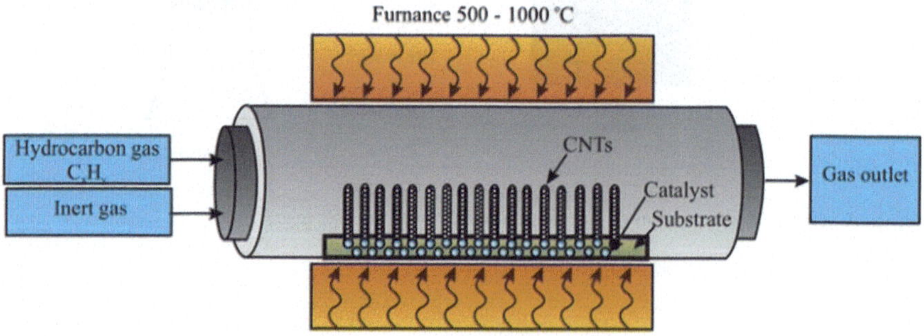

Fig. 2 Schematic diagram of chemical vapor deposition (CVD) process (Zaytseva & Neumann, 2016)

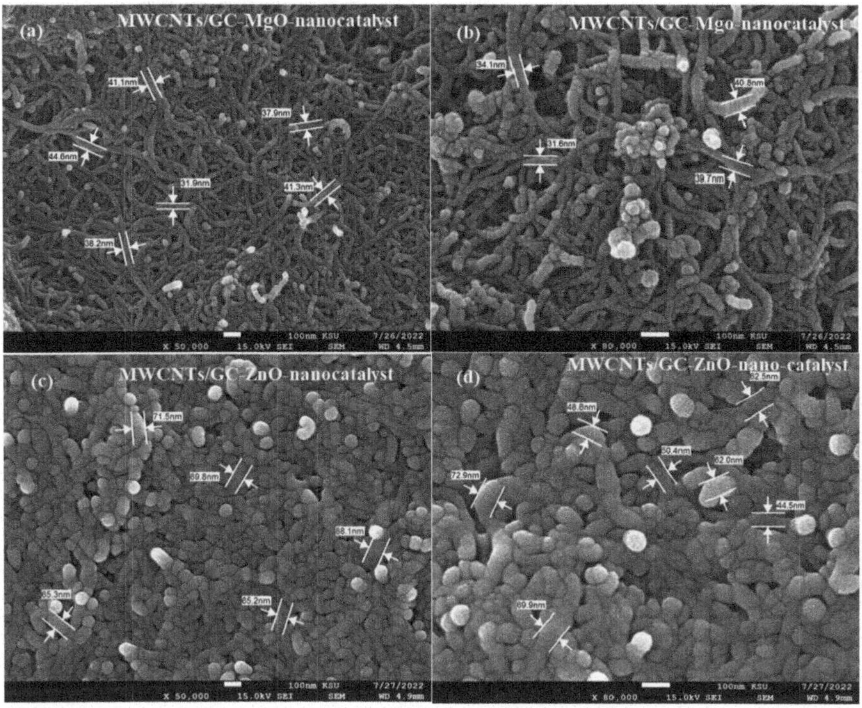

Fig. 3 SEM images of BP membranes synthesized using (**a**, **b**) GC-MgO-nano-catalyst and (**c**, **d**) GC-ZnO-nano-catalyst (Alterary et al., 2022)

Wu et al. (2021) manufactured innovative carbon fiber mat (i-MBP-PCF) by depositing MWCNT BPs on the carbon fiber fabric (CF) surface for improving the interlaminar mechanical and electrical properties of carbon fiber composites. Also, MBP/CF mats were prepared by stacking BPs on the CF surface for the comparison purpose. Both fabrics were employed as reinforcements in the epoxy infiltrated composites. The results revealed that both i-MBP-PCF and MBP/CF mats exhibited significant improvement in the interlaminar shear strength, and the electrical conductivity as compared to CF/epoxy composites. Figure 4 shows the SEM images of various CF fabrics. More gaps between BP layer (MBP) and CF fabric in MBP/PCF hybrid fabric mat are seen compared to i-MBP/PCF hybrid fabric.

Hu et al. (2019a) analyzed the pore structures of various BPs prepared by dispersing MWCNTs in different dispersing mediums including water (BP-A-W), ethanol (BP-A-E), and 1-Methyl-2-pyrrolidone (BP-A-N) based on the SEM images using Image-Pro software (Fig. 5).

Figure 6 shows TEM images of the agglomerates of CNTs and the individual CNTs ultrasonicated in the water. The CNT shortening and peeling, possibly due to the dispersion of the agglomerates in ultrasonication is seen in Fig. 6b (Rojas et al., 2019).

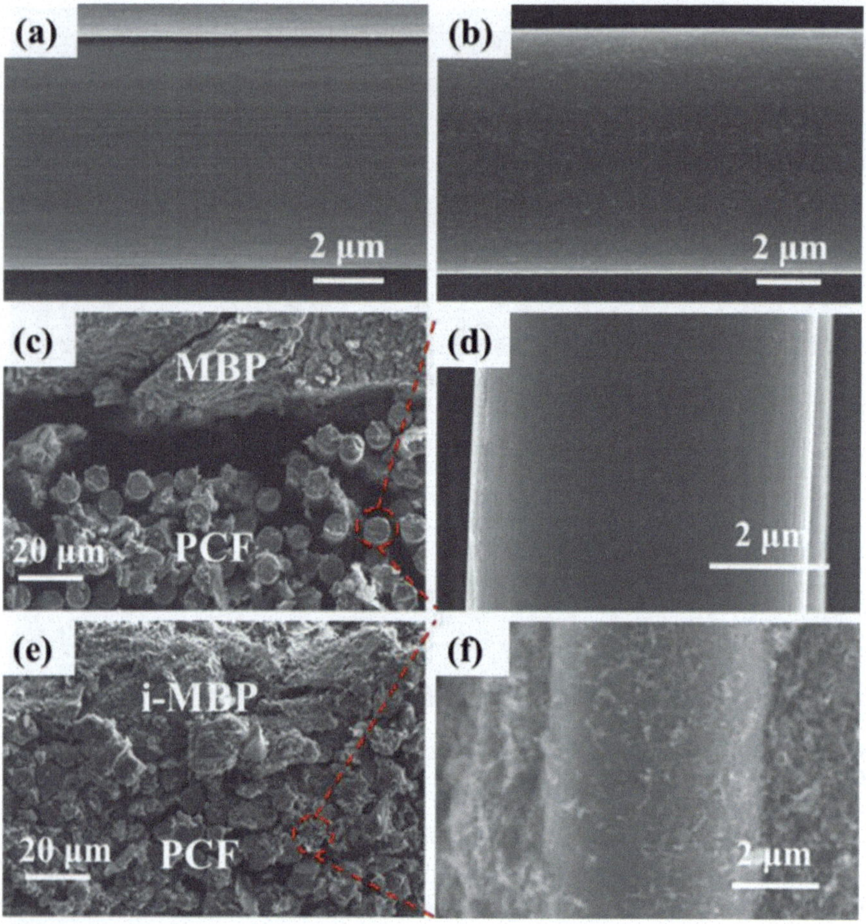

Fig. 4 SEM micrographs of (**a**) carbon fiber (CF), (**b**) functionalized carbon fiber (PCF), (**c**) stacked buckypapers/carbon fiber fabric (MBP/PCF), (**e**) in situ deposited buckypaper/carbon fiber fabric (i-MBP-PCF), (**d**) the detailed image of carbon fiber cut from MBP/PCF and (**f**) the detailed image of carbon fiber CF that cut from i-MBP-PCF (Wu et al., 2021)

Figure 7 shows the optical microscopy image of a multi-wall carbon nanotube (MWCNT) buckypapers prepared by filtering method (Trakakis et al., 2013).

3.1 *Surface Area Assessment*

The pore feature is directly related to the surface area in BPs. The membrane's surface area can be assessed through the application of the Brunauer–Emmett–Teller (BET) method. The characterization of the textural properties of the BPs is done by

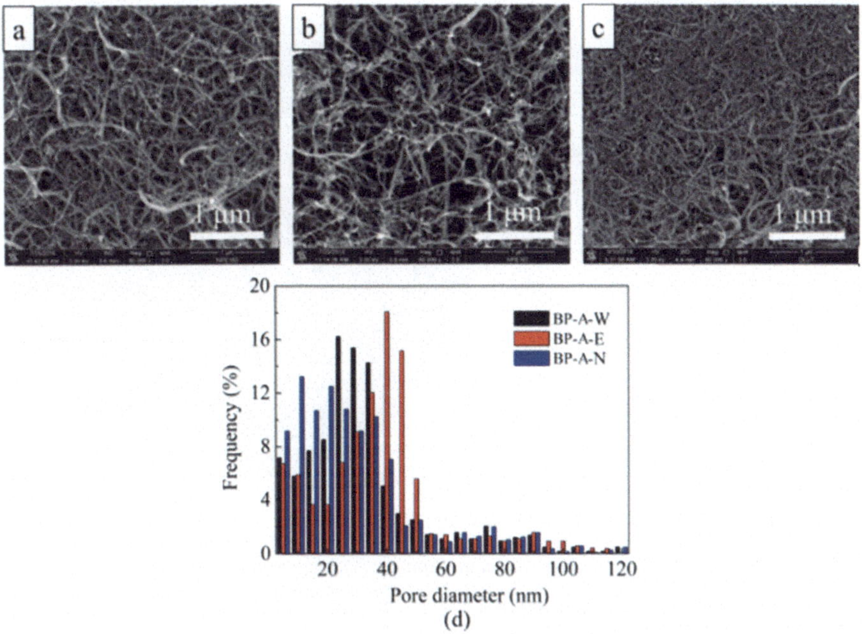

Fig. 5 The SEM images of buckypapers, (**a**) dispersed in water (BP-A-W), (**b**) dispersed in ethanol (BP-A-E), (**c**) dispersed in 1-Methyl-2-pyrrolidone (BP-A-N), (**d**) pore size distribution of BP-A-W, BP-A-E and BP-A-N obtained from data processing of their SEM images (Hu et al., 2019a)

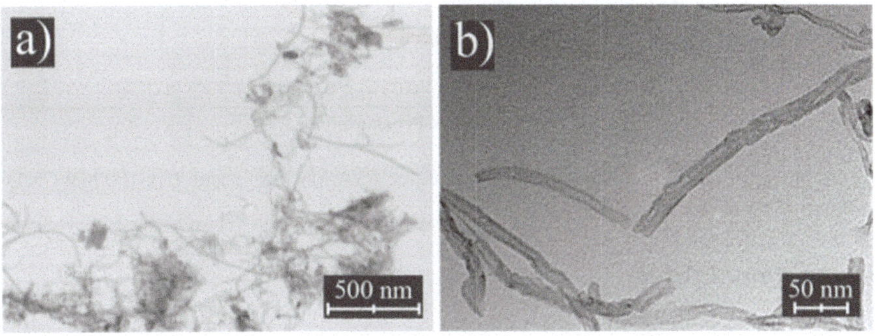

Fig. 6 TEM of (**a**) Primary agglomerates, (**b**) individual of CNTs (Rojas et al., 2019)

nitrogen adsorption-desorption isotherm measurements (Hu et al., 2019a). Figure 8 shows the Nitrogen adsorption-desorption isotherms for BPs prepared in PVA/MWCNTs and Trix-100/MWCNTs solutions (Alshahrani et al., 2022). Using the BET method, the surface area of PVA/MWCNTs BP and Trix-100/MWCNTs BP membranes were calculated as 67.7 m^2/g and 88.5 m^2/g, respectively.

Fig. 7 Optical image of a multi-wall carbon nanotube (MWCNT) buckypapers (Trakakis et al., 2013)

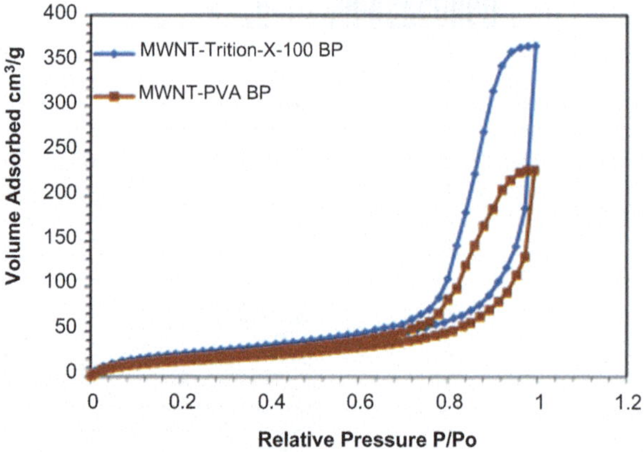

Fig. 8 Nitrogen adsorption-desorption isotherms for PVA/MWCNTs and Trix-100/MWCNTs (Alterary et al., 2022)

3.2 Contact Angle

To obtain desired functional properties in BP/polymer composites, BP films are required to exhibit good wettability characteristic by the resin material. The contact angle measurement is carried out to characterize the resin compatibility of BPs (Kar et al., 2012; Ribeiro et al., 2017). A smaller contact angle between BP and resin indicates greater ease of resin spreading across the BP surface, requiring a lower surface energy. Moreover, a surface exhibits lower hydrophobicity when it possesses a smaller contact angle, and conversely, higher hydrophobicity is associated with a larger contact angle. Given that the hydrophilicity/hydrophobicity of a membrane plays a crucial role in its interaction with water, organic compounds, and inorganic materials, the contact angle can be considered a standard metric for membrane characterization. Alshahrani et al. (Alshahrani et al., 2022) performed contact angle measurements on PVA/MWCNTs BP and Trix-100/MWCNTs BP membranes using Milli-Q water. A 2 μL water droplet was dropped onto the flat surface of BP membrane using a micro-syringe. Digital images of the water droplet were then captured with a digital camera for subsequent contact angle calculations. The contact angles for the PVA/MWCNTs BP and Trix-100/MWCNTs BP membranes were obtained to be 58 ± 3° and 34 ± 3°, respectively.

4 Physical Properties of BPs

The properties of BPs are greatly influenced by the parameters such as dispersant type, solvent type, ultrasonication time, CNT content, CNT type, and filter type utilized in the preparation process.

4.1 Mechanical Properties of BPs

On of the fundamental characteristics governing the usability of BPs are their mechanical properties; poor mechanical properties often result in frequent instances of breakage during application. The microstructural form of CNTs and the preparation techniques employed for BP preparation influence the interplay between CNTs and their state within BPs. As a consequence, significant disparities occur in the mechanical properties across diverse BPs. Generally, the mechanical properties of BPs are weak due to weak van der Waals forces acting between CNTs.

In general, CNTs are randomly distributed in the CNT thin films such as BPs. However, the van der Waals interactions among CNTs lead to bigger CNT aggregations resulting in entangled networks. The entanglement behavior causes BPs to have low mechanical properties (Dai et al., 1999; Ma et al., 2007; Meng et al., 2005; Rinzler et al., 1998). To overcome this problem, CNTs are well-ordered via some alignment methods.

The alignment of carbon nanotubes (CNTs) within BP can be manipulated through the application of a magnetic field and/or compressive force during the synthesis phase. When buckypapers consist of well-oriented CNTs their strength, weight efficiency, and stiffness can be boosted by an order of magnitude compared to BPs of which CNTS are poorly dispersed. This establishes the significant capability of BP to serve as the primary reinforcing element within composite materials.

Chemical functionalization of CNTs is employed to enhance the mechanical characteristics of BPs. Zhang et al. (2004) discovered that subjecting CNTs to nitric acid treatment led to an augmentation in both film tensile strength and tensile modulus; however, this improvement came at the cost of sacrificing electrical conductivity. Alignment is one of the efficient methods to improve the mechanical properties of BPs (Wang et al., 2008; Zhang et al., 2012). Cheng et al. (2011) demonstrated that aligning CNTs and establishing conjugational cross-linkages through chemical functionalization notably elevated the electrical conductivity and mechanical properties of BPs in comparison to their pristine counterparts.

Guo et al. (2013) investigated the influence of surface functionalization of CNTs on the mechanical properties of the BP composites. Two different oxidation methods were employed for the surface functionalization of CNTs. The morphological observations showed that surface modification of CNTs by acidification results in more condensed and uniform carbon nanotube network. Furthermore, it was shown that the modulus and mechanical strength of the functionalized BP composites exhibited notable enhancements, rising by 20–37% and 42–68%, respectively, in comparison to the unmodified BP composite.

Hall et al. (2008) discovered that adjusting the quantity of MWCNTs can alter the Poisson's ratio of the material, transitioning it from positive to negative values. Consequently, CNT BP has the capability to demonstrate auxetic characteristics, which is a crucial quality in the development of synthetic muscles.

Pham et al. (2008) examined the mechanical and electrical attributes of composites comprised of polycarbonate and CNT BP. Their investigation revealed that the incorporation of polycarbonate led to improvement in stiffness and toughness within the BP, at the expense of a decrease of in the electrical conductivity.

Oh et al. (2015) used simple filtration method to develop BP film consisting of self-assembled and well-aligned densely packed CNTs. It was shown that both the alignment and the assembly behavior of CNTs has a strong effect on the concentration of CNT suspension. It was also observed that highly dense aligned BP exhibits significant improvement in packing density, mechanical strength, elasticity modulus, and hardness.

Zhang (2008) introduced a novel manufacturing method of BP membranes via hydroentangling. Hydroentangling represents a straightforward, rapid, cost-effective, and environmentally friendly method for mechanically bonding fibers, resulting in the creation of a durable and uniform nonwoven fabric or membrane. In hydroentangling process, dry CNT powder can be directly applied to the perforated plate support. The drag force exerted by water jets overcomes the van der Waals force, leading to the formation of entangled CNT structures. Schematic illustration of hydroentangling process of BPs is depicted in Fig. 9. Thereby, this method eliminates any pre-processing steps such as using dispersant and chemical functionalization of CNTs to avoid from bundling due to the intertube van der Waals force. Furthermore, the thickness of hydroentangled CNT membranes can be readily controlled by adjusting the quantity of CNTs deposited onto the support. By utilizing a patterned substrate, desired topologies can also be achieved. Hydroentangled CNT membranes possess a high density of entanglements, resulting in improved structural integrity and mechanical properties. The findings indicated a tensile strength of 51 MPa for hydroentangled CNT membranes, surpassing that of filtration-produced CNT buckypapers by a factor of three. Notably, the incorporation of polyethylene oxide (PEO) into hydroentangled CNT membranes led to a 37% strength enhancement. This effect is attributed to PEO chains effectively reducing nanotube slippage and stabilizing entanglements during tensile loading.

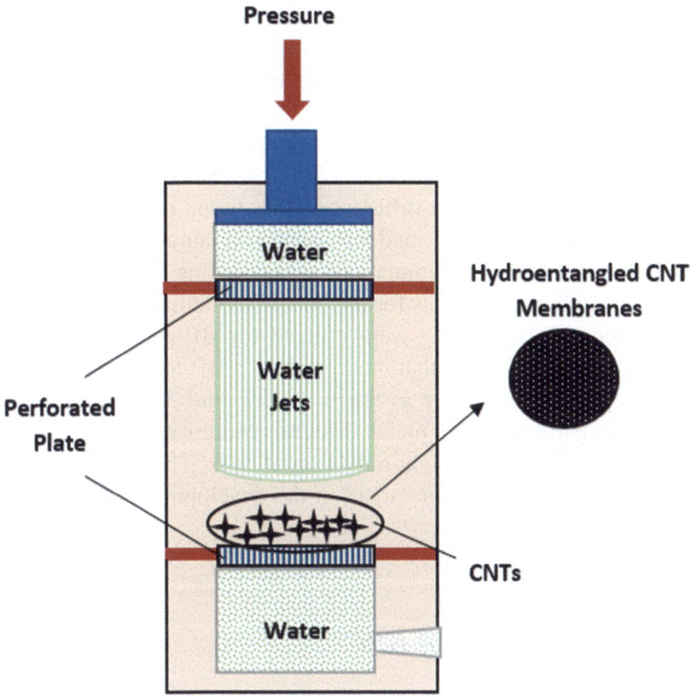

Fig. 9 Schematic illustration of hydroentangling process of BPs

4.1.1 Theoretical Investigations of Mechanical Properties

The numerous studies have been carried out on the molecular-level interactions between adjacent carbon nanotubes (Chen et al., 2003; Liew et al., 2005; Zhou et al., 2007; Zhu et al., 2004). However, it is very challenging to understand the properties of BPs in the atomic scale due to their inherently complex structure.

Buehler and coauthors (Buehler, 2006) have formulated a 'bead-spring' model aiming to understand the mechanical characteristics of ultra-long carbon nanotubes (CNTs) and their process of self-assembly. This model simplifies the CNT structure into a series of uniformly spaced beads, connected by springs. The elasticity of these springs, both in terms of stretching and bending, is adjusted to effectively depict specific mechanical behaviors exhibited by CNTs.

Li and Kroger (2012) numerically investigated the structural formation mechanisms of BPs and their influence on the mechanical properties. They suggested an algorithm using the combination of the molecular dynamics and Monte Carlo methods. The results indicated that both the entanglement and the bundling mechanisms have a significant effect on the mechanical properties of BP. The entanglement network and the pore size were used to quantitatively characterize the effect of both mechanisms. The entanglement level is found to be more dominant in SWCNT BPs than DWCNT BPs due to the fact that SWCNTs have smaller adhesion energy.

Zhigilei and his team (2005) proposed an mesoscopic model for both static and dynamic simulations of CNT nanomechanics. This model adopts a coarse-grained approach, representing CNTs as 'breathing flexible cylinders' that are composed of a variable quantity of segments.

Chandra et al. (2022) employed a modeling approach that combines atomistic and continuum methods to investigate the behavior of BP composites across a range from the nano to macro scale. Six different carbon-based nanostructures including graphene, CNT, CNT bundle, BP, and BP composite automotive components are analyzed in the study. Atomistic simulations for modelling of graphene, CNTs, and CNT bundles, beam representations for the bundle modelling, and continuum solid representation for resin modelling were utilized for BP and its composites. The computational findings indicated that incorporating a BP composite can elevate the natural frequency and stiffness by as much as 25% and 37%, respectively, when compared to conventional solid metal designs. Moreover, this improvement is accompanied by a substantial weight reduction of 57%. These results indicate that BP is a good candidate for reinforcement in the development of future aerospace and automotive structures.

4.2 Electrical Properties of BPs

Enhancing the mechanical load transfer capacity of a randomly arranged carbon nanotube BP through a strategic increase in nanotube interfaces has its advantages but also leads to a consequent decrease in the rate of improvement in its electrical conductivity due to the elevated contact resistance. To overcome this problem, many studies have been conducted.

Ahmadalinezhad et al. (2011) conducted an investigation aimed at characterizing the mechanical, electrical, and thermal properties of two distinct types of buckypapers, namely uncompressed buckypapers (UBPs) and compressed buckypapers (CBPs). These BPs are fabricated using vacuum filtration method. CBPs were found to exhibit enhanced thermal and electrical conductivities due to their denser network of carbon nanotubes, resulting in a greater number of conductive pathways. However, the mechanical properties of both CBPs and UBPs remained nearly identical.

Facilitating high current transport within carbon nanotube (CNT) networks, such as buckypapers (BPs), presents a challenge due to the pronounced contact resistance among CNTs. This challenge becomes even more formidable when incorporating BPs into insulating polymer matrices to create functional nanocomposites. BPs produced via the hydroentangling method have exhibited a promising enhancement in conductivity, attributed to a heightened interconnection between CNTs, in contrast to BPs formed through the filtration method (Zhang, 2008). Zhang's study (2008) revealed that the electrical conductivity of hydroentangled CNT membranes is approximately ten times higher than that of CNT BPs produced through filtration.

Employing covalently bonded cross-linking among CNTs, where inter-tube linkers function as efficient medium for stress and current transfer, represents a promising strategy to amplify both the mechanical and electrical attributes of BP (Cha et al., 2008; Kis et al., 2004; Ventura et al., 2010; Zhang & Jiang, 2011). However, chemical functionalization in this process damages the intrinsic properties of CNTs resulting with reduction in the improvement level of electrical and mechanical properties. To address this issue Zhang et al. (2013) demonstrated in-situ cross-linking technique of CNTs. In this approach, the process involves the initial creation of a pristine BP with minimal CNT aggregations using a pressurized filtration technique. Subsequently, the CNTs present within the BP are subjected to in situ cross-linking by filtering a solution containing 1,4-benzoquinone in an aqueous medium through the pre-fabricated BP. The results of micro-tensile and electrical conductivity tests showed that the in-situ cross-linking method leads to significant improvement in the bulk mechanical and electrical properties of BP.

The alignment of carbon nanotubes (CNTs) results in significant enhancements in both their electrical and mechanical characteristics compared to those with random orientation (Hamasaki et al., 2022; Inoue et al., 2011). The mechanical strength experiences an elevation owing to the efficient load distribution among the nanotubes, facilitated by increased packing density. Simultaneously, the electrical conductivity along the alignment direction undergoes a significant enhancement due to reduced contact resistance (Park et al., 2009b; Zhang et al., 2012).

Alignment can be achieved through various methods, such as applying a magnetic field during the filtration process (Fischer et al., 2003; Hone et al., 2000), depositing an aligned array of CNTs prior to filtration to induce self-alignment (Oh et al., 2015), and utilizing mechanical stretching techniques (Zhang et al., 2017).

Sakurai et al. (2013) investigated the effect of SWCNT length on the electrical and mechanical properties of BPs fabricated by vacuum filtration approach. Results indicated that the increase in the length of SWNCTs results in enhancement in the mechanical strength and electrical conductivity of BPs. This might be attributed to the fact that longer CNTs allows more efficient electron and load transfer via lower number of CNT connections.

Tieu et al. (2018) conducted an investigation into the impact of sonication time and power on the electrical conductivity of BP. Their findings revealed that the optimal electrical conductivity was achieved as the dispersion underwent sonication for 30 minutes at the ultrasonication power of 1000 Watt. The aggressive sonication process with the longer sonication time at high powers led to CNT damage, thereby diminishing the properties of the eventual structure. On the other hand, significantly shorter sonication durations were insufficient to yield a suitable dispersion.

Vahid et al. (2018) investigated the effect of the dispersant type on the electrochemical performance of BPs. The various types of non-ionic surfactants including Triton X-100, Tween 20, Tween 80, Span 20, and saponin were employed to synthesize the BPs. The results showed that the use of Tween 20 surfactant in the preparation of BP yielded the highest electrical conductivity. This outcome was attributed to the rapid electron transfer facilitated by the remarkably uniform structure.

Rani et al. (2023) explored the influence of surfactant content on the electrical characteristics of buckypapers. An alkali-soluble emulsifier was employed as the surfactant to counteract agglomeration. The findings revealed that samples with the least amount of surfactant demonstrated the highest electrical conductivity, measuring at 91.74 S/cm. This conductivity was twice that of BP produced with the highest surfactant content.

4.3 *Thermal Conductivity of BPs*

Examination of thermal conductivity is quite challenging issue due to the number of parameters such as heat losses. Methods for measuring thermal conductivity can be categorized based on the heating process employed. In contact-heating methods samples are in direct contact with the heater. Methods in which the sample is heated using radiation are termed non-contact heating methods. The former category encompasses approaches such as 3-ω, bolometer, and steady-state methods. In contrast, the latter group includes techniques like laser flash analysis (LFA), thermoreflectance method, and Raman spectroscopy, among others. Kumanek and Janas (2019) reports the detail explanation of these methods.

The transmission of thermal energy along a CNT is commonly understood to be primarily governed by the phonon conduction mechanism (Marconnet et al., 2013; Pop et al., 2006). While certain CNTs possess metallic properties, potentially suggesting electron-mediated thermal energy conduction (Z. Han & Fina, 2011), the predominant mode of transport remains rooted in the collective atomic vibrations (Balandin, 2011; Hone, 2001).

Nanotubes exhibit excellent thermal conductivity (e.g., 300 W m^{-1} K^{-1} for CNTs and of 350 W m^{-1} K^{-1} for BNNTs) (Chang et al., 2006). The inherently high thermal conductivity of nanotubes can be enhanced on a macroscopic scale by the creation of interconnected pathways within the BP network. The packing density of CNTs within a film exert a substantial influence on the thermal conductivity value of CNT networks. Enhancements in thermal conductivity can be achieved by increasing the number of junctions between CNTs, which facilitate the movement of heat within the sample. The density of CNT films can be enhanced by either increasing the ordering of CNTs or increasing the packing density of CNTs with pressure.

Zhang et al. (2012) investigated the effect of the CNT density packing on the thermal conductivity values of BPs. The densities ranging from 0.8 to 1.39 g/cm^3 were obtained using various pressures ranging from 20 to 30 MPa. The thermal conductivity measurements conducted using steady-state methods showed an enhancement in the thermal conductivity from 472 to 766 W/mK. This result suggests that as the density increases, thermal conductivity also increases. This relationship stems from the greater packing of CNTs in the sample, leading to a more accessible thermal percolation pathway.

Itkis et al. (2007) employed the bolometric method (Fig. 10) to explore the impact of film preparation techniques on the thermal properties of CNT buckypapers. Two distinct CNT films with varying packing densities were created. The initial film was assembled on a stainless-steel grid via the electric arc discharge

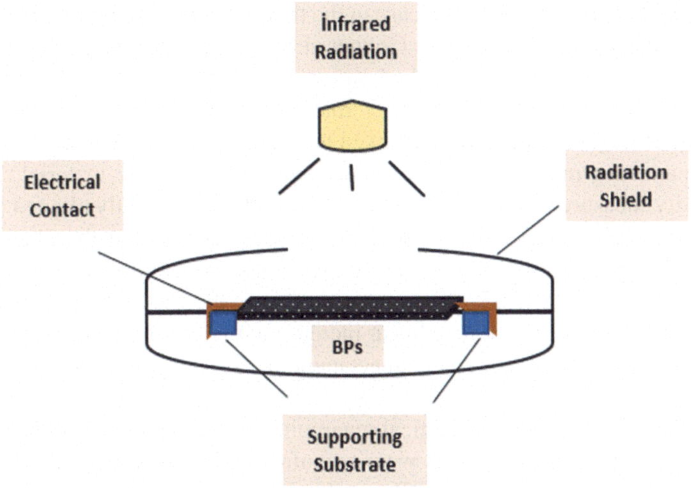

Fig. 10 Schematic view of bolometric method

method. The second film was generated through a vacuum filtration process. The thermal conductivities of the first and second films were measured at 75 and 30 W/mK, respectively. This outcome suggests that the film with lower packing density, the first one, exhibits weaker contact junctions within its networks compared to the second film, which has a higher packing density.

The procedure for acquiring and refining CNTs plays a substantial role in shaping their characteristics. A significant factor influencing these properties is the annealing process applied to the end product which eliminates and reduces the structural defects (Hone et al., 2000, 2002; Lin et al., 2012; Zhang et al., 2005, 2007a). Moreover, annealing alters the material's density, which was previously discussed.

Hone et al. (2000, 2002) investigated the annealing effect on the thermal conductivity of magnetically aligned SWCNT films. Magnetically aligned SWCNT films were exposed to vacuum annealing at 1200 °C. They attributed the enhanced thermal conductivity to the elimination of acidic contaminants from the CNTs, which subsequently altered the conduction mechanism.

Zhang et al. (2007a) characterized bulk MWCNTs synthesized via the CVD method involving the decomposition of propylene and hydrogen on a Ni catalyst. The acquired nanotubes were subsequently purified and subjected to annealing at different temperatures (1600 °C, 1800 °C, 2000 °C). They examined the structural modifications in the bulk MWCNTs post-annealing and observed that higher annealing temperatures correlated with increased density, reaching up to 1.45 g/cm^3 at 2000 °C. Through assessments of the thermal characteristics of disk-shaped MWCNT samples, it was revealed that elevating the annealing temperature (and consequently the density) led to an augmentation in thermal conductivity from 2.8 to 4.2 W/mK.

Alignment of BPs results in increase in the thermal conductivity ranging from 50 to 200 W/mK (Hone et al., 2016). Gonnet et al. (2006) introduced an innovative technique for creating densely packed and well-oriented SWNT nanocomposites. This method involves saturating SWNT mats with a low viscosity resin solution. Alignment of SWNTs in-plane is achieved through magnetic orientation within the BP under a strong magnetic field before loading the composite material. Thermal conductivity measurements were conducted on both magnetically aligned BPs and their corresponding composites. The findings demonstrate that the alignment of nanotubes has a noticeable impact on the thermal conductivities of both BPs and nanocomposites. Additionally, it was observed that the thermal conductivities of both BPs and composites exhibit a linear increase with rising temperature. However, the extent of thermal conductivity enhancement in the nanocomposites was found to be limited.

Wang et al. (2008) used mechanical manipulation to align the CNTs in BPs. The thick BPs with large surface areas were manufactured by this method. The results showed that mechanically aligned BPs exhibit enhanced thermal and electrical conductivity properties.

Numerous researches have been focused on the CNT based BPs due to the large-scale manufacturing advantage of CNTs. Yet, there has been limited documentation related to the fabrication and characteristics of macroscopic films comprised of different nanoparticle materials, such as boron nitride nanotubes (BNNTs).

BNNTs offer specific benefits compared to CNTs. In contrast to CNTs, which exhibit metallic or semiconducting behavior based on their chirality (Saito et al., 1992), BNNTs consistently demonstrate electrical insulating characteristics (Blase et al., 1994). Having the excellent elastic modulus, BNNTs are known as the strongest insulating fibers (Chopra & Zettl, 1998). They also possess the exceptionally uncommon combination of functioning as both a dielectric and a thermal conductor. BNNTs also exhibit superior thermal stability compared to CNTs and can withstand oxidation in ambient air at temperatures ranging from 800 to 900 °C (and up to 1500 °C in an inert atmosphere) (Chen et al., 2004, 2017; Kim et al., 2015). The exceptional characteristics of both h-BN and CNTs can be merged by creating a coaxial heteronanotube hybrid structure, wherein a central CNT serves as the core, enveloped by an outer layer of BNNT. Through this approach, the layers of BNNT contribute to thermal conduction while simultaneously acting as a dielectric barrier by insulating the conductive core CNTs. Therefore, both BNNTs and CNT/BNNTs are ideal candidates for the purpose of heat dissipation in electronics packaging. Jones et al. (2023) reported the synthesis of carbon (core)/boron nitride (shell) heteronanotube (MWCNT/BNNT), boron nitride nanotube (BNNT), and MWCNT/BNNT BPs. MWCNT/BNNTs heteronanotubes were fabricated using atmospheric pressure chemical vapor deposition (APCVD) and the safe hexagonal boron nitride (h-BN) precursor ammonia borane. Then, thermal oxidation of the core carbon in MWCNT/BNNTs was achieved to synthetize the BNNT film. BPs composed of MWCNTs, BNNTs, and MWCNT/BNNTs were fabricated by filtration method. Thermal conductivity of BPs was measured using scanning thermal microscopy. The outcomes showed a 14% improvement in the thermal conductivity of MWCNT/BNNT BPs compared to MWCNT buckypapers (Fig. 11).

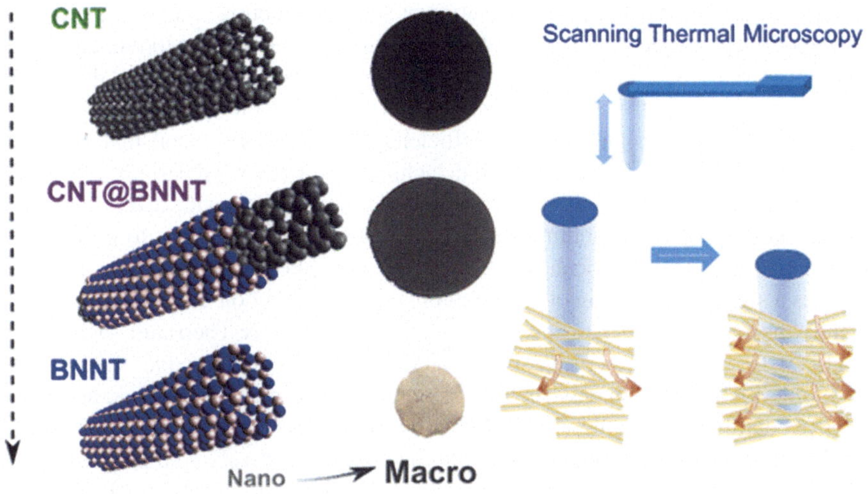

Fig. 11 The main steps of thermal characterization CNT BP, CNT/BNNT BP, and BNNT/BP (Jones et al., 2023)

5 Application of BPs in Composites

The potential uses of BPs and their nanocomposites hold great promise. BP can be impregnated with resin and subsequently fabricated into composite materials, or easily integrated into traditional fiber-reinforced composites (Wang et al., 2008). Notably distinct from composites featuring dispersed carbon nanotubes (CNTs) within the matrix, composites derived from BP offer the advantage of accommodating significantly higher CNT concentrations. This, coupled with the formation of highly conductive networks of tubes, results in the optimization of both electrical conductivity and mechanical properties. This innovative approach introduces a new avenue for developing multifunctional composites that possess combined conductive and structural attributes, potentially applicable within the aerospace industry. To achieve this objective, there is a requirement for BP with enhanced mechanical and electrical characteristics.

BP composites have been investigated for possible structural candidate in aeronautic applications such as lightning strike protection (LSP) (Zhang et al., 2019), de-icing (Chu et al., 2014), electromagnetic interference (EMI) shielding (Xia et al., 2018), flame retardancy (Tang et al., 2011a, b), sensor (Cao & Rogers, 2009; Cheung et al., 2009; Kukovecz et al., 2008; Smajda et al., 2007a) and electrodes of battery and supercapacitor (Faraji & Mohammadzadeh Aydisheh, 2019; Guler et al., 2014a, b; Kim & Park, 2014; Singh et al., 2003; Xu et al., 2015, 2017).

Fiber-reinforced polymers (FRPs), such as carbon-fiber reinforced polymers (CFRP), have found extensive utilization within the aviation sector. This application has the potential to lower fuel consumption and enhance aircraft fatigue resistance as opposed to conventional metallic materials. However, transitioning from electrically conductive metals to insulating or semi-conducting composites presents an important drawback: an increased susceptibility of the aircraft to damage from lightning strikes. The existing protective strategy involves bonding a metallic mesh onto the composite structure's surface. Nonetheless, this increase in weight has a negative effect on the aircraft's fuel efficiency. Furthermore, metal materials are prone to experiencing degradation and detachment over extended periods of use due to their susceptibility to corrosion and limited compatibility with resins. The primary objective of the Lightning Strike Protection (LSP) system is to establish a secure pathway for electrical conductivity on the outer structure. This pathway enables a significant portion of the lightning current to stay on the surface and efficiently dissipate without causing substantial harm (Gagné & Therriault, 2014; Gou et al., 2010). This method of passive protection relies on the principle of a Faraday cage, employing components such as metal airframes, metallic meshes, and conductive coatings. Schematic view of lightning strike protection displayed in Fig. 12.

BP stands out as a promising material that could effectively address the limitations associated with conventional metallic LSP materials. Notably, BP boasts a significantly lower density (approximately 0.8 g/cm^3) compared to Aluminum (Al) (2.7 g/cm^3) and Copper (Cu) (8.9 g/cm^3), as highlighted in reference (Han et al., 2015). Additionally, BP with micron-scale pores exhibits improved compatibility with resins, a feature advantageous over metallic materials within the context of Carbon-Fiber Reinforced Polymer (CFRP) composite modeling (Younes et al., 2016). Numerous studies have been carried out to investigate the LSP protection of airframes by BPs (Kumar et al., 2019; Mall et al., 2009; Rein et al., 2011; Younes et al., 2016).

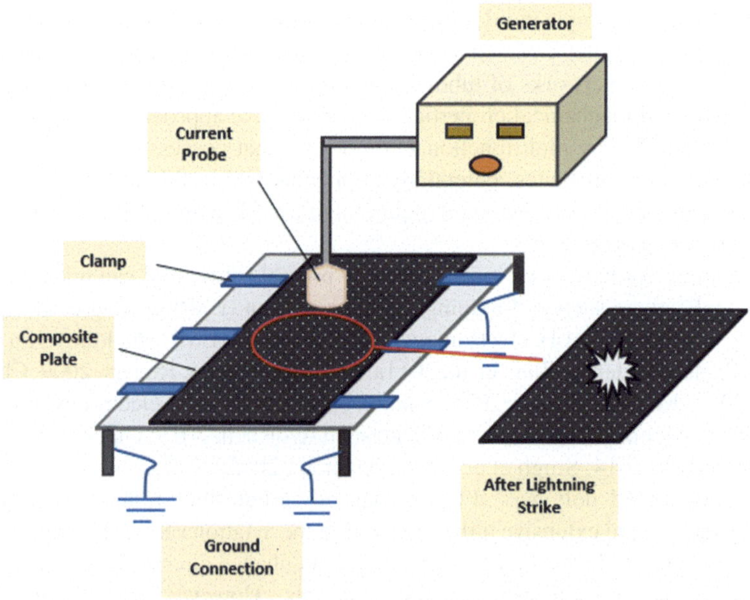

Fig. 12 Schematic view of lightning strike protection (LSP) and after lightning strike

Mall et al. (2009) studied the degradation of compressive strength in different electrically conductive nanocomposites by subjecting them to a simulated lightning strike. In this research, an examination was conducted on five conductive nanocomposites to assess the reduction in their compressive strength following exposure to a simulated lightning strike. Among these systems, one comprises Nickel-coated carbon woven fabric, while the remaining four systems encompass carbon fabric incorporated with a nickel-nano strand veil (NiNS), aligned buckypapers, random buckypapers, and a combination of buckypapers derived from vapor-grown carbon fibers and single-walled carbon nanotubes (SWCNTs). Simulated lightning strike showed that the reduction in the ultimate compressive strength was about 75—30%. The degradation was maximum in the case of using NiNS whereas it was minimum as random BPs were utilized indicating that damage due to lightning strike is strongly dependent on the electrical conductivity. In a pioneering study, Bai et al. (2023) incorporated quartz fibers as an insulating layer for the first time in combination with highly conductive super aligned carbon nanotube films (SA-CNTFs) for Lightning Strike Protection (LSP). Their research revealed that quartz fibers demonstrate superior potential as an alternative insulating material compared to conventional glass fibers, mainly due to their exceptional thermal resistance. The investigation highlighted a synergistic relationship between the conductive and insulating layers. Furthermore, when compared to CNT-based LSP laminate samples constructed with layers of buckypaper/Boron Nitride/epoxy or SA-CNTF/glass fiber, the resulting weight reduction was at least approximately 30% and 38%, respectively. Kumar et al. (2019) illustrated the utilization of interleaved multi-walled carbon nanotube buckypapers (MWCNT BPs) between Carbon-Fiber Reinforced Polymer (CFRP) laminates, aiming to enhance the electrical conductivity across the thickness and diminish the effects of lightning strike damage. The findings revealed a remarkable enhancement of electrical conductivity, with improvements of 697% in the in-plane direction and 643% through the thickness. Moreover, when subjected to simulated lightning tests of 40 kA, both pristine composites and those interleaved with BP were mechanically tested. The results demonstrated the exceptional effectiveness of the BP composite in reducing the impact of lightning strikes.

Electromagnetic Interference (EMI) derives from electromagnetic waves that interfere with electronics in aircraft. The secondary purpose for LSP is to provide the aircraft with EMI shielding.

The fundamental principle underlying EMI shielding is rooted in the concept of the Faraday Cage. In this principle, electromagnetic waves are effectively prevented from infiltrating any enclosed surface encompassed by conductive materials.

Investigations pertaining to electromagnetic interference (EMI) shielding of composites have been conducted on a wide range of materials (K. Chen et al., 2022; Gelves et al., 2011; He et al., 2023; Li et al., 2022; Park et al., 2009a; Zhang et al., 2022).

Tian et al. (2023) presented an extensive review of the recent progress of multifunctional EMI shielding polymer composites. In this review, they also discussed in detail the basic theory, attenuation mechanisms and characterization methods of electromagnetic interference (EMI) shielding. Devi et al. (2022) reviewed the EMI shielding behavior of polymer based composite materials modified by carbon based nanoparticles. It is widely recognized that carbon nanotubes (CNTs), due to their remarkable corrosion resistance, low density, and exceptional electrical conductivity, have assumed a pivotal role in the development of EMI shielding materials. Therefore, the incorporation of a substantial proportion of CNT within polymer matrix composites becomes crucial to achieve optimal electrical conductivity. However, utilizing the high CNT content in the composites is a challenging task due to the fact that agglomerations might severely reduce mechanical properties of the composite at high CNT contents. BPs emerge as promising candidates for EMI shielding due to their remarkably high CNT content, porous architecture, and outstanding compatibility with resins (Choi et al., 2017; Iqbal et al., 2023; Park et al., 2009a; Ribeiro et al., 2023; Xia et al., 2018; Yang et al., 2017). Ribeiro et al. (2023). investigated EMI shielding effectiveness of BP/glass fiber/epoxy for five different sandwich configurations. Iqbal et al. (2023) comprehensively reviewed the Mxene/CNT BPs for EMI shielding applications. It was emphasized that Mxene/CNT films exhibiting superior mechanical properties as well as excellent electrical conductivity compared with their individual films are novel material for EMI shielding.

The accumulation of ice on aircraft structures causes deterioration in aerodynamic configuration, primarily by adding weight (Lynch & Khodadoust, 2001). As ice builds up on these structures, the roughness of the wing surface amplifies. This not only negatively impacts aerodynamic efficiency but also triggers distortions in airfoil shape and leads to heightened energy consumption. Two primary approaches have been employed to address the issue of icing: anti-icing and de-icing strategies. One of them is to use the hot air generated by the engines to heat the metallic pipe networks for the anti-icing purpose (Chu et al., 2014). However, this method reduces the effectiveness of the engines and increases the weight of aircraft (Falzon et al., 2015). De-icing technique involves removing ice from the aircraft surface. De-icing salts are often utilized as a predominant method among de-icing techniques to mitigate the problem, primarily due to their cost-effectiveness and high level of effectiveness. Nonetheless, it is important to note that de-icing salts containing chloride compounds can result in the corrosion of the metal structures in airplanes and contribute to environmental pollution (Novotny et al., 2008). Another deicing method works based on heating strategy including internal and external heating. In the context of internal heating, the process of de-icing the outer surface is accomplished by utilizing components with high thermal conductivity that are embedded within the composite structure. Examples of such components include electric cables or carbon fiber heating wires. On the other hand, the external heating approach involves employing methods like microwave or infrared heating. However, both techniques come with certain limitations. In the case of internal heating, effectively reaching the outer surface proves challenging due to the poor thermal conductivity of CFRP. On the other hand, the external heating method faces complexities in achieving uniform heating across a large surface area.

BP boasts a combination of low density and remarkable thermal conductivity, rendering it suitable for utilization as a heating element within de-icing systems designed for aircraft constructed from CFRPs. When an electric current flows through a BP composite aided by CNTs, rapid heating occurs, and this generated heat is subsequently dispersed into the surrounding environment through processes of radiation and conduction (Chu et al., 2014; Falzon et al., 2015; Tarfaoui et al., 2019; Younes et al., 2016; Zangrossi et al., 2022) used BP in glass fiber/epoxy composites to fabricate self-heating composite for de-icing. The panel underwent heating and cooling cycles, captured by thermal cameras to observe the temperature variations that emerge within the composite. The outcomes revealed that the material holds promise as a viable option for an electric heating element intended for de-icing purposes. Zangrossi et al. (Zangrossi et al., 2022) examined the influence of surface modification on the microstructure of CNT-based BP and its subsequent impact on electrical heating performance when impregnated with an ice-phobic resin. The findings suggested that incorporating surface-modified MWCNTs into composites along with the use of advanced ice-phobic resin presents a potentially promising approach for electrical heating. This strategy could effectively address ice-related challenges while upholding the principles of lightweight design and efficiency.

Structural Health Monitoring (SHM) involves the permanent embedding of lightweight and small sensors within composites. These sensors are used to provide information about the structural integrity of composites such as stress—strain distributions, matrix cracks, fiber failures, and delamination. Therefore, one can monitor and evaluate the safety of composite structure. (Zhao et al., 2009). In general, most common SHM approaches include piezoelectric ceramic sensors (PZT), fibre-optic sensors (FBG), and nano-material-based sensors. Among them FBG sensors have some disadvantages. Since the optical fibres are inherently fragile it is very challenging to embed them in small scale complex structures. In addition, they may act as defects in composite materials due to their larger size than the reinforcing fibers. On tlhe other hand, PZT sensors exclusively respond to dynamic stimuli. In contrast to FBG and PZT, electrical resistance measurement (ERM) is significantly valuable in SHM and exhibits greater sensitivity compared to other traditional methods of damage detection. (Irving & Thiagarajan, 1999; Kostopoulos et al., 2009; Seo & Lee, 2000; Wang & Chung, 1997; Zhang et al., 2007b).

The underlying principle of ERM method is that the electrical resistance composites increases as the electrically conductive networks within the composite decreases due to the damages occurring under the various load during service. The intrinsic piezoresistive nature of CNTs make them good candidates to be used as strain or stress sensors in ERM approach. One of the techniques to make the composites electrically conductive is to disperse the CNTs within the polymer matrix (Zhang et al., 2015a). Since the CNTs have very high aspect ratio, the electrically conductive networks within the composite can be easily formed at low contents. Then the mechanical deformation within the composite becomes the function of electrical resistance change due to piezo-resistive properties of composites. However, it is a challenging task to uniformly disperse the CNTs within the resin due to the tendency of CNTs to form aggregates because of van der Waals interactions between them. This drawback might negatively effect the sensing behavior of damage within the composites. BP film sensors on the other hand eliminates this drawback since they have densely packed CNT networks formed by van der Waals interactions (Li & Kröger, 2012; Zhang et al., 2015b). Wang et al. (2022) used BP sensors to monitor the damage of composite wrapped pressure vessel (COPV) subjected to hydrostatic pressure tests and hydraulic fatigue tests. The electrical resistance change was recorded by BP sensors and resistance change was correlated with the strain data obtained by strain gages. The experimental findings established that BP sensors exhibit greater sensitivity and dependability in detecting the onset and spread of microcracks, as well as subtle stiffness reduction, within the COPV, surpassing the capabilities of strain gages.

Meng et al. (2009) employed BP as a framework to generate composites of CNT/polyaniline. These composites exhibited elevated specific capacitance, reduced internal resistivity, and enhanced stability across varying current loads. This unique combination of properties holds significant potential for the advancement of energy storage devices.

BP stands as a highly promising option for supercapacitors because of its excellent electrochemical capabilities, its lightweight composition, and its compact dimensions. Che et al. (2013) reported a novel technique to fabricate cellular SWNT BP films using conductive polymer and thermosetting resin. This novel manufacturing approach involves wrapping the initially prepared BP with a consistent coaxial layer of polypyrrole (PPy) through a pulsed electrochemical deposition method. This process targets the surfaces of individual SWNTs or bundles of SWNTs. Subsequently, the structure undergoes complete immersion with cyanate ester resin via dip infiltration technique. The resultant material demonstrates a substantial enhancement in mechanical characteristics, with a remarkable improvement of approximately 400% in both tensile modulus and strength compared to unmodified BP. In addition, its electrochemical performance was found to increase, reaching 320 F g^{-1} at a current density of 1 A g^{-1} and maintaining its original electrical and thermal properties.

Xiao et al. (2008) discovered that extremely thin buckypapers (BPs), when subjected to sound frequency electric currents, could emit loud sounds.

Xiao et al. (2008) discovered that extremely thin buckypapers (BPs), when subjected to sound frequency electric currents, could emit loud sounds. The extremely low heat capacity per unit area of BP films allows them to exhibit a broad frequency response range and a high sound pressure level. Since the developed BPs are extremely thin, flexible, and transparent, they can be customized into various shapes and sizes and could introduce innovative applications and manufacturing approaches for speakers and other acoustic devices.

Rechargeable lithium-ion batteries are extensively employed as energy storage devices due to their notably high energy density. These batteries are comprised of two electrodes, segregated by an electrolyte containing dissociated lithium salts, which facilitates the movement of lithium ions between the two electrodes. Zhang (2008) illustrated the benefits of employing hydroentangled CNT membranes in the context of lithium-ion batteries. Zhang employed hydroentangled CNT membranes as the active electrode within lithium-ion half cells. It was found that these membranes have the reversible capacity of approximately 250 mAh g^{-1} at a current density of 5 mA g^{-1}.

Battery electrodes are commonly crafted by binding active materials with polymers like polyvinylidene fluoride (PVDF) or polytetrafluoroethylene (PTFE), often supplemented with carbon black particles to enhance electrical conductivity. However, the inclusion of polymer binders in electrodes leads to a physical separation of numerous active materials from the electrolyte. This diminishes the effective interface area between the electrode and electrolyte, resulting in reductiom reaction kinetics, particularly under high-current density conditions. Remarkably conductive hydroentangled BP membranes offer a direct alternative as electrodes, eliminating the need for binders or carbon black conductors. These unique electrodes exhibit rapid kinetics even at elevated current densities. Zhang's work in 2008 (Zhang, 2008) illustrated that the interfacial resistance of hydroentangled CNT electrodes is approximately 69% lower than that of CNT electrodes bound with PVDF. This finding points to a notably accelerated electrode kinetics facilitated by hydroentangled CNT electrodes.

Many studies have focused on the use of BPs as adsorbent for the removal of contaminants and pathogens from the wate (Brady-Estévez et al., 2008; Holt et al., 2006; Srivastava et al., 2004). Wang et al. (2005) introduced an innovative high-flux filtration medium characterized by a three-layer composite architecture. The hydrophilic nanocomposite coating on the upper layer of this composite consists of a hydrophilic polyether-b-polyamide copolymer or a cross-linked PVA hydrogel and BP and was used for oil/water emulsion separation. Test outcomes demonstrated that the composite ultrafiltration membrane exhibited remarkable efficiency in retaining oil/water emulsions. Esteves et al. (2008) exhibited the successful elimination of bacteria and virus based pathogens from water using SWNT based BP filter, operating at low pressures. This filter was designed by incorporating a poly(vinylidene fluoride) (PVDF)-derived microporous membrane with a pore size of 5 μm, which was then coated with a thin layer of SWNTs.

6 Conclusion

This comprehensive review provides a general concept of the manufacturing techniques, characterization, morphology, and various applications of Boron Phosphides (BPs). BPs have gained considerable prominence in functional composites owing to their impressive electrical and thermal conductivity, resistance to corrosion, and favorable resin compatibility.

BPs can be produced through wet and dry methods. The wet approach involves dispersing CNTs in a solvent followed by filtration. Introducing chemical functionalization using different surfactants often enhances network stability at the expense of electrical conductivity. Aligning CNTs within BPs leads to heightened electrical and thermal conductivity, as well as improved mechanical properties. This alignment can be achieved via mechanical stretching or exposure to a magnetic field.

Hybridizing CNTs with other nanoparticles, such as Mxene and boron nitride, contributes to the enhancement of mechanical, electrical, and thermal properties. BPs exhibit potential as membrane materials due to their porous structures, suitable for adsorbing low-molecular-weight organic contaminants and toxins from water. Highly conductive BP membranes have found application as electrodes in battery technologies like lithium-ion batteries.

The minimal heat capacity per unit area of BP films grants them a broad frequency response range and high sound pressure level, rendering them excellent candidates for use in loudspeakers. BP's remarkable electrochemical capabilities make it a promising contender for supercapacitors. BP sensors have proven effective in structural health monitoring of composite structures due to their exceptional electrical conductivity.

Lastly, BPs have garnered significant utility in aircraft applications, serving purposes such as electromagnetic interference (EMI) shielding, lightning strike protection (LSP), and de-icing of aircraft components. In summary, BPs stand out as exceptional materials with versatile potential for innovative technological applications.

References

Ahmadalinezhad, A., Wu, G., & Chen, A. (2011). Mediator-free electrochemical biosensor based on buckypaper with enhanced stability and sensitivity for glucose detection. *Biosensors and Bioelectronics, 30*(1), 287–293. https://doi.org/10.1016/j.bios.2011.09.030

Alshahrani, A. A., AlQahtani, M., Almushaikeh, A. M., Hassan, H. M. A., Alzaid, M., Alrashidi, A. N., & Alsohaimi, I. H. (2022). Synthesis, characterization, and heavy-ion rejection rate efficiency of PVA/MWCNTs and Triton X-100/MWCNTs Buckypaper membranes. *Journal of Materials Research and Technology, 18*, 2310–2319. https://doi.org/10.1016/j.jmrt.2022.03.099

Alterary, S. S., Alshahrani, A. A., & Alsahli, S. A. (2022). Fabrication of novel buckypaper metal oxide nano-catalysis glycerol carbonate/MWCNTs membrane for efficient removal of heavy metals. *Heliyon, 8*(12), e12633. https://doi.org/10.1016/j.heliyon.2022.e12633

Amrollahi, A., Rashidi, A. M., Emami Meibodi, M., & Kashefi, K. (2009). Conduction heat transfer characteristics and dispersion behaviour of carbon nanofluids as a function of different parameters. *Journal of Experimental Nanoscience, 4*(4), 347–363. https://doi.org/10.1080/17458080902929929

Bai, Y., Zhu, M., Wang, S., Gao, F., Gao, R., Wang, C., Wang, G., Jin, H., Liu, L., Zhang, H., & Zhang, Z. (2023, February). Superaligned carbon nanotube film/quartz fiber composites towards advanced lightweight lightning strike protection. *Composites Part A: Applied Science and Manufacturing, 173*, 107617. https://doi.org/10.1016/j.compositesa.2023.107617

Balandin, A. A. (2011). Thermal properties of graphene and nanostructured carbon materials. *Nature Materials, 10*(8), 569–581. https://doi.org/10.1038/nmat3064

Barisci, J. N., Wallace, G. G., MacFarlane, D. R., & Baughman, R. H. (2004). Investigation of ionic liquids as electrolytes for carbon nanotube electrodes. *Electrochemistry Communications, 6*(1), 22–27. https://doi.org/10.1016/j.elecom.2003.09.015

Blase, X., Rubio, A., Louie, S. G., & Cohen, M. L. (1994). Stability and band gap constancy of boron nitride nanotubes. *EPL, 28*(5), 335–340. https://doi.org/10.1209/0295-5075/28/5/007

Brady-Estévez, A. S., Kang, S., & Elimelech, M. (2008). A single-walled-carbon-nanotube filter for removal of viral and bacterial pathogens. *Small, 4*(4), 481–484. https://doi.org/10.1002/smll.200700863

Buehler, M. J. (2006). Mesoscale modelling of mechanics of carbon nanotubes: Self-assembly, self-folding, and fracture. *Journal of Materials Research, 21*(11), 2855–2869. https://doi.org/10.1557/jmr.2006.0347

Cao, Q., & Rogers, J. A. (2009). Ultrathin films of single-walled carbon nanotubes for electronics and sensors: A review of fundamental and applied aspects. *Advanced Materials, 21*(1), 29–53. https://doi.org/10.1002/adma.200801995

Cha, S. I., Kim, K. T., Lee, K. H., Mo, C. B., Jeong, Y. J., & Hong, S. H. (2008). Mechanical and electrical properties of cross-linked carbon nanotubes. *Carbon, 46*(3), 482–488. https://doi.org/10.1016/j.carbon.2007.12.023

Chandra, Y., Adhikari, S., Mukherjee, S., & Mukhopadhyay, T. (2022). Unfolding the mechanical properties of buckypaper composites: nano- to macro-scale coupled atomistic-continuum simulations. In Engineering with computers (38, 6). Springer . https://doi.org/10.1007/s00366-021-01538-w.

Chang, C. W., Fennimore, A. M., Afanasiev, A., Okawa, D., Ikuno, T., Garcia, H., Li, D., Majumdar, A., & Zettl, A. (2006). Isotope effect on the thermal conductivity of boron nitride nanotubes. *Physical Review Letters, 97*(8), 085901. https://doi.org/10.1103/PhysRevLett.97.085901

Che, J., Chen, P., & Chan-Park, M. B. (2013). High-strength carbon nanotube buckypaper composites as applied to free-standing electrodes for supercapacitors. *Journal of Materials Chemistry A, 1*(12), 4057–4066. https://doi.org/10.1039/c3ta01421e

Chen, B., Gao, M., Zuo, J. M., Qu, S., Liu, B., & Huang, Y. (2003). Binding energy of parallel carbon nanotubes. *Applied Physics Letters, 83*(17), 3570–3571. https://doi.org/10.1063/1.1623013

Chen, Y., Zou, J., Campbell, S. J., & Caer, G. L. (2004). Boron nitride nanotubes: Pronounced resistance to oxidation. *Applied Physics Letters, 84*(13), 2430–2432. https://doi.org/10.1063/1.1667278

Chen, H., Chen, M., Di, J., Xu, G., Li, H., & Li, Q. (2012a). Architecting three-dimensional networks in carbon nanotube buckypapers for thermal interface materials. *Journal of Physical Chemistry C, 116*(6), 3903–3909. https://doi.org/10.1021/jp2086158

Chen, I. W. P., Cottinet, P. J., Tsai, S. Y., Foster, B., Liang, R., Wang, B., & Zhang, C. (2012b). Improved performance of carbon nanotube buckypaper and ionic-liquid-in- Nafion actuators for rapid response and high durability in the open air. *Sensors and Actuators, B: Chemical, 171–172*, 515–521. https://doi.org/10.1016/j.snb.2012.05.022

Chen, X., Dmuchowski, C. M., Park, C., Fay, C. C., & Ke, C. (2017). Quantitative characterization of structural and mechanical properties of boron nitride nanotubes in high temperature environments. *Scientific Reports, 7*(1), 11388. https://doi.org/10.1038/s41598-017-11795-9

Chen, K., Feng, Y., Shi, Y., Wang, H., Fu, L., Liu, M., Lv, Y., Yang, F., Yu, B., Liu, M., & Song, P. (2022). Flexible and fire safe sandwich structured composites with superior electromagnetic interference shielding properties. *Composites Part A: Applied Science and Manufacturing, 160*, 107070. https://doi.org/10.1016/j.compositesa.2022.107070

Cheung, W., Chiu, P. L., Parajuli, R. R., Ma, Y., Ali, S. R., & He, H. (2009). Fabrication of high performance conducting polymer nanocomposites for biosensors and flexible electronics: Summary of the multiple roles of DNA dispersed and functionalized single walled carbon nanotubes. *Journal of Materials Chemistry, 19*(36), 6465–6480. https://doi.org/10.1039/b823065j

Choi, J., Jang, S. H., Oh, J. Y., Yang, S. J., Park, N. H., Jeong, W., & Lim, D. Y. (2017). High-density carbon nanotube wet-laid buckypapers with enhanced strength and conductivity using a high-pressure homogenization process. *Bulletin of the Korean Chemical Society, 38*(4), 438–443. https://doi.org/10.1002/bkcs.11104

Chopra, N. G., & Zettl, A. (1998). Measurement of the elastic modulus of a multi-wall boron nitride nanotube. *Solid State Communications, 105*(5), 297–300. https://doi.org/10.1016/S0038-1098(97)10125-9

Chu, H., Zhang, Z., Liu, Y., & Leng, J. (2014). Self-heating fiber reinforced polymer composite using meso/macropore carbon nanotube paper and its application in deicing. *Carbon, 66*, 154–163. https://doi.org/10.1016/j.carbon.2013.08.053

Coleman, J. N., Blau, W. J., Dalton, A. B., Muñoz, E., Collins, S., Kim, B. G., Razal, J., Selvidge, M., Vieiro, G., & Baughman, R. H. (2003). Improving the mechanical properties of single-walled carbon nanotube sheets by intercalation of polymeric adhesives. *Applied Physics Letters, 82*(11), 1682–1684. https://doi.org/10.1063/1.1559421

Collins, P., & Avouris, P. (2000, December). Nanotubes electronics. *Scientific American, 283*, 62.

Coluci, V. R., Hall, L. J., Kozlov, M. E., Zhang, M., Dantas, S. O., Galvão, D. S., & Baughman, R. H. (2008). Modeling the auxetic transition for carbon nanotube sheets. *Physical Review B – Condensed Matter and Materials Physics, 78*(11), 115408. https://doi.org/10.1103/PhysRevB.78.115408

Cottinet, P. J., Souders, C., Labrador, D., Porter, S., Liang, Z., Wang, B., & Zhang, C. (2011). Nonlinear strain-electric field relationship of carbon nanotube buckypaper/Nafion actuators. *Sensors and Actuators, A: Physical, 170*(1–2), 164–171. https://doi.org/10.1016/j.sna.2011.06.013

Dai, H., Kong, J., Zhou, C., Franklin, N., Tombler, T., Cassell, A., Fan, S., & Chapline, M. (1999). Controlled chemical routes to nanotube architectures, physics, and devices. *Journal of Physical Chemistry B, 103*(51), 11246–11255. https://doi.org/10.1021/jp992328o

Devi, G., Priya, R., Tapas Bapu, B. R., Thandaiah Prabu, R., Sathish Kumar, P. J., & Anusha, N. (2022). Role of carbonaceous fillers in electromagnetic interference shielding behavior of polymeric composites: A review. *Polymer Composites, 43*(11), 7701–7723. https://doi.org/10.1002/pc.27009

Falzon, B. G., Robinson, P., Frenz, S., & Gilbert, B. (2015). Development and evaluation of a novel integrated anti-icing/de-icing technology for carbon fibre composite aerostructures using an electro-conductive textile. *Composites Part A: Applied Science and Manufacturing, 68*, 323–335. https://doi.org/10.1016/j.compositesa.2014.10.023

Faraji, M., & Mohammadzadeh Aydisheh, H. (2019). Facile and scalable preparation of highly porous polyvinyl chloride-multi walled carbon nanotubes-polyaniline composite film for solid-state flexible supercapacitor. *Composites Part B: Engineering, 168*, 432–441. https://doi.org/10.1016/j.compositesb.2019.03.060

Fischer, J. E., Zhou, W., Vavro, J., Llaguno, M. C., Guthy, C., Haggenmueller, R., Casavant, M. J., Walters, D. E., & Smalley, R. E. (2003). Magnetically aligned single wall carbon nanotube films: Preferred orientation and anisotropic transport properties. *Journal of Applied Physics, 93*(4), 2157–2163. https://doi.org/10.1063/1.1536733

Frizzell, C. J., In Het Panhuis, M., Coutinho, D. H., Balkus, K. J., Minett, A. I., Blau, W. J., & Coleman, J. N. (2005). Reinforcement of macroscopic carbon nanotube structures by polymer intercalation: The role of polymer molecular weight and chain conformation. *Physical Review B – Condensed Matter and Materials Physics, 72*(24), 245420. https://doi.org/10.1103/PhysRevB.72.245420

Gagné, M., & Therriault, D. (2014). Lightning strike protection of composites. *Progress in Aerospace Sciences, 64*, 1–16. https://doi.org/10.1016/j.paerosci.2013.07.002

Gelves, G. A., Al-Saleh, M. H., & Sundararaj, U. (2011). Highly electrically conductive and high performance EMI shielding nanowire/polymer nanocomposites by miscible mixing and precipitation. *Journal of Materials Chemistry, 21*(3), 829–836. https://doi.org/10.1039/c0jm02546a

Gonnet, P., Liang, Z., Choi, E. S., Kadambala, R. S., Zhang, C., Brooks, J. S., Wang, B., & Kramer, L. (2006). Thermal conductivity of magnetically aligned carbon nanotube buckypapers and nanocomposites. *Current Applied Physics, 6*(1), 119–122. https://doi.org/10.1016/j.cap.2005.01.053

Gou, J. (2006). Single-walled nanotube bucky paper and nanocomposite. *Polymer International, 55*(11), 1283–1288. https://doi.org/10.1002/pi.2079

Gou, J., Tang, Y., Liang, F., Zhao, Z., Firsich, D., & Fielding, J. (2010). Carbon nanofiber paper for lightning strike protection of composite materials. *Composites Part B: Engineering, 41*(2), 192–198. https://doi.org/10.1016/j.compositesb.2009.06.009

Guler, M. O., Cetinkaya, T., Cevher, O., Tocoglu, U., & Akbulut, H. (2014a). Nanocomposite ZnO:MWCNT thin films for Li-ion batteries prepared via reactive magnetron sputtering. *Acta Physica Polonica A, 125*(2), 319–321. https://doi.org/10.12693/APhysPolA.125.319

Guler, M. O., Cetinkaya, T., Tocoglu, U., & Akbulut, H. (2014b). Electrochemical performance of MWCNT reinforced ZnO anodes for Li-ion batteries. *Microelectronic Engineering, 118*, 54–60. https://doi.org/10.1016/j.mee.2013.12.029

Guo, J., Li, M., Liu, Q., Gu, Y., Li, Y., & Zhang, Z. (2013). Influence of oxidation and distribution of carbon nanotube on mechanical properties of buckypaper/epoxy composites. *Journal of Reinforced Plastics and Composites, 32*(4), 248–257. https://doi.org/10.1177/0731684412468799

Hall, L. J., Coluci, V. R., Galvão, D. S., Kozlov, M. E., Zhang, M., Dantas, S. O., & Baughman, R. H. (2008). Sign change of poisson's ratio for carbon nanotube sheets. *Science, 320*(5875), 504–507. https://doi.org/10.1126/science.1149815

Hamasaki, H., Kawase, T., & Hirahara, K. (2022). Anisotropic temperature distribution in carbon nanotube bundles determined using two types of phase transitions of nanoparticles. *Physical Review Materials, 6*(2), L023001. https://doi.org/10.1103/PhysRevMaterials.6.L023001

Han, Z., & Fina, A. (2011). Thermal conductivity of carbon nanotubes and their polymer nanocomposites: A review. *Progress in Polymer Science (Oxford), 36*(7), 914–944. https://doi.org/10.1016/j.progpolymsci.2010.11.004

Han, J. H., Zhang, H., Chen, M. J., Wang, G. R., & Zhang, Z. (2014). CNT buckypaper/thermoplastic polyurethane composites with enhanced stiffness, strength and toughness. *Composites Science and Technology, 103*, 63–71. https://doi.org/10.1016/j.compscitech.2014.08.015

Han, J. H., Zhang, H., Chen, M. J., Wang, D., Liu, Q., Wu, Q. L., & Zhang, Z. (2015). The combination of carbon nanotube buckypaper and insulating adhesive for lightning strike protection of the carbon fiber/epoxy laminates. *Carbon, 94*, 101–113. https://doi.org/10.1016/j.carbon.2015.06.026

He, J., Han, M., Wen, K., Liu, C., Zhang, W., Liu, Y., Su, X., Zhang, C., & Liang, C. (2023). Absorption-dominated electromagnetic interference shielding assembled composites based on modular design with infrared camouflage and response switching. *Composites Science and Technology, 231*, 109799. https://doi.org/10.1016/J.COMPSCITECH.2022.109799

Holt, J. K., Park, H. G., Wang, Y., Stadermann, M., Artyukhin, A. B., Grigoropoulos, C. P., Noy, A., & Bakajin, O. (2006). Fast mass transport through sub-2-nanometer carbon nanotubes. *Science, 312*(5776), 1034–1037. https://doi.org/10.1126/science.1126298

Hone, J. (2001). Phonons and thermal properties of carbon nanotubes. Carbon nanotubes. *Topics in Applied Physics, 80*, 273–286. https://doi.org/10.1007/3-540-39947-X_11

Hone, J., Llaguno, M. C., Nemes, N. M., Johnson, A. T., Fischer, J. E., Walters, D. A., Casavant, M. J., Schmidt, J., & Smalley, R. E. (2000). Electrical and thermal transport properties of magnetically aligned single wall carbon nanotube films. *Applied Physics Letters, 77*(5), 666–668. https://doi.org/10.1063/1.127079

Hone, J., Llaguno, M. C., Biercuk, M. J., Johnson, A. T., Batlogg, B., Benes, Z., & Fischer, J. E. (2002). Thermal properties of carbon nanotubes and nanotube-based materials. *Applied Physics A: Materials Science and Processing, 74*(3), 339–343. https://doi.org/10.1007/s003390201277

Hone, J., Whitney, M., Piskoti, C., & A. Z. (2016). Thermal conductivity of single-walled carbon nanotubes. *Journal of Experimental and Theoretical Physics, 123*(6), 1084–1089. https://doi.org/10.1103/PhysRevB.59.R2514

Hong, W. T., & Tai, N. H. (2008). Investigations on the thermal conductivity of composites reinforced with carbon nanotubes. *Diamond and Related Materials, 17*(7–10), 1577–1581. https://doi.org/10.1016/j.diamond.2008.03.037
Hu, Y., Li, D., Tang, P., Bin, Y., & Wang, H. (2019a). Comparative study of structure, mechanical and electromagnetic interference shielding properties of carbon nanotube buckypapers prepared by different dispersion media. *Materials and Design, 184*, 108175. https://doi.org/10.1016/j.matdes.2019.108175
Hu, Y., Li, D., Wu, L., Yang, J., Jian, X., & Bin, Y. (2019b, April). Carbon nanotube buckypaper and buckypaper/polypropylene composites for high shielding effectiveness and absorption-dominated shielding material. *Composites Science and Technology, 181*, 107699. https://doi.org/10.1016/j.compscitech.2019.107699
Inoue, Y., Suzuki, Y., Minami, Y., Muramatsu, J., Shimamura, Y., Suzuki, K., Ghemes, A., Okada, M., Sakakibara, S., Mimura, H., & Naito, K. (2011). Anisotropic carbon nanotube papers fabricated from multiwalled carbon nanotube webs. *Carbon, 49*(7), 2437–2443. https://doi.org/10.1016/j.carbon.2011.02.010
Iqbal, A., Hassan, T., Gao, Z., Shahzad, F., & Koo, C. M. (2023). MXene-incorporated 1D/2D nano-carbons for electromagnetic shielding: A review. *Carbon, 203*, 542–560. https://doi.org/10.1016/j.carbon.2022.11.104
Irving, P. E., & Thiagarajan, C. (1999). Fatigue damage characterization in carbon fibre composite materials using an electrical potential technique. *Smart Materials and Structures, 7*(4), 456–466. https://doi.org/10.1088/0964-1726/7/4/004
Itkis, M. E., Borondics, F., Yu, A., & Haddon, R. C. (2007). Thermal conductivity measurements of semitransparent single-walled carbon nanotube films by a bolometric technique. *Nano Letters, 7*(4), 900–904. https://doi.org/10.1021/nl062689x
Jia, X., Chen, Z., Suwarnasarn, A., Rice, L., Wang, X., Sohn, H., Zhang, Q., Wu, B. M., Wei, F., & Lu, Y. (2012). High-performance flexible lithium-ion electrodes based on robust network architecture. *Energy and Environmental Science, 5*(5), 6845–6849. https://doi.org/10.1039/c2ee03110h
Jones, R. S., Gonzalez-Munoz, S., Griffiths, I., Holdway, P., Evers, K., Luanwuthi, S., Maciejewska, B. M., Kolosov, O., & Grobert, N. (2023). Thermal conductivity of carbon/boron nitride heteronanotube and boron nitride nanotube Buckypapers: Implications for thermal management composites. *ACS Applied Nano Materials., 6*(17), 15374–15384. https://doi.org/10.1021/acsanm.3c01147
Kang, I., Schulz, M. J., Kim, J. H., Shanov, V., & Shi, D. (2006). A carbon nanotube strain sensor for structural health monitoring. *Smart Materials and Structures, 15*(3), 737–748. https://doi.org/10.1088/0964-1726/15/3/009
Kar, S., Bindal, R. C., & Tewari, P. K. (2012). Carbon nanotube membranes for desalination and water purification: Challenges and opportunities. *Nano Today, 7*(5), 385–389. https://doi.org/10.1016/j.nantod.2012.09.002
Kessler, M. R., Sottos, N. R., & White, S. R. (2003). Self-healing structural composite materials. *Composites Part A: Applied Science and Manufacturing, 34*(8), 743–753. https://doi.org/10.1016/S1359-835X(03)00138-6
Kim, D. S., & Park, Y. J. (2014). Buckypaper electrode containing carbon nanofiber/Co3O4 composite for enhanced lithium air batteries. *Solid State Ionics, 268*(PB), 216–221. https://doi.org/10.1016/j.ssi.2014.03.010
Kim, Y. A., Muramatsu, H., Hayashi, T., Endo, M., Terrones, M., & Dresselhaus, M. S. (2006). Fabrication of high-purity, double-walled carbon nanotube buckypaper. *Chemical Vapor Deposition, 12*(6), 327–330. https://doi.org/10.1002/cvde.200504217
Kim, K. S., Jakubinek, M. B., Martinez-Rubi, Y., Ashrafi, B., Guan, J., O'Neill, K., Plunkett, M., Hrdina, A., Lin, S., Dénommée, S., Kingston, C., & Simard, B. (2015). Polymer nanocomposites from free-standing, macroscopic boron nitride nanotube assemblies. *RSC Advances, 5*(51), 41186–41191. https://doi.org/10.1039/c5ra02988k
Kis, A., Csányi, G., Salvetat, J. P., Lee, T. N., Couteau, E., Kulik, A. J., Benoit, W., Brugger, J., & Forró, L. (2004). Reinforcement of single-walled carbon nanotube bundles by intertube bridging. *Nature Materials, 3*(3), 153–157. https://doi.org/10.1038/nmat1076

Kostopoulos, V., Vavouliotis, A., Karapappas, P., Tsotra, P., & Paipetis, A. (2009). Damage monitoring of carbon fiber reinforced laminates using resistance measurements. Improving sensitivity using carbon nanotube doped epoxy matrix system. *Journal of Intelligent Material Systems and Structures, 20*(9), 1025–1034. https://doi.org/10.1177/1045389X08099993

Kukovecz, Á., Smajda, R., Oze, M., Schaefer, B., Haspel, H., Kónya, Z., & Kiricsi, I. (2008). Multiwall carbon nanotube films surface-doped with electroceramics for sensor applications. *Physica Status Solidi (B) Basic Research, 245*(10), 2331–2334. https://doi.org/10.1002/pssb.200879556

Kumanek, B., & Janas, D. (2019). Thermal conductivity of carbon nanotube networks: a review. *Journal of Materials Science, 54*(10), 7397–7427. https://doi.org/10.1007/s10853-019-03368-0

Kumar, V., Sharma, S., Pathak, A., Singh, B. P., Dhakate, S. R., Yokozeki, T., Okada, T., & Ogasawara, T. (2019). Interleaved MWCNT buckypaper between CFRP laminates to improve through-thickness electrical conductivity and reducing lightning strike damage. *Composite Structures, 210*(September 2018), 581–589. https://doi.org/10.1016/j.compstruct.2018.11.088

Li, Y., & Kröger, M. (2012). A theoretical evaluation of the effects of carbon nanotube entanglement and bundling on the structural and mechanical properties of buckypaper. *Carbon, 50*(5), 1793–1806. https://doi.org/10.1016/j.carbon.2011.12.027

Li, Y. L., Zhong, X. H., & Windle, A. H. (2008). Structural changes of carbon nanotubes in their macroscopic films and fibers by electric sparking processing. *Carbon, 46*(13), 1751–1756. https://doi.org/10.1016/J.CARBON.2008.07.021

Li, H., Ru, X., Song, Y., Wang, H., Yang, C., Gong, L., Liu, Z., Zhang, Q., & Chen, Y. (2022). Flexible and self-healing 3D MXene/reduced graphene oxide/polyurethane composites for high-performance electromagnetic interference shielding. *Composites Science and Technology, 227*, 109602. https://doi.org/10.1016/j.compscitech.2022.109602

Liew, K. M., Wong, C. H., & Tan, M. J. (2005). Buckling properties of carbon nanotube bundles. *Applied Physics Letters, 87*(4), 041901. https://doi.org/10.1063/1.2001135

Lin, W., Shang, J., Gu, W., & Wong, C. P. (2012). Parametric study of intrinsic thermal transport in vertically aligned multi-walled carbon nanotubes using a laser flash technique. *Carbon, 50*(4), 1591–1603. https://doi.org/10.1016/j.carbon.2011.11.038

Lynch, F. T., & Khodadoust, A. (2001). Effects of ice accretions on aircraft aerodynamics. *Progress in Aerospace Sciences, 37*(8), 669–767. https://doi.org/10.1016/S0376-0421(01)00018-5

Ma, W., Song, L., Yang, R., Zhang, T., Zhao, Y., Sun, L., Ren, Y., Liu, D., Liu, L., Shen, J., Zhang, Z., Xiang, Y., Zhou, W., & Xie, S. S. (2007). Directly synthesized strong, highly conducting, transparent single-walled carbon nanotube films. *Nano Letters, 7*(8), 2307–2311. https://doi.org/10.1021/nl070915c

Malekimoghadam, R., Hosseini, S. A., & Icardi, U. (2023). Bending analysis of carbon nanotube coated–fiber multi-scale composite beams using the refined zigzag theory. *Aerospace Science and Technology, 138*, 108328. https://doi.org/10.1016/j.ast.2023.108328

Mall, S., Ouper, B. L., & Fielding, J. C. (2009). Compression strength degradation of nanocomposites after lightning strike. *Journal of Composite Materials, 43*(24), 2987–3001. https://doi.org/10.1177/0021998309345337

Marconnet, A. M., Panzer, M. A., & Goodson, K. E. (2013). Thermal conduction phenomena in carbon nanotubes and related nanostructured materials. *Reviews of Modern Physics, 85*(3), 1295. https://doi.org/10.1103/RevModPhys.85.1295

Meng, G., Jung, Y. J., Cao, A., Vajtai, R., & Ajayan, P. M. (2005). Controlled fabrication of hierarchically branched nanopores, nanotubes, and nanowires. *Proceedings of the National Academy of Sciences of the United States of America, 102*(20), 7074–7078. https://doi.org/10.1073/pnas.0502098102

Meng, C., Liu, C., & Fan, S. (2009). Flexible carbon nanotube/polyaniline paper-like films and their enhanced electrochemical properties. *Electrochemistry Communications, 11*(1), 186–189. https://doi.org/10.1016/j.elecom.2008.11.005

Muramatsu, H., Hayashi, T., Kim, Y. A., Shimamoto, D., Kim, Y. J., Tantrakarn, K., Endo, M., Terrones, M., & Dresselhaus, M. S. (2005). Pore structure and oxidation stability of double-walled carbon nanotube-derived bucky paper. *Chemical Physics Letters, 414*(4–6), 444–448. https://doi.org/10.1016/j.cplett.2005.08.110

Novotny, E. V., Murphy, D., & Stefan, H. G. (2008). Increase of urban lake salinity by road deicing salt. *Science of the Total Environment, 406*(1–2), 131–144. https://doi.org/10.1016/j.scitotenv.2008.07.037

Oh, J. Y., Yang, S. J., Park, J. Y., Kim, T., Lee, K., Kim, Y. S., Han, H. N., & Park, C. R. (2015). Easy preparation of self-assembled high-density buckypaper with enhanced mechanical properties. *Nano Letters, 15*(1), 190–197. https://doi.org/10.1021/nl5033588

Park, J. G., Louis, J., Cheng, Q., Bao, J., Smithyman, J., Liang, R., Wang, B., Zhang, C., Brooks, J. S., Kramer, L., Fanchasis, P., & Dorough, D. (2009a). Electromagnetic interference shielding properties of carbon nanotube buckypaper composites. *Nanotechnology, 20*(41), 415702. https://doi.org/10.1088/0957-4484/20/41/415702

Park, J. G., Smithyman, J., Lin, C. Y., Cooke, A., Kismarahardja, A. W., Li, S., Liang, R., Brooks, J. S., Zhang, C., & Wang, B. (2009b). Effects of surfactants and alignment on the physical properties of single-walled carbon nanotube buckypaper. *Journal of Applied Physics, 106*(10), 104310. https://doi.org/10.1063/1.3255901

Peter Chen, I. W., Liang, R., Zhao, H., Wang, B., & Zhang, C. (2011). Highly conductive carbon nanotube buckypapers with improved doping stability via conjugational cross-linking. *Nanotechnology, 22*(48), 485708. https://doi.org/10.1088/0957-4484/22/48/485708

Pham, G. T., Park, Y. B., Wang, S., Liang, Z., Wang, B., Zhang, C., Funchess, P., & Kramer, L. (2008). Mechanical and electrical properties of polycarbonate nanotube buckypaper composite sheets. *Nanotechnology, 19*(32), 325705. https://doi.org/10.1088/0957-4484/19/32/325705

Pierson, H. O. (2021). *Handbook of carbon, graphite, diamond and fullerenes properties, processing and applications* (Vol. 1–2). Noyes Publications. https://doi.org/10.1002/9783527674046.ch2

Pop, E., Mann, D., Wang, Q., Goodson, K., & Dai, H. (2006). Thermal conductance of an individual single-wall carbon nanotube above room temperature. *Nano Letters, 6*(1), 96–100. https://doi.org/10.1021/nl052145f

Rani, M., Sehrawat, M., Rani, R., Gahtori, B., & Singh, B. P. (2023, February). Tailoring the physical characteristics of buckypaper via controlling the surfactant concentration. *Surfaces and Interfaces, 37*, 102713. https://doi.org/10.1016/j.surfin.2023.102713

Rein, M. D., Breuer, O., & Wagner, H. D. (2011). Sensors and sensitivity: Carbon nanotube buckypaper films as strain sensing devices. *Composites Science and Technology, 71*(3), 373–381. https://doi.org/10.1016/j.compscitech.2010.12.008

Ribeiro, B., Botelho, E. C., Costa, M. L., & Bandeira, C. F. (2017). Carbon nanotube buckypaper reinforced polymer composites: A review. *Polimeros, 27*(3), 247–255. https://doi.org/10.1590/0104-1428.03916

Ribeiro, B., Gomes, N. A. S., Baldan, M., & Rezende, M. C. (2023). Designing Sandwich architecture towards glass Fiber/epoxy reinforced multi-walled carbon nanotube Buckypaper composites for electromagnetic interference shielding. *Electronic Materials Letters, 19*(2), 184–191. https://doi.org/10.1007/s13391-022-00382-4

Rinzler, A., Liu, J., Dai, H., & Nikolaev, P. (1998). Large-scale purification of single-wall carbon nanotubes: Process, product, and characterization. *Applied Physics A, 37*, 29–37. http://www.springerlink.com/index/mrmex1mghbrrjqmw.pdf

Rojas, J. A., Ardila-Rodríguez, L. A., Diniz, M. F., Gonçalves, M., Ribeiro, B., & Rezende, M. C. (2019). Optimization of Triton X-100 removal and ultrasound probe parameters in the preparation of multiwalled carbon nanotube buckypaper. *Materials and Design, 166*, 107612. https://doi.org/10.1016/j.matdes.2019.107612

Roy, S., Jain, V., Bajpai, R., Ghosh, P., Pente, A. S., Singh, B. P., & Misra, D. S. (2012). Formation of carbon nanotube bucky paper and feasibility study for filtration at the nano and molecular scale. *Journal of Physical Chemistry C, 116*(35), 19025–19031. https://doi.org/10.1021/jp305677h

Saito, R., Fujita, M., Dresselhaus, G., & Dresselhaus, M. S. (1992). Electronic structure of chiral graphene tubules. *Applied Physics Letters, 60*(18), 2204–2206. https://doi.org/10.1063/1.107080

Sakurai, S., Kamada, F., Futaba, D. N., Yumura, M., & Hata, K. (2013). Influence of lengths of millimeter-scale single-walled carbon nanotube on electrical and mechanical properties of buckypaper. *Nanoscale Research Letters, 8*(1), 546. https://doi.org/10.1186/1556-276X-8-546

Sears, K., Dumée, L., Schütz, J., She, M., Huynh, C., Hawkins, S., Duke, M., & Gray, S. (2010). Recent developments in carbon nanotube membranes for water purification and gas separation. *Materials, 3*(1), 127–149. https://doi.org/10.3390/ma3010127

Seo, D., & Lee, J. (2000). Damage detection of CFRP laminates using electrical resistance measurement and neural network. *Composite Structures, 47*(1999), 525–530.

Singh, C., Shaffer, M. S. P., & Windle, A. H. (2003). Production of controlled architectures of aligned carbon nanotubes by an injection chemical vapour deposition method. *Carbon, 41*(2), 359–368. https://doi.org/10.1016/S0008-6223(02)00314-7

Skákalová, V., Kaiser, A. B., Dettlaff-Weglikowska, U., Hrnčariková, K., & Roth, S. (2005). Effect of chemical treatment on electrical conductivity, infrared absorption, and raman spectra of single-walled carbon nanotubes. *Journal of Physical Chemistry B, 109*(15), 7174–7181. https://doi.org/10.1021/jp044741o

Smajda, R., Györi, Z., Sápi, A., Veres, M., Oszkó, A., Kis-Csitári, J., Kukovecz, Á., Kónya, Z., & Kiricsi, I. (2007a). Spectroscopic studies on self-supporting multi-wall carbon nanotube based composite films for sensor applications. *Journal of Molecular Structure, 834–836*(SPEC. ISS), 471–476. https://doi.org/10.1016/J.MOLSTRUC.2006.10.012

Smajda, R., Kukovecz, Á., Kónya, Z., & Kiricsi, I. (2007b). Structure and gas permeability of multi-wall carbon nanotube buckypapers. *Carbon, 45*(6), 1176–1184. https://doi.org/10.1016/j.carbon.2007.02.022

Sreekumar, P. A., & Thomas, S. (2008). *Properties and performance of natural-fibre composites.* CRC Press. https://doi.org/10.1533/9781845694593.1.67

Srivastava, A., Srivastava, O. N., Talapatra, S., Vajtai, R., & Ajayan, P. M. (2004). Carbon nanotube filters. *Nature Materials, 3*(9), 610–614. https://doi.org/10.1038/nmat1192

Suppiger, D., Busato, S., & Ermanni, P. (2008). Characterization of single-walled carbon nanotube mats and their performance as electromechanical actuators. *Carbon, 46*(7), 1085–1090. https://doi.org/10.1016/j.carbon.2008.03.019

Tang, Y., Zhuge, J., Gou, J., Chen, R. H., Ibeh, C., & Hu, Y. (2011a). Morphology, thermal stability, and flammability of polymer matrix composites coated with hybrid nanopapers. *Polymers for Advanced Technologies, 22*(10), 1403–1413. https://doi.org/10.1002/pat.1621

Tang, Y., Zhuge, J., Lawrence, J., McKee, J., Gou, J., Ibeh, C., & Hu, Y. (2011b). Flame retardancy of carbon nanofibre/intumescent hybrid paper based fibre reinforced polymer composites. *Polymer Degradation and Stability, 96*(5), 760–770. https://doi.org/10.1016/j.polymdegradstab.2011.02.013

Tarfaoui, M., El Moumen, A., Boehle, M., Shah, O., & Lafdi, K. (2019). Self-heating and deicing epoxy/glass fiber based carbon nanotubes buckypaper composite. *Journal of Materials Science, 54*(2), 1351–1362. https://doi.org/10.1007/s10853-018-2917-9

Tian, K., Hu, D., Wei, Q., Fu, Q., & Deng, H. (2023). Recent progress on multifunctional electromagnetic interference shielding polymer composites. *Journal of Materials Science and Technology, 134,* 106–131. https://doi.org/10.1016/j.jmst.2022.06.031

Tieu, D. T., Nguyen, M. K., Nguyen, T. T. H., Ngo, T. V. K., & Do, Q. H. (2018). Study to fabricate the large scale Buckypaper based on carbon nanotubes. *Communications Physics, 28*(3), 217–224. https://doi.org/10.15625/0868-3166/28/3/10758

Trakakis, G., Tasis, D., Parthenios, J., Galiotis, C., & Papagelis, K. (2013). Structural properties of chemically functionalized carbon nanotube thin films. *Materials, 6*(6), 2360–2371. https://doi.org/10.3390/ma6062360

Vahid, N. F., Marvi, M. R., Naimi-Jamal, M. R., Naghib, S. M., & Ghaffarinejad, A. (2018). Effect of surfactant type on buckypaper electrochemical performance. *Micro and Nano Letters, 13*(7), 927–930. https://doi.org/10.1049/mnl.2017.0691

Ventura, D. N., Stone, R. A., Chen, K. S., Hariri, H. H., Riddle, K. A., Fellers, T. J., Yun, C. S., Strouse, G. F., Kroto, H. W., & Acquah, S. F. A. (2010). Assembly of cross-linked multi-walled carbon nanotube mats. *Carbon, 48*(4), 987–994. https://doi.org/10.1016/j.carbon.2009.11.016

Vohrer, U., Kolaric, I., Haque, M. H., Roth, S., & Detlaff-Weglikowska, U. (2004). Carbon nanotube sheets for the use as artificial muscles. *Carbon, 42*(5–6), 1159–1164. https://doi.org/10.1016/j.carbon.2003.12.044

Wan Dalina, W. A. D., Mariatti, M., & Tan, S. H. (2019). Multi-walled carbon nanotubes buckypaper/epoxy composites: Effect of loading and pressure on tensile and electrical properties. *Polymer Bulletin, 76*(6), 2801–2817. https://doi.org/10.1007/s00289-018-2530-8

Wang, X. W., & Chung, D. D. L. (1997). Sensing delamination in a carbon fiber polymer-matrix composite during fatigue by electrical resistance measurement. *Polymer Composites, 18*(6), 692–700. https://doi.org/10.1002/pc.10322

Wang, Z., Liang, Z., Wang, B., Zhang, C., & Kramer, L. (2004). Processing and property investigation of single-walled carbon nanotube (SWNT) buckypaper/epoxy resin matrix nanocomposites. *Composites Part A: Applied Science and Manufacturing, 35*(10), 1225–1232. https://doi.org/10.1016/j.compositesa.2003.09.029

Wang, X., Chen, X., Yoon, K., Fang, D., Hsiao, B. S., & Chu, B. (2005). High flux filtration medium based on nanofibrous substrate with hydrophilic nanocomposite coating. *Environmental Science and Technology, 39*(19), 7684–7691. https://doi.org/10.1021/es050512j

Wang, D., Song, P., Liu, C., Wu, W., & Fan, S. (2008). Highly oriented carbon nanotube papers made of aligned carbon nanotubes. *Nanotechnology, 19*(7), 075609. https://doi.org/10.1088/0957-4484/19/7/075609

Wang, G. J., Zu, M., & Qiu, J. (2011). Preparation of amine-modified multiwalled carbon nanotubes and their use in composites. *Xinxing Tan Cailiao/New Carbon Materials, 26*(2), 3705. https://doi.org/10.1016/j.carbon.2011.04.026

Wang, X., Lin, L., Lu, S., Zhang, L., Li, B., Zhang, D., & Luo, Y. (2022). Evaluation of embedded buckypaper sensors in composite overwrappedped pressure vessels for progressive damage monitoring. *Composite Structures, 284*(May 2020), 115223. https://doi.org/10.1016/j.compstruct.2022.115223

Wu, Y., Cheng, X., Chen, S., Qu, B., Wang, R., Zhuo, D., & Wu, L. (2021). In situ formation of a carbon nanotube buckypaper for improving the interlaminar properties of carbon fiber composites. *Materials and Design, 202*, 109535. https://doi.org/10.1016/j.matdes.2021.109535

Xia, Q., Zhang, Z., Chu, H., Liu, Y., & Leng, J. (2018). Research on high electromagnetic interference shielding effectiveness of a foldable buckypaper/polyacrylonitrile composite film via interface reinforcing. *Composites Part A: Applied Science and Manufacturing, 113*, 132–140. https://doi.org/10.1016/j.compositesa.2018.07.019

Xiao, L., Chen, Z., Feng, C., Liu, L., Bai, Z. Q., Wang, Y., Qian, L., Zhang, Y., Li, Q., Jiang, K., & Fan, S. (2008). Flexible, stretchable, transparent carbon nanotube thin film loudspeakers. *Nano Letters, 8*(12), 4539–4545. https://doi.org/10.1021/nl802750z

Xing, H., Sun, L., Song, G., Gou, J., & Hao, Y. W. (2008). Surface coating of carbon nanofibers/nanotubes by electrodeposition for multifunctionalization. *Nanotechnology, 19*(2), 025704. https://doi.org/10.1088/0957-4484/19/02/025704

Xu, G., Yuan, J., Tao, X., Ding, B., Dou, H., Yan, X., Xiao, Y., & Zhang, X. (2015). Absorption mechanism of carbon-nanotube paper-titanium dioxide as a multifunctional barrier material for lithium-sulfur batteries. *Nano Research, 8*(9), 3066–3074. https://doi.org/10.1007/s12274-015-0812-0

Xu, G., Kushima, A., Yuan, J., Dou, H., Xue, W., Zhang, X., Yan, X., & Li, J. (2017). Ad hoc solid electrolyte on acidized carbon nanotube paper improves cycle life of lithium-sulfur batteries. *Energy and Environmental Science, 10*(12), 2544–2551. https://doi.org/10.1039/c7ee01898c

Yang, T., Xie, D., Li, Z., & Zhu, H. (2017). Recent advances in wearable tactile sensors: Materials, sensing mechanisms, and device performance. *Materials Science and Engineering R: Reports, 115*, 1–37. https://doi.org/10.1016/j.mser.2017.02.001

Younes, H., Al-Rub, R. A., Mahfuzur Rahman, M., Dalaq, A., Al Ghaferi, A., & Shah, T. (2016). Processing and property investigation of high-density carbon nanostructured papers with superior conductive and mechanical properties. *Diamond and Related Materials, 68*, 109–117. https://doi.org/10.1016/j.diamond.2016.06.016

Zangrossi, F., Ghosh, B., Xu, F., Warrior, N., & Hou, X. (2022). Improvement in electrical characteristics by surface modification of multi-wall carbon nanotube based buckypaper for de-icing application. *Journal of Composite Materials, 56*(29), 4487–4499. https://doi.org/10.1177/00219983221121876

Zaytseva, O., & Neumann, G. (2016). Carbon nanomaterials: Production, impact on plant development, agricultural and environmental applications. *Chemical and Biological Technologies in Agriculture, 3*(1), 1–26. https://doi.org/10.1186/s40538-016-0070-8

Zhang, X. (2008). Hydroentangling: A novel approach to high-speed fabrication of carbon nanotube membranes. *Advanced Materials, 20*(21), 4140–4144. https://doi.org/10.1002/adma.200801919

Zhang, J., & Jiang, D. (2011). Interconnected multi-walled carbon nanotubes reinforced polymer-matrix composites. *Composites Science and Technology, 71*(4), 466–470. https://doi.org/10.1016/j.compscitech.2010.12.020

Zhang, X., Sreekumar, T. V., Liu, T., & Kumar, S. (2004). Properties and structure of nitric acid oxidized single wall carbon nanotube films. *Journal of Physical Chemistry B, 108*(42), 16435–16440. https://doi.org/10.1021/jp0475988

Zhang, H. L., Li, J. F., Yao, K. F., & Chen, L. D. (2005). Spark plasma sintering and thermal conductivity of carbon nanotube bulk materials. *Journal of Applied Physics, 97*(11), 114310. https://doi.org/10.1063/1.1927286

Zhang, H. L., Li, J. F., Zhang, B. P., Yao, K. F., Liu, W. S., & Wang, H. (2007a). Electrical and thermal properties of carbon nanotube bulk materials: Experimental studies for the 328–958 K temperature range. *Physical Review B – Condensed Matter and Materials Physics, 75*(20), 205407. https://doi.org/10.1103/PhysRevB.75.205407

Zhang, W., Sakalkar, V., & Koratkar, N. (2007b). In situ health monitoring and repair in composites using carbon nanotube additives. *Applied Physics Letters, 91*(13), 1–3. https://doi.org/10.1063/1.2783970

Zhang, L., Zhang, G., Liu, C., & Fan, S. (2012). High-density carbon nanotube buckypapers with superior transport and mechanical properties. *Nano Letters, 12*(9), 4848–4852. https://doi.org/10.1021/nl3023274

Zhang, J., Jiang, D., Peng, H. X., & Qin, F. (2013). Enhanced mechanical and electrical properties of carbon nanotube buckypaper by in situ cross-linking. *Carbon, 63*, 125–132. https://doi.org/10.1016/j.carbon.2013.06.047

Zhang, H., Liu, Y., Kuwata, M., Bilotti, E., & Peijs, T. (2015a). Improved fracture toughness and integrated damage sensing capability by spray coated CNTs on carbon fibre prepreg. *Composites Part A: Applied Science and Manufacturing, 70*, 102–110. https://doi.org/10.1016/j.compositesa.2014.11.029

Zhang, Z., Wei, H., Liu, Y., & Leng, J. (2015b). Self-sensing properties of smart composite based on embedded buckypaper layer. *Structural Health Monitoring, 14*(2), 127–136. https://doi.org/10.1177/1475921714568405

Zhang, S., Park, J. G., Nguyen, N., Jolowsky, C., Hao, A., & Liang, R. (2017). Ultra-high conductivity and metallic conduction mechanism of scale-up continuous carbon nanotube sheets by mechanical stretching and stable chemical doping. *Carbon, 125*, 649–658. https://doi.org/10.1016/j.carbon.2017.09.089

Zhang, J., Zhang, X., Cheng, X., Hei, Y., Xing, L., & Li, Z. (2019). Lightning strike damage on the composite laminates with carbon nanotube films: Protection effect and damage mechanism. *Composites Part B: Engineering, 168*(1–4), 342–352. https://doi.org/10.1016/j.compositesb.2019.03.054

Zhang, P., Tian, R., Zhang, X., Ding, X., Wang, Y., Xiao, C., Zheng, K., Liu, X., Chen, L., & Tian, X. (2022). Electromagnetic interference shielding epoxy composites with satisfactory thermal conductivity and electrical insulation performance enabled by low-melting-point alloy layered structure. *Composites Part B: Engineering, 232*, 109611. https://doi.org/10.1016/j.compositesb.2022.109611

Zhao, J., Wang, R., He, X., & Liu, W. (2009). Strain monitoring of composite pressure vessel with thin metal liner using fiber Bragg grating. *Second International Conference on Smart Materials and Nanotechnology in Engineering, 7493*, 74930F. https://doi.org/10.1117/12.840654

Zhigilei, L. V., Wei, C., & Srivastava, D. (2005). Mesoscopic model for dynamic simulations of carbon nanotubes. *Physical Review B – Condensed Matter and Materials Physics, 71*(16), 165417. https://doi.org/10.1103/PhysRevB.71.165417

Zhou, W., Huang, Y., Liu, B., Hwang, K. C., Zuo, J. M., Buehler, M. J., & Gao, H. (2007). Self-folding of single- and multiwall carbon nanotubes. *Applied Physics Letters, 90*(7), 073107. https://doi.org/10.1063/1.2535874

Zhu, H., & Wei, B. (2008). Assembly and applications of carbon nanotube thin films. *Journal of Materials Science and Technology, 24*(4), 447–456.

Zhu, C., Xie, Z., & Guo, K. (2004). Formation of close-packed multi-wall carbon nanotube bundles. *Diamond and Related Materials, 13*(1), 180–183. https://doi.org/10.1016/J.DIAMOND.2003.10.034

Nanoparticles for Diagnosis and Treatment of Infectious Diseases

Sujit Mukherjee, Debmalya Pal, Arunava Bhattacharyya, Pritam Mondal, and Subhasis Roy

1 Introduction

In modern healthcare, the convergence of nanotechnology and medical science has paved the way for groundbreaking approaches to diagnosing and treating infectious diseases. Nanoparticles, with their miniature dimensions and remarkable physicochemical properties, have emerged as veritable champions in this endeavor (Moyez et al., 2016; Sundar & Kumar, 2012; Datta et al., 2023). These infinitesimal agents hold the potential to revolutionize the landscape of infectious disease management, offering innovative solutions that are both precise and effective. In diagnosis, nanoparticles exhibit an incredible aptitude for detection and identification. By harnessing their unique surface characteristics, nanoparticles can be engineered to bind selectively with specific pathogens or biomarkers associated with infectious diseases. This interaction can trigger detectable signals, providing rapid and highly sensitive diagnostic tools.

Moreover, nanoparticles can play a pivotal role in imaging, enhancing contrast in various modalities such as MRI, CT, and ultrasound (Roy et al., 2015a). This newfound clarity in visualizing infected tissues empowers healthcare providers with unprecedented insights for accurate disease detection and monitoring. Beyond diagnosis, nanoparticles unleash a new frontier in treating infectious diseases. As versatile drug delivery vehicles, they enable precise targeting of therapeutic agents to infected sites. By encapsulating antibiotics, antiviral drugs, or immune-modulating

Sujit Mukherjee, Debmalya Pal, Arunava Bhattacharyya, Pritam Mondal and Subhasis Roy contributed equally with all other contributors.

S. Mukherjee · D. Pal · A. Bhattacharyya · P. Mondal · S. Roy (✉)
Department of Chemical Engineering, University of Calcutta, Kolkata, India

S. A. Bansal et al. (eds.), *Emerging Applications of Novel Nanoparticles*, Lecture Notes in Nanoscale Science and Technology 37, https://doi.org/10.1007/978-3-031-57843-4_4

substances, nanoparticles ensure enhanced drug stability and bioavailability while minimizing off-target effects and drug resistance.

Furthermore, certain nanoparticles' remarkable photothermal and photodynamic properties open the door to innovative therapies. These nanoparticles can harness light energy to selectively destroy pathogens or infected cells, offering a targeted and minimally invasive approach to treatment. The potential of nanoparticles extends even to genetic manipulation and immune modulation. They carry genetic materials like siRNA and mRNA, enabling gene silencing and vaccine development (Zazo et al., 2016; Roy et al., 2015a). This tailored genetic intervention can target specific components of pathogens or enhance the host's immune responses.

Additionally, nanoparticles can be harnessed to interact with the immune system, bolstering its defenses against infections. However, while the promise of nanoparticles is tantalizing, challenges remain on the path to their widespread implementation. Ensuring biocompatibility, long-term safety, and large-scale production are critical considerations. As researchers continue to push the boundaries of scientific exploration, the integration of nanoparticles into infectious disease diagnosis and treatment holds immense potential to reshape the future of healthcare, offering precision, efficiency, and hope in the battle against some of humanity's most formidable adversaries (Moyez et al., 2016; Datta et al., 2023).

Additionally, nanomaterials can be tailored through functionalization, expanding their potential and enabling customization based on specific objectives (Sundar & Kumar, 2012). A promising research area is infectious diseases, where nanoparticle-based systems show great potential in the diagnosis and treatment of the infections like bacterial, viral, and fungal, as well as in vaccine development. In bacterial infections, nanoparticle-based systems offer a novel approach to combat biofilms and bacteria resistant to multi-drugs. Nanoparticles can be functionalized with specific molecules like proteins, DNA, or antibodies to modulate their targeting capabilities and pharmacokinetics, leading to increased efficacy of anti-infective drugs.

Furthermore, certain types of nanoparticles can be tailored for diagnostic analysis, miniaturizing, and simplifying pathogen detection techniques (Moyez et al., 2016). For viral infections like HIV and hepatitis, as well as fungal infections, nanoparticles hold promise in diagnostics and treatment (Burlage & Tillmann, 2017; Agrawal & Prajapati, 2012). Their functionalization with specific molecules allows for enhanced delivery of drugs and targeted therapy. Additionally, nanoparticles can be useful in theranostic applications, combining therapeutic agents with imaging capabilities to enable simultaneous treatment and visualization of target organs or tissues (Zazo et al., 2016; Chen et al., 2014; El Zowalaty et al., 2015). One particularly intriguing application of nanoparticles is that they can be designed for biosensors. These analytical devices encompass biological materials or biomimetics with physicochemical transducers or microsystems (Perfezou et al., 2012). Biosensors can detect specific analytes by using receptors fixed on a transducer surface to recognize the analyte, generating a detectable signal. Though it was for developing cancer biomarkers first (Perfezou et al., 2012; Bohunicky & Mousa, 2010), this concept can be adapted for other pathological conditions if appropriate biomarkers are identified (Panagiotopoulou et al., 2017; Li et al., 2015; Fang et al.,

2014). Nanoparticles include various proteins and peptides in biosensors (Roy et al., 2015b; Chen et al., 2015; Bera et al., 2022; Moyez & Roy, 2018), such as enzymes or specific ligands. This capability helps identify and count the interested molecules, expanding the scope of diagnostic applications.

Moreover, advanced multilayered nanoparticles that serve as biosensors can be designed to detect surrogate molecules present in the infected cells before the nanoparticle cargo delivery. This approach facilitates precise drug delivery and gives feedback to control the administered drug amount (Prow et al., 2004). Another area where nanoparticle-based systems show promise is in combating bacterial pathogenicity by conflicting with quorum sensing. Bacteria use this chemical process to manage the formation of biofilms. Natural or synthetic quorum-sensing inhibitors are valuable in treating bacteria resistant to multi-drug because these can perform as anti-biofilm agents.

Overall, the diverse applications of nanoparticles in infectious disease diagnosis, treatment, and biosensing demonstrate their enormous potential in revolutionizing healthcare and combating challenging infections. Ongoing research in this field is vital to unlocking further advancements in medical science and therapeutic strategies.

2 Brief About Nanoparticles

Nanotechnology primarily focuses on objects at the nanometer scale. In living organisms, cells contain nano-sized components. The subject area of nanotechnology involves the designing, production, and characterization of particles at this tiny scale. These particles, which can be as small as 1–100 nanometers (ultrafine particles) or within the range of 100–2500 nanometers (fine particles), function as cohesive units and possess unique transport and properties. Nanoparticles (NPs) exhibit a complex structure comprising three distinct layers: (i) Surface layer that can be implemented with various tiny molecules, polymers, and metal ions. (ii) The shell layer, which differs from the core, consists of distinctive chemical material. (iii) Core represents the central part of the nanoparticles. The nanoparticle is often referred to by the core itself (Shin et al., 2016). Due to these extraordinary features, researchers from diverse disciplines have shown immense interest in exploring the potential of these materials. As example scanning electron microscopy images (SEM) of SiO_2 nanoparticles are shown in Fig. 1.

One significant application of nanotechnology is enhancing drugs' pharmaceutical and therapeutic effects. By leveraging the high surface area of nano-sized particles, various functional groups can be attached to them, allowing targeted interactions with tumor cells (Pandey & Khuller, 2006). Consequently, they have emerged as promising alternatives to traditional radiation and chemotherapy methods, as they can efficiently accumulate in the microenvironment of tumors. Also, nanoparticles offer numerous advantages over other carrier systems, making them highly efficient

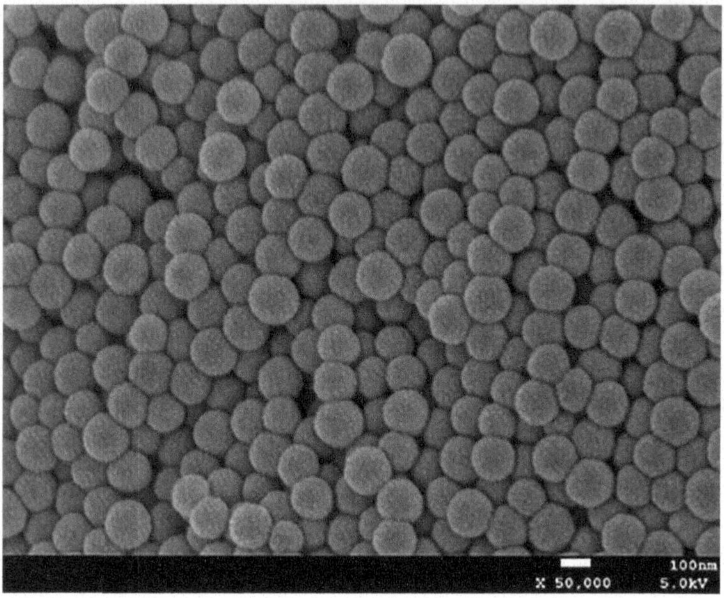

Fig. 1 SEM images of SiO_2 nanoparticle. (Adapted with permission from (Fu et al., 2018). Copyright (2018) Scientific Research Publishing, Creative Commons Attribution 4.0 International License)

delivery systems. One major advantage is their submicron size, enabling them to extravasate and occlude terminal blood vessels (Mosqueira et al., 2001).

Additionally, nanoparticles can encapsulate, disperse, or dissolve high densities of therapeutic agents, with different preparation processes yielding variation of properties and release manner of the entrapped agent. Liposomes are potential carriers that offer benefits such as drug protection, targeted delivery, and reduced toxicity. Still, they suffer from technical limitations such as poor reproducibility and stability issues, leading to drug leakage (Allen & Cullis, 2004). However, polymeric nanoparticles provide certain advantages, increasing drug/protein stability and offering controlled release properties. They require fewer excipients for the formulation, have simple preparation procedures, exhibit high physical stability, and allow the feasible release of drugs, making them suitable for treating persistent diseases. A controlled variation in release characteristics can be achieved by adjusting the polymer composition, enabling moderate constant doses over extended periods. Studies have shown that drug availability within nanoparticles is 77% higher compared to a similar formation containing microspheres (Alex & Bodmeier, 1990). These nanoparticles can be administered via several routes, such as nasal, parenteral, oral, and ophthalmic applications. Oral delivery becomes feasible for small drug molecules, which might otherwise be limited to injectable forms, such as anticancer agents. Nanoparticles also hold promise as adjuvants for vaccines, maximizing drug absorption by enhancing interaction with epithelial cells (Boudad et al., 2001).

3 Nanoparticles for Diagnosis of Infectious Diseases

In recent times, nanoparticles have shown great promise in diagnosing infectious diseases, offering unique advantages and capabilities. With their distinct characteristics and wide range of applications, nanoparticles have become attractive instruments to detect and diagnose infectious diseases. Nanomaterials and nanostructures have special properties that make it possible to create nanodiagnostic platforms to quickly and accurately diagnose illnesses using tiny amounts of patient samples. Nanoparticles possess a significant quality of higher surface area to volume ratio, which enables thick covering of molecules, which is a key benefit. Because of this characteristic, disease biomarkers have more places to bind, resulting in an effect that boosts the specificity and sensitivity of bioassays. Figure 2 shows a schematic of several nanomaterials in infectious disease treatment.

3.1 Point-of-Care Tests (POCT)

A contemporary method of diagnosing that is done right at the patient's point of care (i.e., near-patient testing) is called point-of-care testing (POCT) (Chen et al., 2019). It seeks to deliver immediate, accurate, and real-time medical condition detection. In recent years, POC diagnostics have drawn more and more attention, especially in clinical care, where on-site detection is important, especially in places with few resources. This prompted the creation of several techniques, tools, and

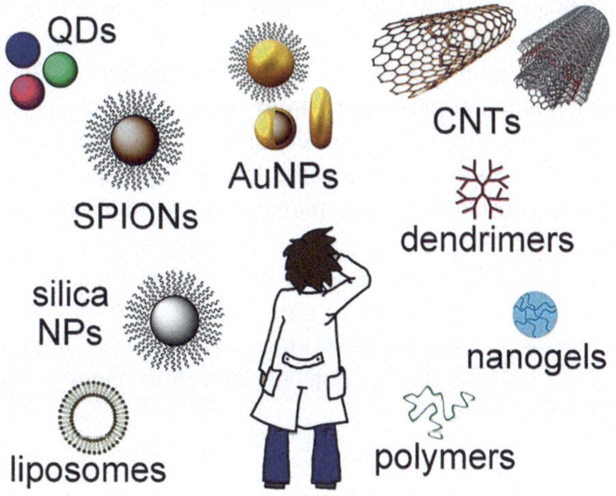

Fig. 2 Several nanomaterials in the field of infectious disease treatment (Marchesan & Prato, 2013). (Adapted with permission) (Copyright (2012) American Chemical Society)

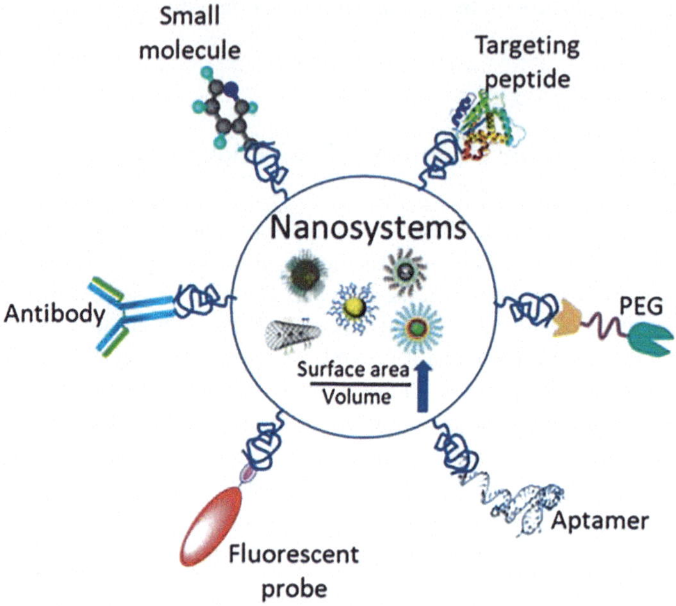

Fig. 3 Biomolecule detection nanosystems in diagnostics with their functional parts (Thwala et al., 2023). (Creative Commons Attribution (CC BY) license CC BY 4.0 DEED)

biosensors that can fulfill the POC diagnostic tool requirement. Clinical applications, including tissue engineering, medication transport, imaging, and diagnostics have all advanced thanks to nanotechnology. Interests in nano-diagnostics are increasing rapidly for infectious disease treatment because of their unique qualities, which include fast detection, sensitivity enhancement, and probable point-of-care testing. Innovative and efficient nano-diagnostics have been created for infectious illness, with the promise of providing convenient, reliable, and cost-effective POC diagnostic tools to identify infectious diseases in underdeveloped nations. POC diagnostic methods must be disposable, affordable, simple to use, and portable to succeed. Figure 3 describes biomolecule detection nanosystems in diagnostics with their functional parts. They must be able to examine tiny amounts of body fluids, including saliva, urine, and blood. Applications in global health must be affordable, thus, it is important to use the fewest expensive reagents feasible, mass-produce products using low-cost manufacturing methods, and ensure product quality. Miniaturization is also essential for creating portable POC diagnostic instruments (Liu et al., 2020). In addition, while using medical diagnostic instruments in clinical settings, it is important to consider the environmental aspects in areas with few resources, such as water shortage, erratic power, high temperatures, and humidity. Figure 3 shows the biomolecule detection nanosystems in diagnostics with their functional parts.

3.2 *Nanoparticles in the Field of Diagnostics*

Many nanoparticles are employed in diagnostic procedures. The most widely utilized are magnetic nanoparticles, gold nanoparticles, and quantum dots.

3.2.1 Gold Nanoparticles

Gold nanoparticles (AuNPs) have garnered significant interest in infectious disease diagnosis due to their unique optical, chemical, and physical properties. Their ability to interact with biomolecules and produce detectable signals has led to the development of innovative diagnostic techniques. Here's how gold nanoparticles are used for the diagnosis of infectious diseases (Thwala et al., 2023; Jain, 2012):

(a) Immunoassays: Gold nanoparticles can be functionalized with antibodies, antigens, or other biomolecules that specifically bind to target pathogens or disease biomarkers. When the target is present, the nanoparticles aggregate, causing a change in their optical properties, such as color or absorbance. This aggregation can be visually detected or quantified using spectroscopic methods. This approach is commonly used in lateral flow assays, which are simple and rapid diagnostic tests for infections like HIV, malaria, and COVID-19 (Fig. 4).

(b) Surface-Enhanced Raman Scattering (SERS): Gold nanoparticles can greatly enhance Raman scattering signals from molecules adsorbed on their surfaces. In SERS-based assays, target molecules bind to functionalized AuNPs, leading to enhanced Raman signals that can be detected and quantified. SERS has

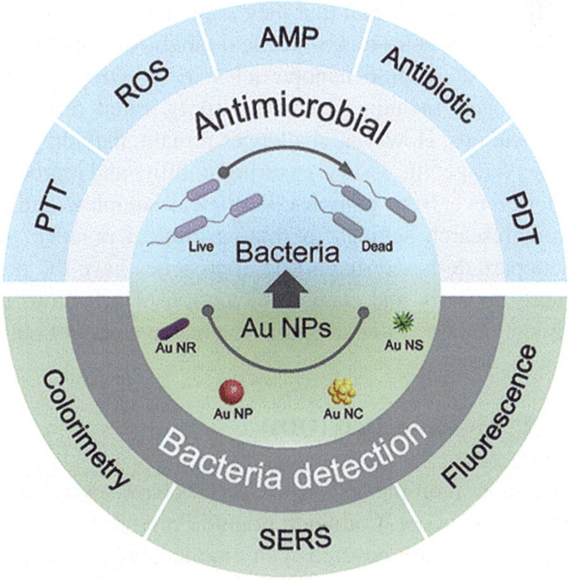

Fig. 4 Gold nanoparticles for the detection of bacterial diseases (Ge et al., 2023). (Adapted with permission)

shown promise in detecting pathogens like bacteria and viruses at low concentrations.

(c) DNA/RNA Detection: Functionalized gold nanoparticles can detect specific DNA or RNA sequences related to infectious agents. This can be achieved by hybridizing the target nucleic acids and complementary DNA probes attached to the nanoparticles. Changes in the optical properties of the AuNPs upon binding enable the detection of nucleic acids associated with various infectious diseases.

(d) Plasmonic Biosensors: Gold nanoparticles exhibit a phenomenon known as localized surface plasmon resonance (LSPR), which causes a shift in their absorption spectrum based on the surrounding environment. Plasmonic biosensors utilize this property to detect molecular interactions between target molecules and immobilized receptors on the AuNP surface. This approach has been employed for sensitive detection of viruses and bacteria.

(e) Photoacoustic Imaging: Gold nanoparticles can convert absorbed light energy into heat, leading to localized temperature increases that generate ultrasound waves detectable by photoacoustic imaging. This technique allows non-invasive imaging of infected tissues, aiding in disease diagnosis and monitoring.

(f) Biosensing Platforms: AuNPs can be integrated into various biosensing platforms, such as microarrays and lab-on-a-chip devices, to simultaneously enable multiplexed detection of pathogens or biomarkers.

Tiny DNA fragments may be coupled with gold particles having diameters less than 13 mm. Only in the presence of a similar target do gold nanoparticles congregate on a sensor surface. The technology can distinguish millions of distinct DNA sequences at once if a patterned sensor surface of several DNA strands is utilized. The current detection limit of this technique is 20 fmol/L which is not well optimized yet (Jain, 2012). Due to the fact that they may be detected using several analytical methods, gold nanoparticles make excellent labels for sensors.

The use of gold nanoparticles in infectious disease diagnosis offers several advantages, including high sensitivity, rapid results, and potential point-of-care applications. However, challenges remain, including optimizing nanoparticle properties for specific assays, ensuring stability and reproducibility, addressing potential interference from complex biological samples, and advancing these techniques from research settings to routine clinical practice. As research continues, gold nanoparticle-based diagnostic approaches have the potential to contribute significantly to early and accurate detection of infectious diseases, thereby improving patient outcomes and aiding in the management of outbreaks.

3.2.2 Quantum Dots (QDs)

Inorganic fluorophores known as QDs have several benefits over fluorescent markers already in use. Cadmium selenide crystals that are 200–10,000 atoms broad that are inorganically covered with zinc sulphide. When exposed to low-energy light, they produce fluorescent light. Emitted light frequency, or colour, is determined by the dot size (10 nm) (Singh et al., 2020). The dots are commonly connected to microbeads or covered in a multivalent bioconjugate polymer. A particular spectral bar code, or collection of different-sized dots affixed on a certain microbead, emits a distinctive spectrum of colours. Theoretically, a detection method using an intensity level of 10 and 6 different colours may produce 106 unique codes. CdSe-ZnS nanocrystals, a quantum dot, do not emit in the near-infrared (Dey et al., 2017). Therefore, infrared can't be utilized with them. They don't require lasers, and have stable fluorescence with easy excitation, great sensitivity, and broad excitation spectra. Complete blood tests are made possible by their infrared colours (Fig. 5).

QDs have several applications in molecular diagnostics and genetic constitution. QDs also enable multidinous diagnostics and the merging of diagnostic and therapeutic methods. The most significant prospective uses of QDs are in treating and detecting cancer. QD bioconjugates that are stable and luminescent make it possible to see cancer cells in still alive animals. Fluorescence microscopy and QDs can be used together to observe cells in living animals at high resolution. To designate the breast cancer marker Her2 immuno-fluorescently, QDs were covalently joined to antibodies and covered with polyacrylate caps. Cancer imaging is made possible by carbohydrate-encapsulated QDs with observable luminous characteristics. For biological experiments, QDs can be employed in multicolour optical coding.

3.2.3 Nanobarcodes

By electrochemically reducing the essential metals with dimensions ranging from 12 nm to 15 μm in width and 1–50 μm in length, independent, cylindrical nanoparticles with the quality of submicron type metal ions, are created. Under the light,

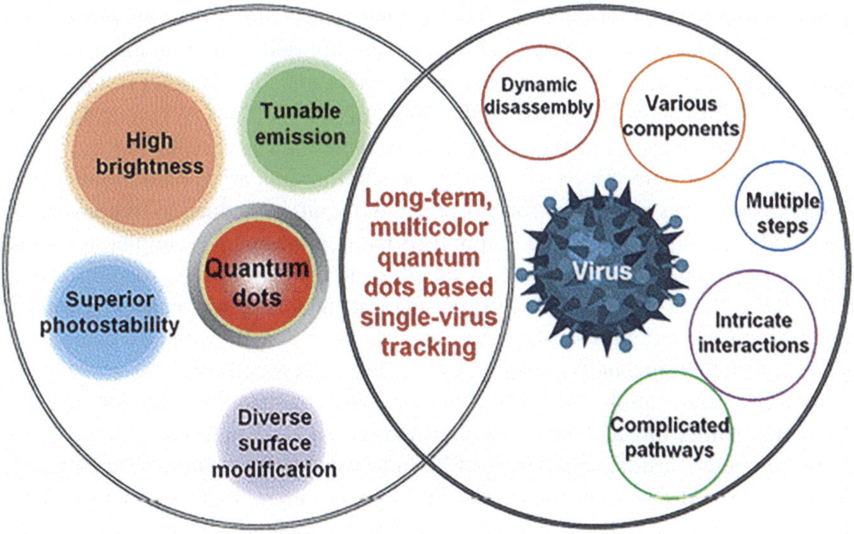

Fig. 5 QD based virus tracking (Wang et al., 2021). (Adapted with permission)

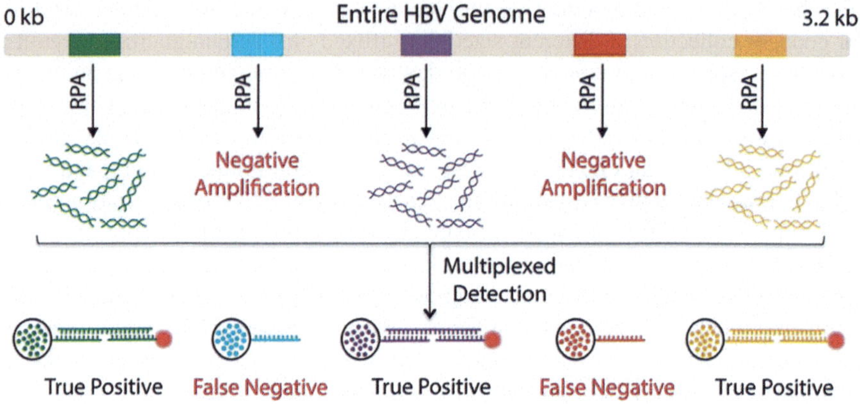

Fig. 6 A representation of QD multiplexed assay for Hepatitis B (HBV) diagnostics (Kim et al., 2016). (Adapted with permission)

fluorescence microscopy, or mass spectrometry, the contrast black and white stripes created by the variable reflectance of the neighbouring metals make them recognizable (like traditional barcodes). Figure 6 represents QD multiplexed assay for Hepatitis B (HBV) diagnostics. This nanobarcodes are helpful in point-of-care mobile devices, multiplexed proteomics experiments, SNP mapping, and population diagnostics (Jain, 2012). Detection of proteins using either fluorescence or mass spectrometry (after the immobilization of proteins on a metal surface).

3.2.4 Magnetic Nanoparticles

In order to generate polyGMA, we seed polymerized copolymer beads that consist of glycidyl methacrylate (GMA) and styrene and inserted saturation magnetic iron nanoparticles that are 15–20 nm in size inside the beads. The diameter of the resulting beads Fe/St-GMA/GMA was in the range of 100–200 nm. When coated with polyGMA, the beads' zeta potential value became 54.8 mV replacing the initial value of 93.7 mV, determined by electrophoresis (Rajasundari & Ilamurugu, 2011). Using nanoparticles as carriers for bioscreening requires a reduction in nonspecific protein adsorption, which gel electrophoresis has demonstrated is simpler using this technique. Nanoparticles are used as labelling molecules in bioscreening (Deng et al., 2021). Nanoparticles made of super-paramagnetic material are convenient for tracking cells and calcium sensing. A polymeric layer with an antibody-coated surface in ferrofluids surrounds a magnetic core that attracts cells (Fig. 7).

MRI has been combined with superparamagnetic nanoparticles to measure 2–3 nm to identify tiny and otherwise invisible lymph node metastases. To visualize cerebral ischemia lesions, ultrasmall SPIO improves MRI. For more than 24 hours, a dextran layered iron oxide nanoparticle improves brain visibility of tumors on MRI. Nanoparticles as Parts of Biosensors The components of biosensors can be

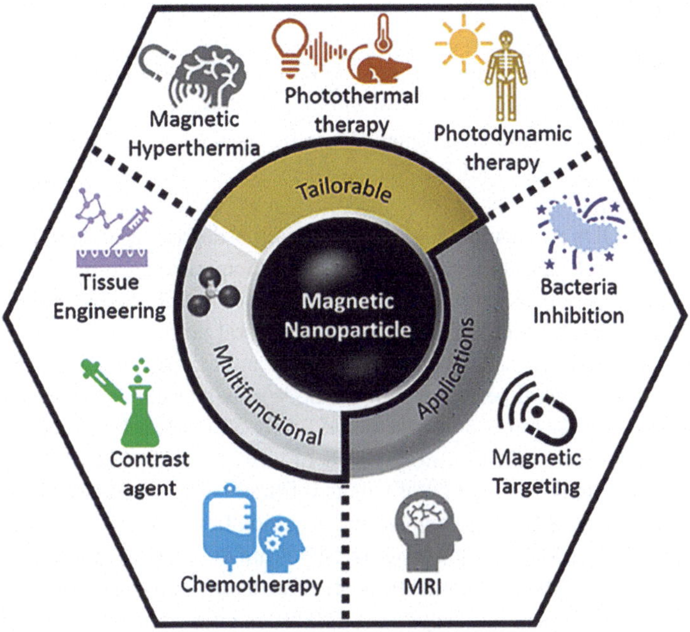

Fig. 7 Various aspects of magnetic nanoparticles in biomedical science (Martins et al., 2021). (Adapted with permission)

made of various nanoparticle types. Most of them function as investigators to detect and distinguish a specimen of interest for the applications of screening and diagnostics. A specific modification process is used in these applications to attach biological molecular species to nanoparticles. The probes link to targets in samples and communicate their existence using physical attributes such as mass or colour (Crnković et al., 2021). One area of biological surface science is the research of molecule binding, which is closely related to the study of how to adjust the characteristics of nanostructures by modifying their surface and structure at the nanoscale. Both topics span a wide range, where one may discover research on endothelization and cell adhesion to nanostructure surfaces or the application of a nanostructured platinum-lipid compound acting as a biosensor. Examples of nanoprobes are Quantum dot-based biosensors, metallic nanoparticles, silica nanoparticles, magnetic beads, carbon nanotubes, and nano barcodes. The other biosensors use different ways of using nanoparticles. In an electrical field, they serve as sieves, allowing charged molecules to pass through. It makes use of particles having artificial nanopores. However, other resources are available for nanoparticle creation than solid-state ones. For instance, DNA nanopores formed of a lipid bilayer and a protein channel called haemolysin with a diameter of around 2 nm are composed of organic material (Crnković et al., 2021). These nanopores can distinguish between separate DNA strands with a length of approximately 30 nucleotides and differ by a single base change. They include native α-haemolysin sensor applications, modified α-haemolysin sensors, and detailing various other organic pores and applications.

3.2.5 Nanodiagnostics

At the point of care, it is crucial to identify harmful germs quickly and accurately. The lack of ultrasensitivity and the wait time for findings are the main drawbacks of most traditional diagnostic techniques. A single bacterium may be found using a bioconjugated bioassay based on nanoparticles for in situ pathogen detection in under 20 minutes (Singh et al., 2016). Using multicolour oligonucleotide-functionalized QDs as nanoprobes, a hybridization-detection technique has been developed to detect lone-molecule hybridization. Sequence-specific spectrum coding is produced by self-assembly of the nanoprobes through the hybridization processes that are independent in the presence of several target sequences. Using this technique, it is possible to analyse the genetics of anthrax pathogenicity by simultaneously detecting several pertinent sequences. With silver nanorods, which greatly boost the signal, a spectroscopic test powered by SERS has been developed that quickly and sensitively detects minute levels of viruses. As viral DNA or RNA scatters, an alteration in the frequency of a near-infrared laser is measured by this technique. It could be possible to recognize the variation in frequency like a fingerprint. This cutting-edge SERS test can recognize spectrum differences in viruses and viral strains (Kuhn et al., 2014). The technique generates repeatable spectra without viral modification and offers quick diagnostics (60 s) for virus identification and characterization. This technique is very affordable and simple to duplicate.

4 Challenges and Limitations of Nanoparticles in Diagnosis of Infectious Diseases

While nanoparticles hold great promise for the diagnosis of infectious diseases, there are several challenges and limitations that researchers and healthcare professionals need to address (Singh et al., 2016; Kuhn et al., 2014; Kelf et al., 2010; Chen et al., 2010; Yu et al., 2012; Albanese & Chan, 2011; Liu & Hurt, 2010; Nel et al., 2006):

(a) Biocompatibility and Toxicity: The potential toxicity of nanoparticles is a significant concern. Some nanoparticles may induce immune responses or cause cellular damage, limiting their safe use in diagnostic applications. Rigorous biocompatibility studies are essential to ensure that nanoparticles do not adversely affect human health.
(b) Clearance and Biodistribution: Nanoparticles can accumulate in various organs and tissues, affecting their clearance from the body. This could lead to unintended side effects or long-term accumulation, especially if the nanoparticles are not efficiently excreted.
(c) Standardization: The development of nanoparticles for diagnostic purposes requires consistent manufacturing processes to ensure reproducibility and reliability. A lack of standardized protocols can lead to variations in nanoparticle properties, affecting their performance in diagnostic assays.

(d) Specificity and False Positives/Negatives: Achieving high specificity in nanoparticle-based diagnostic assays is challenging. Cross-reactivity with similar biomolecules or non-specific binding can lead to false-positive results, while low binding affinity can result in false negatives.
(e) Sample Interference: Biological samples, such as blood or tissue, can contain various components that may interfere with nanoparticle-based assays, affecting their accuracy and reliability.
(f) Nonspecific Targeting: Designing nanoparticles specifically targeting pathogens or biomarkers without binding to healthy cells or interfering substances can be complex. Achieving precise targeting while avoiding nonspecific interactions is a significant challenge.
(g) Regulatory Approval: The regulatory approval process for nanoparticle-based diagnostic tools can be lengthy and complex due to concerns about safety, efficacy, and potential long-term effects.
(h) Cost and Scalability: The production of nanoparticles for diagnostics may be expensive and challenging to scale up for widespread use. High production costs could limit accessibility, especially in resource-limited settings.
(i) Limited Clinical Validation: Many nanoparticle-based diagnostic assays are still in the research and development phase, with limited clinical validation. Robust clinical studies are needed to demonstrate their real-world effectiveness and reliability.
(j) Stability and Storage: Nanoparticles can be sensitive to environmental factors, such as temperature and pH, which may affect their stability during storage and transportation. Ensuring the stability of nanoparticles under different conditions is crucial for their practical use. Ethical and Regulatory Concerns: Using nanoparticles in diagnostics raises ethical and regulatory considerations, including patient privacy, data security, and informed consent.

Addressing these challenges requires interdisciplinary collaboration, rigorous testing, and ongoing research to optimize the design, manufacturing, and implementation of nanoparticle-based diagnostic tools for infectious diseases. While these limitations exist, continued advancements in nanotechnology and biomedicine hold the potential to overcome these hurdles and bring nanoparticle-based diagnostics to the forefront of infectious disease management.

It was observed that the unique properties and potential applications of nanoparticles have made them promising in diagnosing infectious diseases. However, some challenges, barriers, and considerations remain like any new technology. One of the primary concerns with nanoparticles is their potential toxicity and adverse effects on living organisms. Before nanoparticles can be used for diagnostic purposes, thorough biocompatibility testing must be conducted to ensure they are safe for human use. The stability and performance of nanoparticles may be affected by environmental factors like temperature, light, pH etc. Ensuring the stability of nanoparticles during storage and transportation is crucial for their successful application in diagnostics. Biological samples, such as blood or urine, can contain various components that may interfere with nanoparticle-based diagnostics, leading to inaccurate results.

Pre-treatment or purification of samples may be required to minimize interference. Poor permeability of Biological membranes can pose significant challenges in using essential therapeutic agents (Deng et al., 2021). Additionally, certain cell types may lack the necessary machinery for various endocytotic pathways such as macropinocytosis, caveolar endocytosis, phagocytosis and clathrin- and caveolin-1-independent endocytosis) (Kuhn et al., 2014), which limits the intake and applicability of nanoparticles in medical treatments. The non-specific cellular intake process refers to incorporating exterior materials and lacks material selectivity (Kelf et al., 2010). Nanoparticles can be non-specifically taken up by macrophages, a vital immune system constituent and organs like liver and spleen, which belong to the reticuloendothelial system. This shows a notable limitation for using nanoparticles as therapeutic agents since they may be removed from circulation before the intended targets are reached, resulting in a lack of effectiveness in treatment (Chen et al., 2010; Yu et al., 2012). A common approach to get away with these non-specific interactions is to introduce Polyethylene Glycol (PEG) molecules with optimal molecular weight on the surface of nanoparticles and utilize the active targeting ligands (Kelf et al., 2010). Aggregation of nanoparticles can occur due to some physical processes where nanoparticle surfaces contact each other. This results in the formation of clusters, and the particles become larger than their conventional nanometre range. Such aggregation can have significant implications for nanoparticle intake (Albanese & Chan, 2011), endurance (Liu & Hurt, 2010), toxicity (Nel et al., 2006; Zook et al., 2011), effect, and mobility (Saleh et al., 2008). However, this aggregation in aqueous suspensions can be mitigated by employing polymers that perform as steric stabilizers on the surface of nanoparticles. Systemically administered nanoparticle-based therapeutics face several technical limitations, including intake by macrophages and the reticuloendothelial system (as described earlier), clearance of kidney and bile, and any present proteins degraded by enzymes (Huang & Liu, 2011; Poon et al., 2011). The importance of toxicity studies for evaluating human safety concerns increase because of the widespread applications of nanoparticles in biomedicine. Vital parameters are to be considered in these studies, such as nanoparticle types and their surface modifications, the administered dosage, and both cellular and organismal level biodistribution. Toxicity concerns encompass some effects, such as aggregation of nanoparticles, the circulation time, and slow rate of clearance or elimination. Nanoparticle toxicity can be harmful, eventually leading to hepatotoxicity, neurotoxicity, renal toxicity, pulmonary toxicity, and spermatoxicity (Yoshida et al., 2009). Thoroughly assessing these aspects is crucial to ensure nanoparticles for safe and effective use in medicine. Despite these challenges, nanoparticles hold immense potential for revolutionizing infectious disease diagnosis. Continued research, collaboration between scientists, clinicians, and regulatory bodies, and addressing these considerations is crucial for harnessing the full benefits of nanoparticles in this field.

5 Future of Nanoparticles in Medicinal Sciences

The future of nanoparticles in medicinal sciences holds immense promise and potential for transforming healthcare in unprecedented ways. As technology advances and our understanding of nanomaterials deepens, nanoparticles are poised to play a pivotal role in various aspects of medicine. Here are some key directions that the future of nanoparticles in medicinal sciences may take (Salta, 2004; Logothetidis, 2006; Zhang & An, 2007; Cheon & Lee, 2008):

(a) Personalized Medicine: Nanoparticles have the potential to revolutionize personalized medicine by enabling targeted and tailored therapies. Through precise drug delivery, nanoparticles can deliver medications directly to specific cells or tissues, minimizing side effects and optimizing treatment outcomes based on an individual's unique genetic makeup and disease profile.
(b) Advanced Drug Delivery: Nanoparticles will continue evolving as efficient drug delivery carriers. Researchers are exploring ways to engineer nanoparticles to respond to specific stimuli, such as changes in pH or temperature, enabling on-demand drug release. This could improve therapeutic efficacy, reduce dosing frequency, and enhance patient compliance.
(c) Gene Editing and Therapy: Nanoparticles could play a crucial role in gene editing and therapy, allowing precise delivery of gene-editing tools like CRISPR-Cas9 to target cells. This holds the potential to correct genetic mutations and treat inherited diseases at the molecular level.
(d) Imaging and Diagnostics: Nanoparticles are likely to advance medical imaging techniques further. They could be developed to provide enhanced contrast and improved visualization of tissues, organs, and cellular structures, aiding in early and accurate disease diagnosis.
(e) Regenerative Medicine: Nanoparticles could facilitate regenerative medicine approaches by delivering growth factors, stem cells, or other therapeutic agents to promote tissue repair and regeneration. This could have profound implications for treating injuries, degenerative diseases, and organ damage.
(f) Immunotherapy Enhancement: Nanoparticles might enhance the effectiveness of immunotherapies by delivering immune-modulating agents directly to immune cells or tumor sites. This could boost the body's immune response against cancer and infectious diseases.
(g) Minimally Invasive Treatments: Nanoparticles may enable minimally invasive treatments by offering precise targeting and localized therapy delivery. This could reduce the need for invasive surgeries and improve patient recovery times.
(h) Combating Antimicrobial Resistance: Nanoparticles with antimicrobial properties could be used as alternatives to traditional antibiotics, helping to combat antimicrobial resistance by disrupting microbial growth and biofilm formation.
(i) Neural Health and Neurological Disorders: Nanoparticles could facilitate drug delivery across the blood-brain barrier, opening new avenues for treating neurological disorders, such as Alzheimer's, Parkinson's, and brain tumors.

(j) Environmental Monitoring and Drug Safety: Nanoparticles may find applications in monitoring environmental pollutants and toxins and drug safety assessments, enhancing our ability to detect and mitigate potential health hazards.

While the potential of nanoparticles in medicinal sciences is exciting, it's important to address safety, regulatory considerations, and ethical implications. Collaborative efforts between scientists, clinicians, regulatory agencies, and policymakers will be crucial to harnessing the full potential of nanoparticles for the betterment of human health. As research and innovation progress, the future undoubtedly holds groundbreaking discoveries and transformative applications of nanoparticles in medicine (Fig. 8).

Nanoparticles have become an intense research and development-specific subject in medicinal sciences because of their exclusive properties and potential applications. These tiny particles offer exciting opportunities to revolutionize various aspects of medicine, such as the delivery of drugs, diagnostics, imaging, and regenerative medicine (Salta, 2004). The overarching objective of nanomedicine research is to study and analyze the minute comprehensively, molecular-scale structures termed nanomachinery. By achieving accurate control and manipulation of these nanomachinery components within cells, we can gain deeper insights into the intricate workings of cellular mechanisms in living organisms. This profound understanding holds the potential to revolutionize the field, leading to cutting-edge

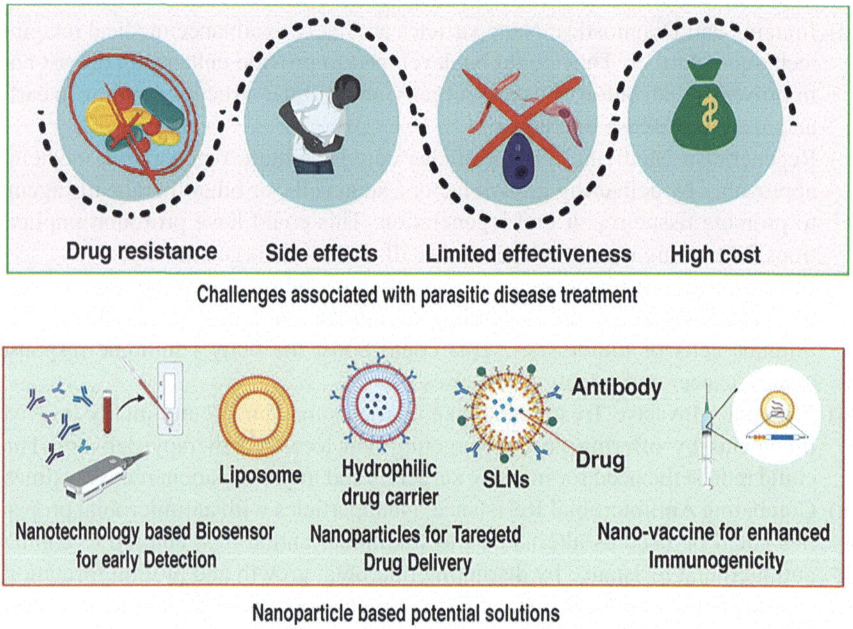

Fig. 8 Utility of Nanoparticles in the treatment of infectious diseases (Tiwari et al., 2023). (LicenseCC BY-NC-ND 4.0)

technologies that can detect diseases early and result in better treatment. This research is important to develop a versatile technology that significantly impacts nanoscale imaging methods, facilitating investigations into the molecular intricacies within living cells (Logothetidis, 2006). One of the most significant applications of nanoparticles in medicinal sciences is the targeted delivery of drugs.

Conventional drug delivery methods often lack specificity, leading to systemic distribution and potential side effects. Nanoparticles can be engineered to carry therapeutic agents, like drugs or nucleic acids, exactly to the region of interest, such as tumor cells or inflamed tissues. Lipid-based nanoparticles, polymeric nanoparticles, and metallic nanoparticles are among the commonly used carriers. These nanoparticles can be functional with ligands or antibodies that recognize specific receptors overexpressed on target cells. By leveraging this targeting approach, nanoparticles can improve drug efficacy, reduce toxicity, and enhance patient outcomes. The future of targeted drug delivery might involve even more sophisticated nanoparticle designs. Researchers may explore advanced surface modifications, allowing nanoparticles to evade the immune system, cross biological barriers efficiently, and accumulate at precise target sites with enhanced precision. Cancer treatment is an area that stands to benefit significantly from nanoparticle research. Nanoparticles can enhance the effectiveness of chemotherapy by delivering drugs directly to the tumor cells. This targeted approach reduces damage to healthy tissues and enhances the therapeutic index of anticancer drugs. Additionally, nanoparticles can play a role in combination therapy, where multiple therapeutic agents can be loaded into a single nanoparticle carrier. This approach allows for synergistic effects and combats drug resistance, a significant challenge in cancer treatment.

Moreover, researchers may explore using nanoparticles in photodynamic therapy, a technique where the light-sensitive nanoparticles become active by light, producing reactive oxygen species that help destroy the cancer cells selectively. This non-invasive and targeted approach holds promise for treating localized tumors. Nanoparticles can be critical in immunotherapy by delivering immune-modulating agents, such as cytokines or antigens (Zhang & An, 2007), directly to immune cells or tumor microenvironments. This approach can help stimulate the immune response, overcome immunosuppression, and improve the effectiveness of immunotherapies. In the future, nanoparticle-based vaccines might be developed to target specific antigens, leading to more potent and tailored immune responses against infectious diseases or cancer.

Nanoparticle probes have revolutionized imaging techniques by offering improved signal sensitivity and enhanced decisiveness and can study biological systems at cellular and molecular levels. Normal magnetic nanoparticles are employed in magnetic resonance imaging (MRI). However, they can also be a versatile platform for integrating various functional parts, such as fluorescence tags, biomolecules, and radionuclides. This enables multimodal imaging, delivery of genes, and tracking of cellular trafficking. By combining magnetic nanoparticles with adenovirus, an MRI system can detect target cells and also helps monitor the delivery of genes and expression of green effulgent proteins optically (Cheon & Lee, 2008).

Other techniques, like positron-emission tomography (PET), have even higher detection sensitivities, allowing nanoparticle use at lower concentrations than typically feasible with routine MRI. In hybrid imaging, the combination of PET with the somatic details obtained from computed tomography (CT) can map signals precisely to atherosclerosis of the vascular system (Nahrendorf et al., 2008). For successful molecular imaging, it is essential to accumulate the contrast media efficiently at the target region. Nanoparticles can be engineered with targeting groups to access molecules hidden behind tissue barriers to achieve this. This targeting approach enhances the accumulation of contrast agents at the desired location, particularly useful for low sensitivity imaging processes.

Moreover, nanoparticles with various contrast groups enhance signal, further improving detection capabilities. Interestingly, these nanoparticles can also serve a dual purpose as carriers for both the contrast medium and therapeutic drugs. This innovative concept, known as theranostics, can simultaneously monitor therapeutic activity and biodistribution (Debbage & Jaschke, 2008). In practical applications, this offers valuable insights into the efficacy of treatments and enhances precision medicine. Furthermore, nanofiber-based scaffolds have been developed with a variation in pore size distribution, high porosity, and greater surface area-to-volume ratio (Roy et al., 2008). These scaffolds provide an excellent environment for cell growth, attachment, and expansion, making them ideal for tissue-engineering applications. The ability to optimize electrospun nanofibrous scaffolds further opens up new possibilities for future advancements in renovative medicine and tissue engineering.

6 Conclusive Remarks

In conclusion, using nanoparticles to diagnose and treat infectious diseases stands at the forefront of scientific innovation and medical advancement. These microscopic agents hold immense potential to revolutionize the way we approach infectious disease management, offering a range of precise and effective strategies. Nanoparticles' remarkable ability to bind with pathogens and biomarkers and their capacity to generate detectable signals has paved the way for rapid and sensitive diagnostic tools. This promises earlier detection, improved patient outcomes, and the potential to curb the spread of infections. In the realm of treatment, nanoparticles shine as versatile carriers for therapeutic agents, enhancing drug delivery precision and minimizing adverse effects. Their photothermal and photodynamic properties introduce novel avenues for targeted therapy, presenting a promising alternative to conventional treatments.

Moreover, nanoparticles enable genetic manipulation and immune modulation, opening doors to tailored interventions that enhance the body's natural defenses against infectious agents. These innovations hold potential not only for treating infections but also for advancing the field of personalized medicine. However, it's essential to acknowledge the challenges that accompany these promising advancements. Striking a balance between innovation and safety, ensuring production scalability, and addressing potential long-term effects remain vital considerations. As the realm of nanotechnology continues to evolve, collaboration between researchers, healthcare professionals, and regulatory bodies will be crucial to navigate these challenges and unlock the full potential of nanoparticles in infectious disease diagnosis and treatment. In essence, nanoparticles are ushering in a new era of precision medicine, where the convergence of cutting-edge technology and medical expertise holds the promise of transforming our ability to combat infectious diseases. The journey ahead is exciting and complex, but it offers hope in the ongoing quest to safeguard global health. With further research and refinement, the management of infectious diseases may be potentially revolutionized by nanoparticles with their versatile properties and applications and contribute to improved global health outcomes.

Acknowledgments Author (S. Roy) would like to acknowledge 'Scheme for Transformational and Advanced Research in Sciences (STARS)' (MoE-STARS/STARS-2/2023-0175) by the Ministry of Education, Govt. of India for promoting translational India-centric research in sciences implemented and managed by Indian Institute of Science (IISc), Bangalore, for their support.

References

Agrawal, S., & Prajapati, R. (2012). Nanosensors and their pharmaceutical applications: A review. *International Journal of Pharmaceutical Sciences and Nanotechnology, 4*, 1528–1535.

Albanese, A., & Chan, W. C. (2011). Effect of gold nanoparticle aggregation on cell uptake and toxicity. *ACS Nano, 5*, 5478–5489.

Alex, R., & Bodmeier, R. (1990). Encapsulation of water-soluble drugs by a modified solvent evaporation method. I. Effect of process and formulation variables on drug entrapment. *Journal of Microencapsulation, 7*(3), 347–355.

Allen, T. M., & Cullis, P. R. (2004). Drug delivery systems: Entering the mainstream. *Science, 303*(5665), 1818–1822.

Bera, S., Saha, A., Mondal, S., Biswas, A., Mallick, S., Chatterjee, R., & Roy, S. (2022). Review of defect engineering in perovskites for photovoltaic application. *Materials Advances, 3*, 5234.

Bohunicky, B., & Mousa, S. A. (2010). Biosensors: The new wave in cancer diagnosis. *Nanotechnology, Science and Applications, 4*, 1–10.

Boudad, H., Legrand, P., Lebas, G., Cheron, M., Duchêne, D., & Ponchel, G. (2001). Combined hydroxypropyl-beta-cyclodextrin and poly(alkylcyanoacrylate) nanoparticles intended for oral administration of saquinavir. *International Journal of Pharmaceutics, 218*(1–2), 113–124.

Burlage, R. S., & Tillmann, J. (2017). Biosensors of bacterial cells. *Journal of Microbiological Methods, 138*, 2–11.

Chen, H., Wang, L., Yeh, J., et al. (2010). Reducing non-specific binding and uptake of nanoparticles and improving cell targeting with an antifouling PEO-b-PgammaMPS copolymer coating. *Biomaterials, 31*, 5397–5407.

Chen, F., Ehlerding, E. B., & Cai, W. (2014). Theranostic nanoparticles. *Journal of Nuclear Medicine, 55*, 1919–1922. https://doi.org/10.2967/jnumed.114.146019

Chen, Q., Sun, Q., Molino, N. M., Wang, S. W., Boder, E. T., & Chen, W. (2015). Sortase A-mediated multi-functionalization of protein nanoparticles. *Chemical Communications, 51*, 12107–12110.

Chen, H., Liu, K., Li, Z., & Wang, P. (2019). Point of care testing for infectious diseases. *Clinica Chimica Acta, 493*, 138–147. https://doi.org/10.1016/j.cca.2019.03.008. Epub 2019 Mar 8. PMID: 30853460; PMCID: PMC6462423.

Cheon, J., & Lee, J. H. (2008). Synergistically integrated nanoparticles as multimodal probes for nanobiotechnology. *Accounts of Chemical Research, 41*, 1630–1640.

Crnković, A., Srnko, M., & Anderluh, G. (2021). Biological nanopores: Engineering on demand. *Life (Basel), 11*(1), 27. https://doi.org/10.3390/life11010027. PMID: 33466427; PMCID: PMC7824896.

Datta, S., Mitra, M., & Roy, S. (2023). Morphological tuning and defect-free lead halide perovskite by surface passivation for solar cell fabrication. *Ionics, 29*, 4397–4405.

Debbage, P., & Jaschke, W. (2008). Molecular imaging with nanoparticles: Giant roles for dwarf actors. *Histochemistry and Cell Biology, 130*, 845–875.

Deng, J., Zhao, S., Liu, Y., Liu, C., & Sun, J. (2021). Nanosensors for diagnosis of infectious diseases. *ACS Applied Bio Materials, 4*(5), 3863–3879. https://doi.org/10.1021/acsabm.0c01247

Dey, A., Karan, P., Sengupta, A., Moyez, S. A., Sarkar, P., Majumder, S. B., Pradhan, D., & Roy, S. (2017). Enhanced charge carrier generation by dielectric nanomaterials for quantum dots solar cells based on CdS-TiO2 photoanode. *Solar Energy, 158*, 83–88.

El Zowalaty, M. E., Hussein Al Ali, S. H., Husseiny, M. I., Geilich, B. M., Webster, T. J., & Hussein, M. Z. (2015). The ability of streptomycin-loaded chitosan-coated magnetic nanocomposites to possess antimicrobial and antituberculosis activities. *International Journal of Nanomedicine, 10*, 3269–3274.

Fang, Y.-S., Wang, H.-Y., Wang, L.-S., & Wang, J.-F. (2014). Electrochemical immunoassay for procalcitonin antigen detection based on signal amplification strategy of multiple nanocomposites. *Biosensors & Bioelectronics, 51*, 310–316.

Fu, X., Song, B., Chen, X. Z., Wang, A., & Wang, C. (2018). Highly-controllable imprinted polymer nanoshell on the surface of silica nanoparticles for selective adsorption of 17β-Estradiol. *Journal of Encapsulation and Adsorption Sciences, 8*, 210–224. https://doi.org/10.4236/jeas.2018.84011

Ge, P., Zhang, J., Ding, T., & Xianyu, Y. (2023). Surface chemistry of gold nanoparticles for bacterial detection and antimicrobial applications. *ACS Materials Letters, 5*(3), 638–655.

Huang, L., & Liu, Y. (2011). *In vivo* delivery of RNAi with lipid-based nanoparticles. *Annual Review of Biomedical Engineering, 13*, 507–530.

Jain, K. K. (2012). Nanomolecular diagnostics. In *The handbook of nanomedicine* (Vol. 12, pp. 113–170). Springer. https://doi.org/10.1007/978-1-61779-983-9_4. PMCID: PMC7123705.

Kelf, T., Sreenivasan, V., Sun, J., et al. (2010). Non-specific cellular uptake of surface-functionalized quantum dots. *Nanotechnology, 21*, 285105.

Kim, J., Biondi, M., Feld, J., & Chan, W. (2016). Clinical validation of quantum dot barcode diagnostic technology. *ACS Nano, 10*(4), 4742–4753.

Kuhn, D. A., Vanhecke, D., Michen, B., et al. (2014). Different endocytotic uptake mechanisms for nanoparticles in epithelial cells and macrophages. *Beilstein Journal of Nanotechnology, 5*, 1625–1636.

Li, Z., Su, W., Liu, S., & Ding, X. (2015). An electrochemical biosensor based on DNA tetrahedron/graphene composite film for highly sensitive detection of NADH. *Biosensors & Bioelectronics, 69*, 287–293.

Liu, J., & Hurt, R. H. (2010). Ion release kinetics and particle persistence in aqueous nano-silver colloids. *Environmental Science & Technology, 44*, 2169–2175.

Liu, D., Wang, J., Lingling, W., Huang, Y., Zhang, Y., Zhu, M., Wang, Y., Zhu, Z., & Yang, C. (2020). Trends in miniaturized biosensors for point-of-care testing. *TrAC Trends in Analytical Chemistry, 122*, 115701.

Logothetidis, S. (2006). Nanotechnology in medicine: The medicine of tomorrow and nanomedicine. *Hippokratia, 10*, 7–21.

Marchesan, S., & Prato, M. (2013). Nanomaterials for (Nano)medicine. *ACS Medicinal Chemistry Letters, 4*(2), 147–149.

Martins, P., Lima, A., Ribeiro, S., Lanceros-Mendez, S., & Martins, P. (2021). Magnetic nanoparticles for biomedical applications: From the soul of the earth to the deep history of ourselves. *ACS Applied Bio Materials, 4*(8), 5839–5870.

Mosqueira, V. C., Legrand, P., Morgat, J. L., Vert, M., Mysiakine, E., Gref, R., Devissaguet, J. P., & Barratt, G. (2001). Biodistribution of long-circulating PEG-grafted nanocapsules in mice: Effects of PEG chain length and density. *Pharmaceutical Research, 18*(10), 1411–1429.

Moyez, S. A., & Roy, S. (2018). Dual-step thermal engineering technique: A new approach for fabrication of efficient CH3NH3PbI3-based perovskite solar cell in open air condition. *Solar Energy Materials and Solar Cells, 185*, 145.

Moyez, A., Dhar, A., Sarkar, P., Jung, H. S., & Roy, S. (2016). A review of the multiple exciton generation in photovoltaics. *Reviews in Advanced Sciences and Engineering, 5*, 51–64.

Nahrendorf, M., Zhang, H., Hembrador, S., Panizzi, P., Sosnovik, D. E., Aikawa, E., et al. (2008). Nanoparticle PET-CT imaging of macrophages in inflammatory atherosclerosis. *Circulation, 117*, 379–387.

Nel, A., Xia, T., Mädler, L., et al. (2006). Toxic potential of materials at the nanolevel. *Science, 311*, 622–627.

Panagiotopoulou, M., Kunath, S., Medina-Rangel, P. X., Haupt, K., Bui, T. S., & B. (2017). Fluorescent molecularly imprinted polymers as plastic antibodies for selective labeling and imaging of hyaluronan and sialic acid on fixed and living cells. *Biosensors & Bioelectronics, 88*, 85–93.

Pandey, R., & Khuller, G. K. (2006). Nanotechnology based drug delivery system(s) for the management of tuberculosis. *Indian Journal of Experimental Biology, 44*(5), 357–366.

Perfezou, M., Turner, A., & Merkoci, A. (2012). Cancer detection using nanoparticle-based sensors. *Chemical Society Reviews, 41*, 2606–2622.

Poon, Z., Lee, J. B., Morton, S. W., et al. (2011). Controlling *in vivo* stability and biodistribution in electrostatically assembled nanoparticles for systemic delivery. *Nano Letters, 11*, 2096–2103.

Prow, T. W., Kotov, N. A., Lvov, Y. M., Rijnbrand, R., & Leary, J. F. (2004). Nanoparticles, molecular biosensors, and multispectral confocal microscopy. *Journal of Molecular Histology, 35*, 555–564. https://doi.org/10.1007/s10735-004-2196-4

Rajasundari, K., & Ilamurugu, K. (2011). Nanotechnology and its applications in medical diagnosis. *Journal of Basic and Applied Chemistry, 1*(2), 26–32.

Roy, S., Kargupta, K., Chakraborty, S., & Ganguly, S. (2008). Preparation of polyaniline nanofibers and nanoparticles via simultaneous doping and electro-deposition. *Materials Letters, 62*(16), 2535–2538, . ISSN 0167-577X. https://doi.org/10.1016/j.matlet.2007.12.066

Roy, U., Rodriguez, J., Barber, P., das Neves, J., Sarmento, B., & Nair, M. (2015a). The potential of HIV-1 nanotherapeutics: From in vitro studies to clinical trials. *Nanomedicine, 10*, 3597–3609.

Roy, S., Han, G. S., Shin, H., Lee, J. W., Mun, J., Shin, H., & Jung, H. S. (2015b). Low temperature synthesis of Rutile TiO nanocrystals and their photovoltaic and photocatalytic properties. *Journal of Nanoscience and Nanotechnology, 15*(6), 4516–4521. American Scientific Publishers. https://doi.org/10.1166/jnn.2015.9778

Saleh, N. B., Pfefferle, L. D., & Elimelech, M. (2008). Aggregation kinetics of multiwalled carbon nanotubes in aquatic systems: Measurements and environmental implications. *Environmental Science & Technology, 42*, 7963–7969.

Salta, O. V. (2004). Applications of nanoparticles in biology and medicine. *Journal of Nanobiotechnology, 2*, 3.

Shin, W.-K., Cho, J., Kannan, A. G., Lee, Y.-S., & Kim, D.-W. (2016). Cross-linked composite gel polymer electrolyte using mesoporous methacrylate-functionalized SiO2 nanoparticles for lithium-ion polymer batteries. *Scientific Reports, 6*, 26332.

Singh, L., Parboosing, R., Kruger, H. G., et al. (2016). Intracellular localization of gold nanoparticles with targeted delivery in MT-4 lymphocytes. *Advances in Natural Sciences: Nanoscience and Nanotechnology, 7*, 045013.

Singh, S., Dhawan, A., Karhana, S., Bhat, M., & Dinda, A. K. (2020). Quantum dots: An emerging tool for point-of-care testing. *Micromachines (Basel)., 11*(12), 1058. https://doi.org/10.3390/mi11121058. PMID: 33260478; PMCID: PMC7761335.

Sundar, S., & Kumar, P. V. (2012). Drug targeting to infectious diseases by nanoparticles surface functionalized with special biomolecules. *Current Medicinal Chemistry, 19*, 3196–3202.

Thwala, L. N., Ndlovu, S. C., Mpofu, K. T., Lugongolo, M. Y., & Mthunzi-Kufa, P. (2023). Nanotechnology-based diagnostics for diseases prevalent in developing countries: Current advances in point-of-care tests. *Nanomaterials, 13*(7), 1247.

Tiwari, R., Gupta, R., Singh, V., Kumar, A., Rajneesh, Madhukar, P., Sundar, S., Gautam, V., & Kumar, R. (2023). Nanotechnology-based strategies in parasitic disease management: From prevention to diagnosis and treatment. *ACS Omega, 8*(45), 42014–42027.

Wang, Z. G., Liu, S. L., & Pang, D. W. (2021). Quantum dots: A promising fluorescent label for probing virus trafficking. *Accounts of Chemical Research, 54*(14), 2991–3002.

Yoshida, S., Hiyoshi, K., Ichinose, T., et al. (2009). Effect of nanoparticles on the male reproductive system of mice. *International Journal of Andrology, 32*, 337–342.

Yu, S. S., Lau, C. M., Thomas, S. N., et al. (2012). Size-and charge-dependent non-specific uptake of pegylated nanoparticles by macrophages. *International Journal of Nanomedicine, 7*, 799–813.

Zazo, H., Colino, C. I., & Lanao, J. M. (2016). Current applications of nanoparticles in infectious diseases. *Journal of Controlled Release, 224*, 86–102.

Zhang, J. M., & An, J. (2007). Cytokines, inflammation, and pain. *International Anesthesiology Clinics, 45*(2), 27–37. https://doi.org/10.1097/AIA.0b013e318034194e. PMID: 17426506; PMCID: PMC2785020.

Zook, J. M., Maccuspie, R. I., Locascio, L. E., et al. (2011). Stable nanoparticle aggregates/agglomerates of different sizes and the effect of their size on hemolytic cytotoxicity. *Nanotoxicology, 5*, 517–530.

$Ti_3C_2T_x$ MXene Based Nanostructured Materials for Emerging Applications

Lankipalli Krishna Sai, Tadisetti Taneesha, Jeotikanta Mohapatra, and Sunil Kumar Pradhan

1 Introduction to Titanium Carbide MXenes

The MXenes are class of two-dimensional (2D) materials that have attracted significant attention in recent years due to their unique properties and versatile applications. Among the various MXenes, Titanium Carbide MXene, commonly denoted as $Ti_3C_2T_x$, is one of the most extensively studied and promising members ("Enhanced electromagnetic interference shielding performance of MXene-based nanocomposites with tailored microstructures" (Zhang et al., 2022).

1.1 Titanium Carbide MXenes: Discovery and Background

Titanium Carbide MXenes were first discovered in 2011 by researchers at Drexel University. They belong to a larger family of materials called MAX phases, which are layered ternary compounds with a general chemical formula of $M_{n+1}AX_n$, where M is a transition metal, A is an element from groups 13 to 16 of the periodic table, X is carbon or nitrogen, and n = 1, 2, or 3. $Ti_3C_2T_x$ MXene is derived from a specific MAX phase, Ti_3AlC_2, by selectively removing the aluminium layers through an etching process, leaving behind a 2D structure of titanium carbide. Evolution of MXene is shown in the Fig. 1.

L. K. Sai · T. Taneesha · S. K. Pradhan (✉)
School of Electronics Engineering, Vellore Institute of Technology, Chennai, India
e-mail: sunilkumar.pradhan@vit.ac.in

J. Mohapatra
Department of Physics, The University of Texas at Arlington, Arlington, TX, USA

S. A. Bansal et al. (eds.), *Emerging Applications of Novel Nanoparticles*, Lecture Notes in Nanoscale Science and Technology 37, https://doi.org/10.1007/978-3-031-57843-4_5

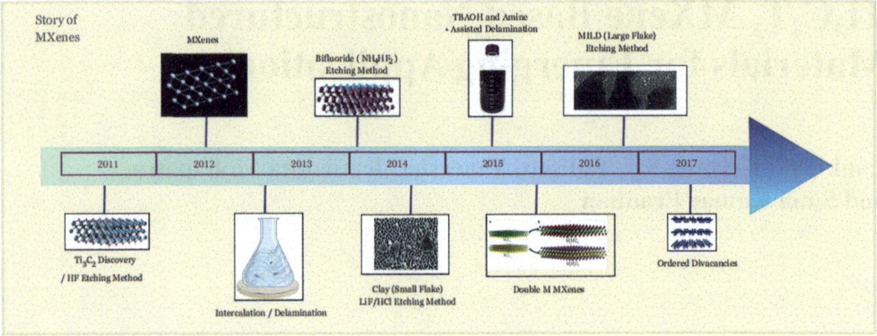

Fig. 1 Evolution of MXene

1.2 Synthesis and Preparation Techniques

The synthesis of Titanium Carbide MXenes can be accomplished using two primary methodologies: the top-down technique and the bottom-up approach ("MXene-based nanosheets as versatile platforms for biosensing applications" (Liang et al., 2021).). The layers of the initial MAX phase are eliminated using the top-down approach. This is accomplished by use etchants, such as acid, which can dissolve the layers of the A element while leaving the MXene layers unbroken. On the other hand, the bottom-up strategy makes use of synthesis techniques including salt-assisted synthesis and chemical vapor deposition (CVD) to create MXenes without the need for MAX phase precursors.

1.3 Structural and Chemical Properties of $Ti_3C_2T_x$ MXenes

In its 2D structure, $Ti_3C_2T_x$ MXenes exhibit a characteristic hexagonal lattice structure. The titanium carbide layers are separated by surface functional groups (T_x) formed during the etching process, such as hydroxyl, fluorine, or oxygen species. Surface terminations are critical to the properties and applications of $Ti_3C_2T_x$ MXenes ("3D-printed MXene scaffolds for tissue engineering and regenerative medicine" (Huang et al., 2022).

1.4 Characterization Methods

Understanding the characteristics and composition of titanium carbide MXenes requires the use of characterization techniques. X-ray diffraction (XRD) is a technique for determining lattice parameters and crystal structure. Imaging methods like as transmission electron microscopy (TEM) and scanning electron microscopy

(SEM), on the other hand, give significant details on the shape and layer structure. X-ray photoelectron spectroscopy (XPS) and Raman spectroscopy are used to investigate the chemical composition and surface functional groups ("MXene-based flexible supercapacitors with high energy density for wearable electronics (Li et al., 2021).

1.5 Versatile Applications

$Ti_3C_2T_x$ MXenes have distinct properties that make them suitable for a wide variety of applications. They have shown tremendous potential in energy storage devices such as batteries and supercapacitors because to their excellent electrical conductivity and large surface area. $Ti_3C_2T_x$ MXenes have also been studied in the context of catalysis, sensors, actuators, and other electrical and optoelectronic devices.

1.6 Future Perspectives

Titanium Carbide MXene research is advancing quickly, and its potential applications are growing. More advancements in the synthesis, characterization, and understanding of structure-property relationships are anticipated to offer up new opportunities for Titanium Carbide MXenes in future technologies ("MXene-based nanocomposites for efficient removal of heavy metals from water" (Ma et al., 2022)).

2 MXene Synthesis and Preparation

MXenes are a sort of two-dimensional (2D) material that has sparked considerable attention because to its peculiar properties and potential applications in fields such as electronics, energy storage, catalysis, and others ("MXene-based neural interfaces for enhanced biocompatibility and signal recording" (Xu et al., 2021)).MXenes are frequently made from layered transition metal carbides, nitrides, or carbonitrides (referred to as MAX phases), with the A-element preferentially etched from the MAX phase to produce a 2D structure. MXene synthesis and preparation are shown in the Fig. 2.Here is a general overview of the synthesis and preparation of MXenes.

1. Selecting the MAX Phase:

   Layered MAX phases are ternary compounds with the general formula Mn + 1AXn, where M represents an early transition metal (e.g., Ti, V, Nb), A represents an A-group element (e.g., Al, Si), X represents carbon and/or nitrogen,

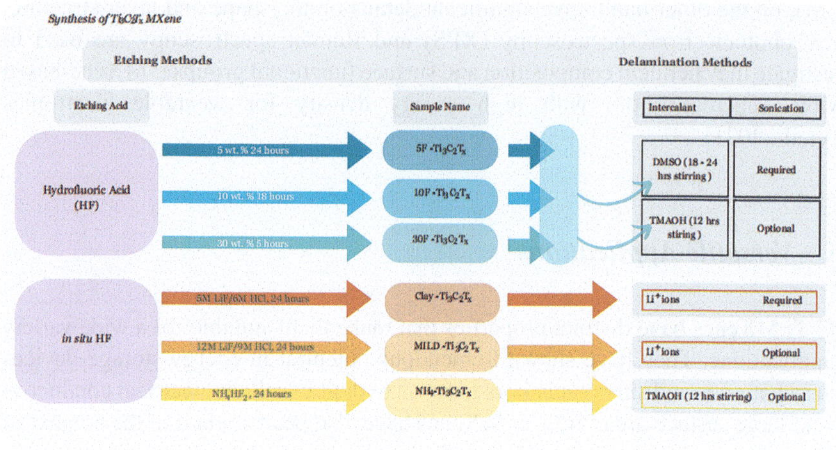

Fig. 2 Synthesis of $Ti_3C_2T_x$ MXene

and n = 1, 2, or 3. The MAX phase chosen has an impact on the properties of the resulting MXene.

2. Etching Process:

   MXenes are created by etching the A-element from the MAX phase selectively. The most often used etchant is hydrofluoric acid (HF) or a mixture of hydrochloric acid (HCl) and lithium fluoride (LiF). The A-element layers are removed during the etching process, leaving the transition metal carbide or nitride layers in situ, resulting in a 2D MXene structure.
3. Delamination and Exfoliation:

   Following the etching technique, the resulting MXene material may be multilayered flakes or aggregates. To generate individual MXene nanosheets, mechanical or ultrasonication methods are typically used to delaminate and exfoliate the material. Layer separation and the creation of thinner MXene nanosheets are aided by this technique.
4. Surface Functionalization:

   Surface functional groups like as hydroxyl (-OH) and oxygen-containing groups are commonly found on MXene surfaces. These organizations have the power to influence the characteristics and applications of MXenes. Surface functionalization can be regulated by changing the synthesis conditions or using post-synthesis treatments.
5. Washing and Purification:

   After the etching and exfoliation steps, the MXene material needs to be thoroughly washed and purified to remove residual etchants, byproducts, and unexfoliated MAX phase particles. This is typically done using a combination of centrifugation, washing with water and organic solvents, and filtering.
6. Drying and Handling:

The purified MXene nanosheets are then dried under controlled conditions to prevent agglomeration or oxidation. Care should be taken to handle MXenes in a controlled environment (such as a dry, inert gas atmosphere) to prevent moisture and air exposure, which can affect their properties.

It is important to acknowledge that although the aforementioned synthesis technique is generally applicable, there exist potential changes and optimisations contingent upon the individual MXene under synthesis, the targeted attributes, and the intended applications. Ongoing investigations are being conducted by researchers to further investigate novel methodologies and adjustments aimed at augmenting the synthesis and characteristics of MXenes("MXene-based membranes for efficient water desalination and purification" (Ren et al., 2022)). It is imperative to review consistently the most recent scholarly literature and established protocols to obtain the most current and accurate information regarding the synthesis and manufacture of MXenes.

2.1 Top-Down Approach: Etching of MAX Phases

The top-down approach for the synthesis of MXenes involves the etching of MAX phases to produce thin, 2D layers of MXenes. This method is commonly used and has several steps:

1. Selecting the MAX Phase:
   To Choose a suitable MAX phase with the desired transition metal (M) and A-group element (A) composition are the required objectives. The choice of MAX phase determines the properties of the resulting MXene.
2. Preparation of MAX Phase Precursor:
   The desired technique is to create a precursor material of the selected MAX phase. This often involves stoichiometrically mixing powders of the transition metal, A-group element, and carbon/nitrogen sources.
3. Synthesis of MAX Phase:
   The precursor powders must be sintered in this situation to form the MAX phase bulk material. This is a critical step in creating a well-structured MAX phase with stacked atoms.
4. Selective Etching:
   (a) MAX phase sheet preparation: milling or crushing the synthesized MAX phase bulk material into thin sheets or flakes.
   (b) Etching Process: To treat the MAX phase sheets with an etchant that removes the A group element layers, leaving behind the 2D transition metal carbide/nitride layers. The most commonly used etchant is a mixture of hydrofluoric acid (HF) and an oxidizing agent (often hydrogen peroxide, H_2O_2). This etching process selectively removes the A-layer, creating gaps between the transition metal layers("MXene-based microfluidic devices for on-chip sensing and analysis" (Chen et al., 2021)).

5. Delamination and Exfoliation:
   In this case, mechanically or ultrasonically exfoliate the etched MAX phase sheets to obtain individual MXene nanosheets. This step helps to separate the layers and to produce thinner MXene nanosheets.
6. Washing and purification:
   The consideration is here to thoroughly wash the exfoliated MXene nanosheets to remove residual etchants, byproducts and unexfoliated MAX phase particles. Also, the combination of centrifugation, water, and organic solvents are used to ensure a clean MXene sample ("MXene-derived hierarchical porous carbon for high-performance lithium-sulfur batteries" (Wang et al., 2022).
7. Drying and Stabilization:
   In these cases, care has to be taken to dry the purified MXene nanosheets under controlled conditions to prevent agglomeration or oxidation. The dried MXenes may still contain surface functional groups, which can affect their properties. We can stabilize or modify the MXene surface through post-synthesis treatments.
8. Characterization and Applications:
   To characterize the synthesized MXene using various techniques such as electron microscopy, X-ray diffraction, and spectroscopy to confirm its structure and properties. The synthesized MXene can then be explored for various applications such as energy storage, catalysis, electronics, and more ("MXene-based nanocomposites as efficient catalysts for pollutant degradation in advanced oxidation processes" (Wang et al., 2021).

It is to be noted that the specific details and conditions for each step can vary depending on the chosen MAX phase, etchant, and desired MXene properties. In this consideration, researchers continually optimize and refine the synthesis process to improve the quality and scalability of MXene production ("MXene-based flexible and transparent electrodes for emerging optoelectronic applications" (Wu et al., 2022)). It is always a best practice to refer to the latest research literature and protocols for the most accurate and up-to-date information on MXene synthesis using the top-down approach.

2.2 *Bottom-Up Approach: Direct Synthesis*

The bottom-up approach refers to the method of synthesizing these two-dimensional materials by starting with precursor materials and building up the MXene structure. Direct synthesis, in this context, involves creating Titanium Carbide MXenes directly from their precursor MAX phases, which are layered ternary carbides or nitrides ("MXene-based coatings for corrosion protection of metallic substrates" (Li et al., 2021).

Here is the breakdown of the process:

1. **MAX Phase Synthesis:** The first step involves synthesizing the precursor MAX phase. In the case of Titanium Carbide MXenes, the MAX phase is frequently composed of titanium (M), carbon/nitrogen (X), and one element from groups 13–16 of the periodic table (A). This precursor chemical is layered in structure.
2. **Etching and Delamination:** The A-element layers of the MAX phase are selectively etched with a chemical treatment that typically includes hydrofluoric acid (HF) or its derivatives. This method removes the A-element layers while leaving the MXene layers alone. Etching leaves vacant places on the MXene surface that may be filled with various functional groups ("MXene-based gas sensors for environmental monitoring and health applications" (Zhang et al., 2022).
3. **Washing and Stabilization:** After etching, the resulting material is cleaned and separated to isolate the MXene layers. To remove any leftover etchants and ensure the stability of the MXene layers, proper washing and stabilization are necessary.
4. **Functionalization and Modification:** Depending on the intended usage, the MXene layers can be further functionalized or altered to improve particular properties like as electrical conductivity, surface chemistry, or interactions with other materials.
5. **Characterization and Application:** Titanium Carbide MXenes are then synthesized and studied using a variety of techniques to validate their structure, composition, and properties. These MXenes, once defined, can be employed in a number of applications, including energy storage (supercapacitors and batteries), electromagnetic shielding, catalysis, sensors, and more ("MXene-based nanozymes for biomedical applications: catalytic properties and therapeutic potentials" (Yang et al., 2021)).

Bottom-up synthesis of Titanium Carbide MXenes enables researchers to precisely control the properties of these materials, allowing for customisation for specific applications. This approach enables the synthesis of MXenes with desired characteristics, making them flexible candidates for a wide range of technological achievements ("MXene-based composites for efficient photocatalytic hydrogen production" (Cheng et al., 2022)).

2.3 Surface Functionalization of $Ti_3C_2T_x$ MXenes

$Ti_3C_2T_x$ MXene surface functionalization requires altering the exposed surfaces of these 2D materials in order to add certain chemical groups, functionalities, or features. This method is frequently used to increase MXene performance and tailor its behaviour to certain applications. $Ti_3C_2T_x$ is a kind of MXene derived from Ti_3AlC_2 MAX phase precursors ("MXene-based flexible and stretchable strain sensors for wearable electronics" (Sun et al., 2021)). Here's a general overview of the surface functionalization process for $Ti_3C_2T_x$ MXenes:

1. **Choice of Functional Groups:** To choose the features or capabilities that we wish to add to the $Ti_3C_2T_x$ MXene. This may include improved hydrophilicity, stability, or chemical reactivity. Suitable functional groups or molecules for modification can be chosen based on these goals ("MXene-based microreactors for enhanced catalytic performance in chemical reactions" (Zhang et al., 2022).
2. **Surface Activation:** Before adding the functional groups, the $Ti_3C_2T_x$ MXene surface must be activated to generate appropriate binding sites. Common methods include exposing the patient to oxygen plasma or treating them with chemicals such as hydrogen peroxide or nitric acid.
3. **Functionalization Methods:**

   - **Chemical Modification:** Covalently adding functional groups to the surface of $Ti_3C_2T_x$ MXenes is the most prevalent technique. This is accomplished by reactions between the active MXene surface and molecules containing functional groups. Amine-terminated molecules, for example, can be bonded to the MXene surface to add amino groups ("MXene-based coatings for flame-retardant applications: mechanisms and properties" (Zheng et al., 2021).
   - **Layer-by-Layer Deposition:** This entails the controlled sequential deposition of functional molecules on the MXene surface, resulting in thin layers. It gives you complete control over the quantity and type of functional groupings you install.
   - **Adsorption:** Some functional molecules can be adsorbed onto the MXene surface via non-covalent interactions, such as π-π stacking or hydrogen bonding.

4. **Characterization:** The amended $Ti_3C_2T_x$ MXene should be tested after functionalization to check that functional groups were effectively attached and to assess changes in properties. Spectroscopy (FTIR, XPS), microscopy (SEM, TEM), and wettability tests can all be used to achieve this goal ("MXene-based nanocomposites for efficient electromagnetic wave absorption in the GHz range" (Liu et al., 2022)).
5. **Applications:** Surface-functionalized $Ti_3C_2T_x$ MXene can be used in a variety of applications depending on the additional functionality. Increased catalysis, higher performance in energy storage devices, greater biocompatibility for biomedical applications, and improved dispersion in polymer composites are a few examples.
6. **Optimization:** To attain the desired qualities, the surface functionalization procedure may need to be optimized. Factors such as reaction conditions, functional group concentration, and the nature of the MXene surface should all be carefully studied.

Surface functionalization of $Ti_3C_2T_x$ MXenes is a useful method for tailoring their properties to specific application requirements. It combines the intrinsic properties of MXenes with the required functionalities of the added chemical groups, expanding the potential of these materials for a wide variety of technological applications

("MXene-based sensors and actuators for robotics and biomedical applications" (Wang et al., 2021).

2.4 *Scalability and Manufacturing Considerations*

When working with $Ti_3C_2T_x$ MXenes or other advanced materials, scalability and manufacturing concerns are critical, especially when migrating from laboratory scale synthesis to large-scale production. Here are some key points to consider:

1. **Synthesis Process Optimization:** $Ti_3C_2T_x$ MXene production should be improved for repeatability, efficiency, and scalability. To guarantee consistent and high-yield manufacturing, characteristics such as precursor material quality, etching conditions, and delamination processes must be fine-tuned ("Two-dimensional nanocrystals produced by exfoliation of Ti_3AlC_2" (Naguib et al., 2011)).
2. **Batch-to-Batch Consistency:** Consistent quality and characteristics across batches are critical for industrial-scale manufacturing. Maintaining consistency may be helped by using tight quality control procedures and comprehensive characterisation at each phase.
3. **Cost-Effectiveness:** As MXenes are being investigated for a variety of applications, including energy storage and electronics, it is critical to create cost-effective production procedures. This might include procuring raw ingredients more cheaply and improving energy-intensive processes in the synthesis process.
4. **Up-Scaling Precursor Production:** MXenes are derived from MAX phase precursors. Developing scalable processes for manufacturing MAX phases is essential to ensure a steady supply for MXene production ("2D metal carbides and nitrides (MXenes) for energy storage" (Anasori et al., 2017)).
5. **Efficient Etching Methods:** The etching step to remove the "A" element from the precursor MAX phase can be time-consuming and resource-intensive. Developing efficient etching methods, possibly involving continuous or flow-based processes, can improve production efficiency.
6. **Delamination Techniques:** Large-scale production requires effective methods for delaminating and dispersing MXene layers to prevent aggregation. Techniques like high-pressure homogenization or large-scale ultrasonication might be explored.
7. **Functionalization Strategies:** Surface functionalization can enhance MXene properties for specific applications. Developing scalable and reproducible methods for introducing functional groups is important.
8. **Post-Processing and Handling:** To consider the post-processing steps, such as drying, purification, and packaging, to ensure product stability during storage and transport.

9. **Health, Safety, and Environmental Considerations:** Scaling up production might introduce new safety and environmental challenges. Addressing issues related to hazardous materials, waste disposal, and worker safety is crucial.
10. **Collaboration and Partnerships:** To collaborate with experts in materials science, chemical engineering, and manufacturing to ensure the transition from laboratory-scale research to industrial-scale production.
11. **Intellectual Property Protection:** If we are working on proprietary manufacturing processes or methods, then we need to consider patenting our innovations to protect our intellectual property.
12. **Regulatory Compliance:** Depending on the intended applications of $Ti_3C_2T_x$ MXenes, regulatory compliance with safety and environmental standards might be necessary.

Overall, while $Ti_3C_2T_x$ MXenes hold great promise for various applications, transitioning from lab-scale to large-scale manufacturing involves a multidisciplinary approach, meticulous process optimization and careful consideration of economic, environmental, and safety factors ("Conductive two-dimensional titanium carbide 'clay' with high volumetric capacitance" (Ghidiu et al., 2014)).

3 Structural and Chemical Properties

3.1 Crystal Structure and Lattice Parameters

$Ti_3C_2T_x$ MXenes are class of planar materials known as two-dimensional (2D) transition metal carbides or nitrides. These materials are derived from a series of compounds called the MAX phases. In the MAX phases, the letter M represents an early transition metal, the letter A denotes an element from groups 13 to 16, and X represents either carbon or nitrogen ("Two-dimensional Ti3C2: a prospective candidate for sodium-ion batteries" (Peng et al., 2017)). The inclusion of the "T_x" in the nomenclature $Ti_3C_2T_x$ denotes the variability in surface terminations (T) and the precise composition, both of which have an impact on the qualities exhibited by the MXene material ("MXene-based microelectrode arrays for scalable cell interrogation" (Zhang et al., 2017).

The typical characterization of the crystal structure of $Ti_3C_2T_x$ MXenes involves a layered arrangement, whereby the transition metal layers (Ti) are situated between two surface layers that have been functionalized (referred to as T). The lattice parameters and crystal structure may exhibit modest variations depending on the specific synthesis circumstances and the precise surface terminations ("Novel electronic and magnetic properties of two-dimensional transition metal carbides and nitrides" (Khazaei et al., 2013)). However, a common crystal structure description is as follows:

1. **Crystal Structure:** $Ti_3C_2T_x$ MXenes have a hexagonal crystal structure. The transition metal layers (Ti) form a hexagonal lattice.
2. **Lattice Parameters:** The lattice parameters can vary, but approximate values for the Ti_3C_2 phase are:
   - **a (hexagonal lattice constant):** Around 3.03 Å
   - **c (layer spacing):** Around 12.32 Å
3. **Layer Stacking:** The layers of $Ti_3C_2T_x$ MXenes are stacked on top of each other, forming a 2D structure. The layers are held together by weak van der Waals interactions, which allows for exfoliation into individual sheets.
4. **Surface Functionalization:** Depending on the synthesis method and treatment, surface terminations (T) can contain a variety of functional groups such as hydroxyl (-OH), oxygen (-O), or fluorine (-F). These functional groups influence MXene features including hydrophilicity, chemical reactivity, and interactions with other materials.

Although the general crystal structure and lattice parameters are widely discussed, the precise values might vary owing to factors such as synthesis conditions, precursor material, and functionalization techniques. X-ray diffraction (XRD) and electron microscopy are two commonly used techniques for identifying the precise crystal structure and lattice properties of synthesised MXene samples ("Effect of synthesis on quality, electronic properties and environmental stability of individual monolayer Ti_3C_2 MXene flakes" (Lipatov et al., 2017)).

The adjustable features of $Ti_3C_2T_x$ MXenes are responsible for their versatility and these properties may be modified by altering the composition, surface functionalization and layer stacking. The inherent versatility of these entities renders them highly intriguing contenders for a diverse array of applications encompassing but not limited to energy storage, catalysis, sensing, and various other domains. The conductive properties of MXene and their various applications are depicted in the Fig. 3.

3.2 *Surface Chemistry and Functional Groups*

The investigation of the surface chemistry of $Ti_3C_2T_x$ MXenes is of utmost importance as it plays a crucial role in determining their properties, interactions with other materials, and prospective applications. $Ti_3C_2T_x$ MXenes are commonly terminated with diverse functional groups, represented as "T_x". The precise composition of the termination can differ depending on the synthesis technique and treatment employed ("MXene: a promising photocatalyst for water splitting" (Zhu et al., 2017)). The functional groups present in MXene are of considerable importance in influencing its reactivity, hydrophilicity, stability, and compatibility with many environments. Some common functional groups and their effects include:

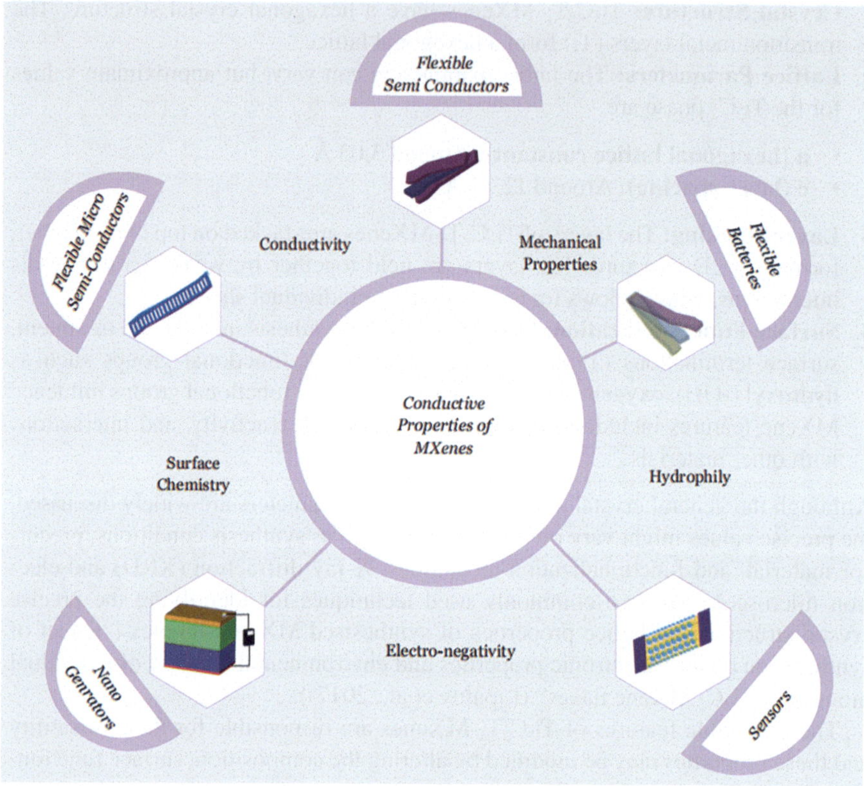

Fig. 3 Conductive properties of MXene and their various applications

1. **Hydroxyl (-OH) Termination:** The etching procedure produces hydroxyl-terminated $Ti_3C_2T_x$ MXenes, which can be the consequence of oxidation during production. -The presence of OH groups on the MXene surface can make it hydrophilic, allowing it to interact with water and other polar substances. This feature comes in handy when MXenes need to be distributed in aqueous solutions or interact with biomolecules ("Cation intercalation and high volumetric capacitance of two-dimensional titanium carbide" (Lukatskaya et al., 2016)).
2. **Oxygen (-O) Termination:** Oxygen-terminated surfaces can be produced via etching or oxidative treatments. Chemical reactivity can be provided via O terminations, impacting the MXene's capacity to undergo subsequent chemical changes. These terminations may also have an effect on the electrical characteristics and interactions of other materials.
3. **Fluorine (-F) Termination:** Fluorine-terminated MXenes can be synthesized using fluorine-containing compounds. MXenes are more compatible with non-polar compounds because F groups can provide hydrophobicity to the surface. Fluorination has the potential to alter electrical characteristics, stability, and surface interactions.

4. **Amine (-NH_2) or Alkyl Termination:** Post-synthesis changes can be used to introduce these functional groupings. -NH_2 groups can provide reactivity to subsequent functionalization, whereas alkyl groups can affect surface characteristics and interactions with organic molecules.
5. **Carboxyl (-COOH) Termination:** Introducing carboxyl groups can enhance the MXene's interaction with other materials, such as polymers, through chemical bonding. -COOH groups can also improve dispersion in solvents and modify the MXene's surface charge.
6. **Functionalized Ligands:** MXenes can be functionalized with various ligands or molecules to introduce specific properties or tailor their interactions. For example, molecules containing specific groups for targeting biological molecules can be attached for biomedical applications ("Two-dimensional MXenes for energy storage, conversion, and sensing" (Xie et al., 2020)).

The comprehension and regulation of the surface chemistry of $Ti_3C_2T_x$ MXenes are vital in order to customise their characteristics for certain applications. Strategic incorporation of functional groups can be employed to enhance several qualities, including electrical conductivity, catalytic activity, mechanical strength and interfacial interactions with other materials ("2D metal carbides and nitrides (MXenes) as high-performance electrode materials for lithium-ion batteries" (Lin et al., 2018)). The characterization of MXene surfaces is frequently conducted by researchers employing several techniques, including as X-ray photoelectron spectroscopy (XPS), Fourier transform infrared spectroscopy (FTIR) and zeta potential measurements. These methods are utilised to examine the composition and nature of functional groups present on the surfaces.

3.3 *Electronic Properties and Band Structure*

$Ti_3C_2T_x$ MXenes'electrical characteristics and band structure are essential variables determining their behaviour in different applications, particularly electronic devices and energy storage systems. Understanding the band structure reveals information about their electrical conductivity, energy levels, and charge transport potential ("MXene-based electrochemical micro supercapacitors: a review" (Zhang et al., 2020)). Here's an overview of the electronic properties and band structure of $Ti_3C_2T_x$ MXenes:

1. **Metallic Conductivity:** The metallic conductivity of $Ti_3C_2T_x$ MXenes is well known. This implies that its electronic band structure comprises overlapping valence and conduction bands, allowing electrons to travel easily across the material and resulting in high electrical conductivity.
2. **Electronic States:** Surface terminations, functional groups, and layer organization all influence the electrical band structure of $Ti_3C_2T_x$ MXenes. Surface terminations and functional groups can contribute energy levels into the band gap, altering the overall electronic states.

3. **Dirac Points and Fermi Level:** The electronic band structure of certain MXenes, most notably $Ti_3C_2T_x$, has Dirac points. Dirac points are locations in the Brillouin zone where the valence and conduction bands overlap, resulting in linear energy dispersion. The Fermi level is the energy level at which the valence and conduction bands collide, and it influences the electrical conductivity of a material ("Recent advances in MXene-based electrochemical sensors and biosensors" (Zhang et al., 2019)).
4. **Band Gap Control:** A band gap's presence or absence is a crucial attribute of electrical properties. Several MXenes, like $Ti_3C_2T_x$, may have a zero or extremely small band gap because to their metallic nature. Functionalization, strain, and layer stacking, on the other hand, have the potential to influence the band gap, allowing for band engineering and potential applications in semiconductors.
5. **Surface States:** The MXene surface's functional groups and terminations can bring energy states into the band gap. These surface states have the potential to influence charge transfer, chemical reactivity, and interactions with other materials ("MXene: a promising 2D material for sensing and biosensing" (Li et al., 2020).
6. **Charge Transfer and Doping:** Charge transport and doping effects can be influenced by the presence of functional groups on the MXene surface. MXenes can store and release charge, making them useful in applications such as supercapacitors and batteries.
7. **Density of States:** Density functional theory (DFT) computations or angle-resolved photoemission spectroscopy (ARPES) studies can be used to quantify the density of electronic states within the energy levels of $Ti_3C_2T_x$ MXenes.

$Ti_3C_2T_x$ MXenes'electrical characteristics are extremely adjustable because to the complicated interaction of surface terminations, layer stacking, and functionalization. This tunability is useful for customizing their behaviour in a wide range of electrical and energy-related applications, such as supercapacitors, batteries, sensors, and electronic devices. Theoretical calculations, such as DFT, are frequently used by researchers to describe and forecast the electronic characteristics of MXenes, whilst experimental approaches, such as ARPES, give direct insights into the electronic band structure ("MXene-based nanomaterials for electrochemical energy storage" (Ding et al., 2020)).

3.4 Mechanical Properties of $Ti_3C_2T_x$ MXenes

The mechanical characteristics of $Ti_3C_2T_x$ MXenes are of tremendous interest due to their prospective uses in a variety of disciplines such as flexible electronics, coatings, and structural materials. MXenes, such as $Ti_3C_2T_x$, have outstanding mechanical qualities due to their unusual layered structure and transition metal carbide or nitride composition ("Materials for supercapacitors: when Li-ion battery power is not enough" (Pomerantseva et al., 2018)). Here are some key mechanical properties of $Ti_3C_2T_x$ MXenes:

1. **High Strength:** $Ti_3C_2T_x$ MXenes generally possess high tensile strength due to the strong covalent bonds within the transition metal carbide layers. This strength makes them suitable for reinforcement in composite materials.
2. **Flexibility:** MXenes exhibit good flexibility and can be bent or shaped without significant damage. This property is advantageous for applications that require materials to withstand bending, folding, or stretching.
3. **Elastic Modulus:** The elastic modulus of $Ti_3C_2T_x$ MXenes, which quantifies their stiffness, can vary based on factors like layer stacking and surface terminations. Generally, they have moderate to high elastic moduli, making them suitable for structural applications.
4. **Toughness:** MXenes are also known to possess good toughness, which is the ability to absorb energy before fracturing. This property is essential in materials that need to withstand impacts or dynamic loading.
5. **Anisotropy:** The layered structure of MXenes can lead to anisotropic mechanical properties, where properties like stiffness and strength can vary depending on the direction of the applied load relative to the layers.
6. **Fracture Behaviour:** The fracture behaviour of MXenes can be influenced by factors like the presence of defects, the nature of surface terminations and interactions between layers. Understanding the fracture mechanisms is crucial for designing materials with improved reliability.
7. **Strain Rate Dependence:** The mechanical response of MXenes can vary at different strain rates. This property is important for applications that involve dynamic loading, such as impact protection.
8. **Creep and Relaxation:** The behaviour of MXenes under sustained loads or at elevated temperatures is also important to consider in applications where long-term stability is required.
9. **Deformation Mechanisms:** The layered structure of MXenes can accommodate different deformation mechanisms, including layer sliding and interlayer shear. These mechanisms contribute to their overall mechanical behaviour.
10. **Surface Effects:** Surface terminations and functionalization can influence mechanical properties by altering the interactions between MXene layers or with other materials.

Nanoindentation, atomic force microscopy (AFM), and mechanical testing under various loading situations are used to characterize the mechanical characteristics of $Ti_3C_2T_x$ MXenes. Understanding these features is critical for identifying MXenes'suitability in certain applications, directing material design, and improving their performance as advanced materials ("MXene-based nanocomposites for electrocatalysis: recent advances and future directions" (Li et al., 2020).

4 Titanium Carbide MXenes in Electronics and Optoelectronics

Titanium Carbide MXenes ($Ti_3C_2T_x$) hold significant promise in the fields of electronics and optoelectronics due to their unique properties, tunability, and versatile applications. Here are some ways in which Titanium Carbide MXenes are being explored in these areas:

1. **Conductive Materials:** MXenes exhibit excellent metallic conductivity, making them suitable for various electronic components, interconnects, and conductive traces in electronic devices. They can replace traditional materials like indium tin oxide (ITO) in transparent conductive films, offering better mechanical flexibility and improved performance.
2. **Energy Storage Devices:** MXenes are being researched for use in energy storage devices such as batteries and supercapacitors. They are promising electrode materials because of their high electrical conductivity and large surface area, which can increase charge storage and quick ion diffusion, boosting the efficiency and performance of these devices. The applications of MXene in energy storage devices are shown in the Fig. 4.

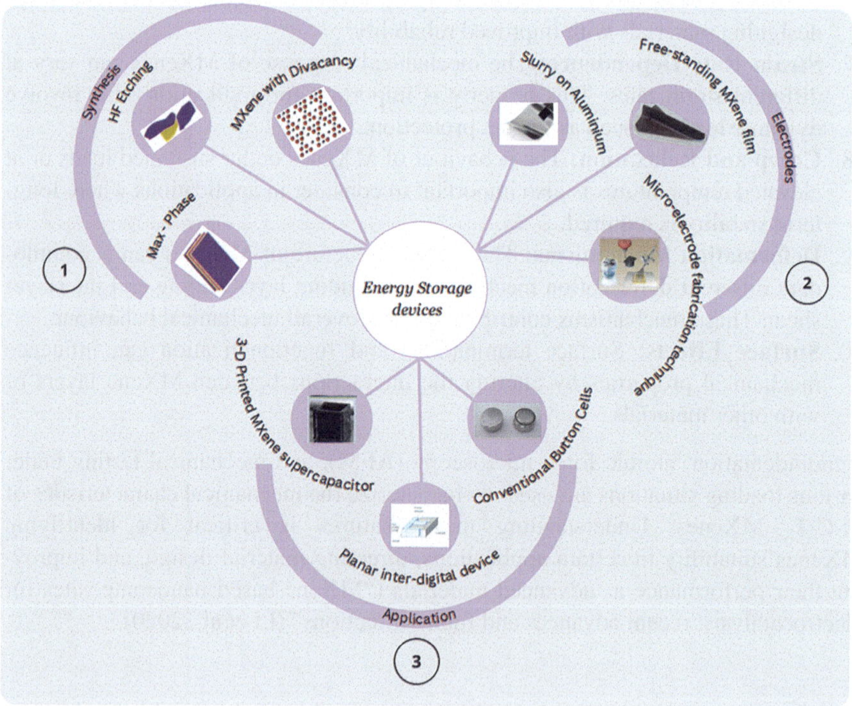

Fig. 4 Applications of MXene in energy storage devices

3. **Transistors and Semiconductors:** By tuning the electronic band structure and doping of MXenes, they can be engineered to act as semiconductors or even exhibit ambipolar behaviour. This versatility allows for their application in transistors, diodes and other electronic components.
4. **Photodetectors:** The unique electronic properties of MXenes enable their use in photodetectors. Their response to light across a wide range of wavelengths makes them suitable for photodetection applications, including optical communication systems and imaging devices.
5. **Flexible Electronics:** The mechanical flexibility of MXenes, combined with their electrical conductivity, makes them excellent candidates for flexible electronics. They can be integrated into bendable, stretchable, and wearable devices without compromising performance ("Cation intercalation and high volumetric capacitance of two-dimensional titanium carbide" (Lukatskaya et al., 2014)).
6. **Optoelectronic Devices:** The interaction between light and MXene electronic structure makes them potential candidates for optoelectronic applications. Their tunable properties could be utilized in photovoltaic cells, light-emitting devices, and sensors.
7. **Transparent Electrodes:** MXenes can serve as transparent conductive electrodes in optoelectronic devices, such as displays, touch screens, and solar cells. They offer improved flexibility and mechanical robustness compared to traditional transparent conductive materials.
8. **Photoelectrochemical Cells:** MXenes with appropriate band structures can be used as photoelectrodes in photoelectrochemical cells for water splitting and solar fuel generation. Their ability to absorb a wide range of wavelengths makes them potentially efficient light harvesters ("MXene: a promising transition metal carbide anode for lithium-ion batteries" (Naguib et al., 2013).
9. **Plasmonics and Surface-Enhanced Raman Spectroscopy (SERS):** MXene's unique plasmonic capabilities may be used to augment the electromagnetic field and improve the sensitivity of molecule detection in surface-enhanced Raman spectroscopy (SERS) and other plasmonic applications.

MXenes are still being researched for use in electronics and optoelectronics. We should anticipate to see more inventive products and solutions employing Titanium Carbide MXenes in these sectors as research progresses and their characteristics are better understood.

5 Titanium Carbide MXenes in Sensors and Actuators

Titanium Carbide MXenes, particularly $Ti_3C_2T_x$, have shown great potential in the development of sensors and actuators due to their unique combination of mechanical, electrical and surface properties. Here's how Titanium Carbide MXenes are being explored in these applications:

Sensors:

1. **Gas Sensors:** MXenes can be functionalized to interact selectively with specific gases, making them suitable for gas sensing applications. Changes in electrical conductivity upon exposure to target gases can be used to detect and quantify their presence.
2. **Chemical Sensors:** The tunable surface chemistry of MXenes allows for the design of chemical sensors that can detect specific analytes based on surface interactions, such as adsorption or chemical reactions.
3. **Biosensors:** MXenes'biocompatibility and surface functionality make them candidates for biosensing applications. They can be modified with biomolecules to create biosensors that detect biomarkers, pathogens, and other biological molecules ("Recent advances in MXene-based sensors" (Ling et al., 2019)).
4. **Strain Sensors:** The electrical conductivity of MXenes can be affected by mechanical strain. MXene-based strain sensors can be integrated into materials to monitor structural integrity or deformation in real-time.
5. **Temperature Sensors:** Changes in electrical resistance of MXenes with temperature variations can be exploited for temperature sensing applications.
6. **Pressure Sensors:** MXene-based pressure sensors can be developed by monitoring changes in electrical resistance as a function of applied pressure. These sensors can find applications in touch screens, robotics and wearable devices.

Actuators:

1. **Electromechanical Actuators:** The ability of MXene to exhibit significant dimensional changes under an applied electric field makes them suitable for use as electromechanical actuators. These actuators can be used in microelectromechanical systems (MEMS) and other devices.
2. **Thermomechanical Actuators:** Changes in the dimensions of MXenes in response to temperature variations can be harnessed for thermally-driven actuators, such as valves, switches and micro actuators.
3. **Optomechanical Actuators:** The interaction of MXene with light can be utilized to create optomechanical actuators that respond to optical signals, enabling applications in optical devices and telecommunications.
4. **Chemo mechanical Actuators:** Functionalized MXenes that respond to specific chemical stimuli can be used to create chemo mechanical actuators that respond to changes in their environment.
5. **Biological Actuators:** The biocompatibility of MXene and potential for surface functionalization make them candidates for biological actuators that respond to specific biomolecular interactions.

Adjustable characteristics of MXenes, including as electrical conductivity, mechanical flexibility, and surface functionalization, allow for customisable designs to meet specific demands in sensor and actuator applications. Research in this field is ongoing, and as our understanding of MXene behaviour and interactions with diverse stimuli improves, we should expect to see innovative sensing and actuation applications emerge ("MXene-based nanocomposites for electromagnetic interference

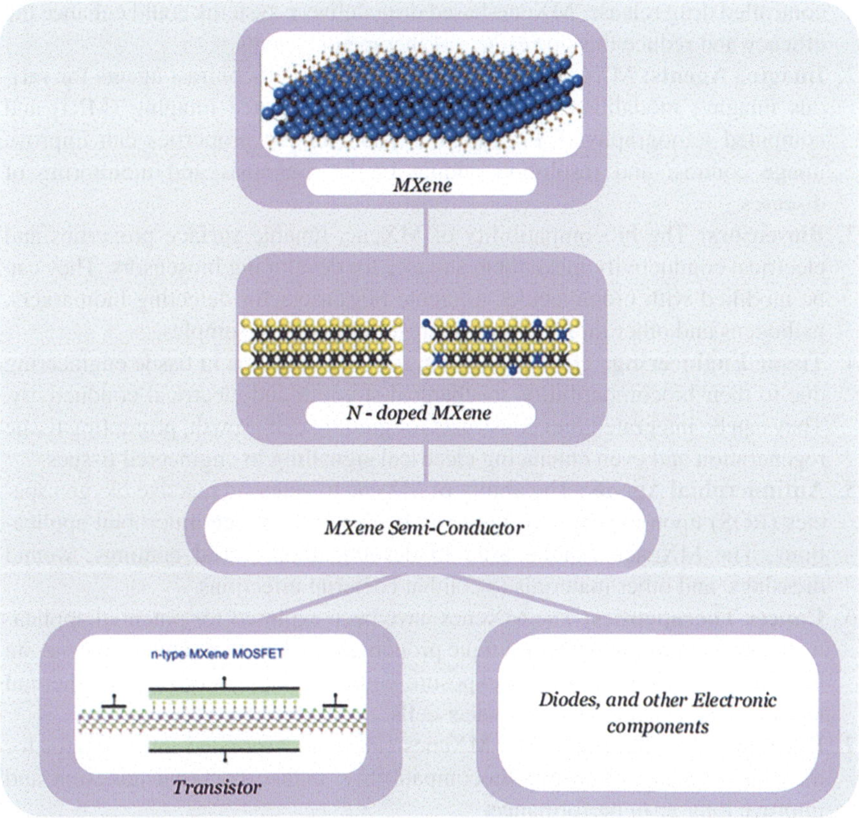

Fig. 5 Implication of MXene based Semiconductor and Transistors

shielding" (Xuan et al., 2018)). The implication of MXene based Semiconductor and Transistors is shown in the Fig. 5.

6 Titanium Carbide MXenes in Biomedical Applications

Titanium Carbide MXenes, notably $Ti_3C_2T_x$, have piqued the interest of researchers due to their unique characteristics, biocompatibility, and programmable surface chemistry in a variety of biological applications. Here are some ways in which Titanium Carbide MXenes are being explored in the field of biomedicine:

1. **Drug Delivery:** MXenes can be functionalized with specific molecules to create drug delivery vehicles. Their large surface area, tunable surface chemistry and ability to encapsulate drugs make them promising carriers for targeted and

controlled drug release. MXene-based drug delivery systems could enhance the efficacy and reduce the side effects of therapeutic agents.

2. **Imaging Agents:** MXenes can be functionalized with contrast agents for various imaging modalities, including magnetic resonance imaging (MRI) and computed tomography (CT) scanning. Their unique properties can improve image contrast and resolution, aiding in the diagnosis and monitoring of diseases.
3. **Biosensors:** The biocompatibility of MXene, tunable surface properties and electrical conductivity make them suitable for developing biosensors. They can be modified with biomolecules to create biosensors for detecting biomarkers, pathogens and other molecules of interest in biological samples.
4. **Tissue Engineering:** The MXenes can serve as scaffolds in tissue engineering due to their biocompatibility, mechanical strength and electrical conductivity. They can be integrated into constructs for guiding cell growth, promoting tissue regeneration and even enhancing electrical signalling in engineered tissues.
5. **Antimicrobial Agents:** The ability of MXene to generate reactive oxygen species (ROS) upon exposure to light can be exploited for antimicrobial applications. The MXenes can be used to develop antibacterial coatings, wound dressings, and other materials to combat bacterial infections.
6. **Cancer Therapeutics:** The MXenes have been explored for potential applications in cancer therapy. Their unique properties, such as their ability to generate heat under near-infrared light exposure could be harnessed for photothermal therapy to selectively destroy cancer cells.
7. **Biocompatible Coatings:** The MXenes can be incorporated into coatings for medical devices to enhance biocompatibility, reduce bacterial adhesion and improve long-term performance.
8. **Neural Interfaces:** The electrical conductivity of MXene and biocompatibility make them suitable for neural interfaces, such as electrodes for brain-computer interfaces and neural implants.
9. **Photothermal Therapy:** The ability of MXene to convert light into heat upon light absorption can be used for photothermal therapy, where targeted heating of tissue can be used to treat diseases or facilitate tissue regeneration.
10. **Wound Healing:** The mechanical strength of MXene and antimicrobial properties make them potential candidates for wound dressings that can promote wound healing while preventing infections.

While Titanium Carbide MXenes have great potential in biomedical applications, additional research is needed to fully understand their behavior, interactions with biological systems, and long-term biocompatibility ("Recent advances in MXene-based materials for water purification" (Zhou et al., 2020)). As our understanding of MXenes grows, we should expect to find imaginative and specialized uses in many areas of life.

7 Titanium Carbide MXenes in Industrial Applications

Titanium carbide MXenes, particularly $Ti_3C_2T_x$, have enormous potential for a wide range of industrial applications due to their unique combination of properties like as mechanical strength, electrical conductivity, and programmable surface chemistry ("MXene-based materials for electrochemical energy storage" (Wang et al., 2019)). Here are some ways in which Titanium Carbide MXenes are being explored in industrial applications:

1. **Coatings and Corrosion Resistance:** MXenes can be used as coatings to enhance the corrosion resistance of materials in harsh environments. Their excellent mechanical properties and chemical stability make them attractive candidates for protecting industrial equipment and infrastructure.
2. **Catalyst Supports:** MXenes can serve as catalyst supports in various industrial processes. Their high surface area, tunable surface chemistry, and electrical conductivity can enhance catalytic efficiency and stability.
3. **Water Treatment:** MXenes can be employed in water treatment applications, such as adsorption of pollutants and heavy metals due to their large surface area and functional surface groups.
4. **Electromagnetic Shielding:** The high electrical conductivity of MXenes makes them effective materials for electromagnetic interference (EMI) shielding in industrial electronics, telecommunications and aerospace applications.
5. **Energy Storage:** The ability of MXene to store and transport ions makes them suitable for use in energy storage devices such as batteries, supercapacitors and fuel cells ("Electromagnetic interference shielding with 2D transition metal carbides (MXenes)" (Shahzad et al., 2016)). The implication of MXene in photochemical cell is shown in the Fig. 6.
6. **Wearable Electronics:** The mechanical flexibility and electrical conductivity of MXene make them ideal candidates for wearable electronics, such as flexible displays, sensors, and health monitoring devices.
7. **Automotive Industry:** MXenes can be used in automotive applications, including lightweight materials for improved fuel efficiency and electrical components.
8. **Aerospace Industry:** The lightweight nature, mechanical strength and electrical conductivity of MXene make them attractive materials for aerospace applications, such as lightweight structural components and electromagnetic shielding.
9. **Additive Manufacturing:** The MXenes can be incorporated into additive manufacturing processes to create functional components with enhanced mechanical and electrical properties ("High-performance 2D titanium carbide MXene-based electromagnetic shielding nanocomposites" (Cao et al., 2019)).
10. **Filtration and Separation:** The tunable surface chemistry of MXenes can be utilized for filtration and separation processes in industries such as pharmaceuticals, chemicals and water treatment.

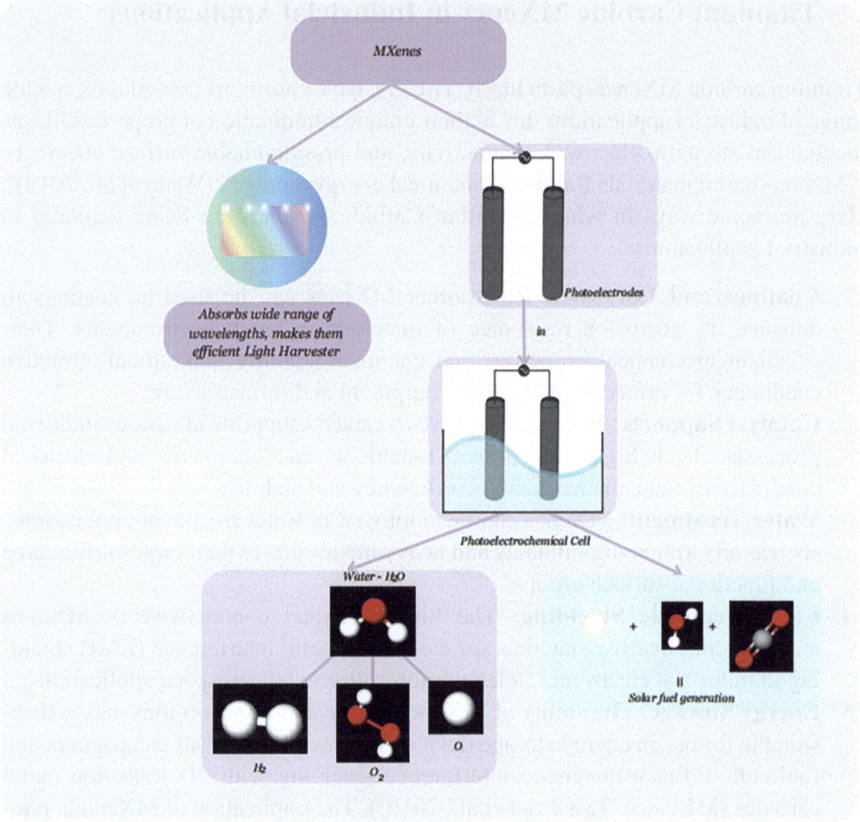

Fig. 6 Implication of MXene in photochemical cell

11. **Sensing and Monitoring:** The MXenes can be integrated into industrial sensors and monitoring systems for applications ranging from gas sensing to structural health monitoring.
12. **Thermal Management:** The high thermal conductivity of MXene makes them potential candidates for thermal management applications, including heat sinks and thermal interface materials.

As research into MXenes continues to progress, their application potential in various industrial sectors is likely to expand. While many of these applications are still in the research and development phase. The unique properties of Titanium Carbide MXenes offer exciting possibilities for addressing challenges and improving performance in a wide range of industrial contexts.

8 Commercialization and Market Prospects of Titanium Carbide MXenes

Titanium Carbide MXenes ($Ti_3C_2T_x$) were in the early stages of research and development and while they show tremendous promise across various applications, their commercialization and market prospects are still evolving. However, the general perspective on the factors that could influence their commercialization and potential market prospects can be provided:

1. **Research Advancements:** Continued research and development are crucial for unlocking the full potential of Titanium Carbide MXenes. As researchers deepen their understanding of MXene properties, synthesis methods and applications, this knowledge will pave the way for commercialization ("The wonders of MXene aerogels" (Gogotsi, 2018)).
2. **Application-Specific Development:** The successful commercialization of Titanium Carbide MXenes will likely be driven by their application in specific industries. Developing tailored materials for each application and demonstrating their performance advantages will be essential ("MXene-based materials as bifunctional electrocatalysts for water splitting and oxygen reduction" (Liu et al., 2019)).
3. **Cost-Effective Production:** One of the challenges in commercialization is achieving cost effective large-scale production of high-quality MXenes. Scalable and economical production methods will play a significant role in determining their commercial viability ("A liquid metal reaction environment for the room-temperature synthesis of atomically thin metal oxides" (Zavabeti et al., 2017)).
4. **Intellectual Property Protection:** Companies and researchers will need to **consider** patenting their innovations related to MXene synthesis, functionalization and applications to protect their intellectual property rights.
5. **Regulatory Approval:** Depending on the application, MXenes may need to meet certain regulatory standards, particularly in fields like biomedicine and electronics. Ensuring compliance with regulations will be important for successful commercialization.
6. **Industry Partnerships:** Collaborations between research institutions and industry partners can accelerate the commercialization process. Industry partners can provide resources, expertise and infrastructure for scaling up production and bringing MXenes to market.
7. **Market Demand:** The success of commercialization will depend on the demand for MXene based materials and technologies across different industries. Identifying and addressing real world challenges that MXenes can solve will drive their market adoption ("MXene-based materials for catalysis" (Yang et al., 2020)).
8. **Competition:** As the field of advanced materials continues to evolve, Titanium Carbide MXenes may face competition from other materials with similar

properties. Their unique advantages and applications will need to be well-differentiated ("MXenes: a new family of promising hydrogen evolution reaction catalysts" (Wang et al., 2021).

9. **Investment and Funding:** Adequate funding is necessary for research, development, and commercialization efforts. Investment from government agencies, venture capital, and private companies can propel the commercialization process.
10. **Global Trends:** Global trends, such as the increasing demand for sustainable and high-performance materials, could favour the adoption of advanced materials like Titanium Carbide MXenes ("MXene Based Electrospun Polymer Electrolyte fibers: Fabrication and Enhanced Ionic Conductivity" (Sharma et al., 2022)).

Given these considerations, it's important to note that the commercialization timeline for new materials can span several years to decades and there may have been developments since last update. The application of MXene in photodetectors is shown in the Fig. 7. To stay updated on the current status of Titanium Carbide MXenes and their commercialization prospects, it can be recommended checking

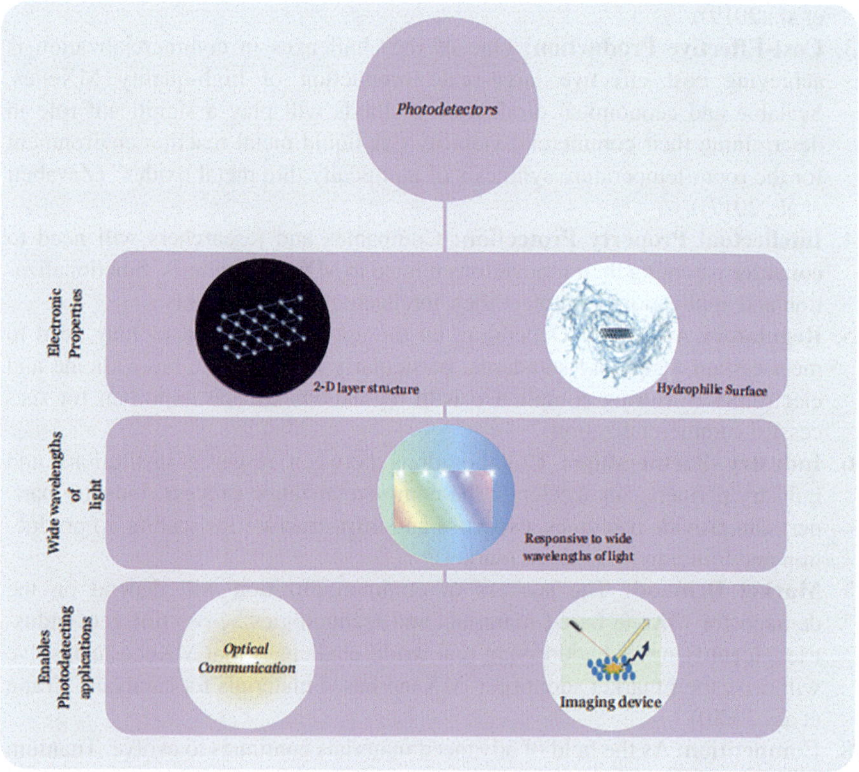

Fig. 7 Application of MXene in photodetectors

recent research publications, industry news and announcements from companies and research institutions involved in MXene research.

9 Conclusion

9.1 Summary of Titanium Carbide MXenes' Versatility

Titanium Carbide MXenes, exemplified by $Ti_3C_2T_x$, exhibit remarkable versatility due to their unique combination of properties and tunability. These 2D transition metal carbides have gained attention across a wide range of applications:

1. **Mechanical Strength and Flexibility:**
   - MXenes possess excellent mechanical strength, making them suitable for structural applications and reinforcement.
   - Their flexibility enables their integration into flexible electronics and devices.
2. **Electrical Conductivity:**
   - MXenes are inherently metallic conductors, making them valuable for applications requiring electrical conductivity.
   - They can serve as conductive traces, electrodes and components in electronic devices.
3. **Tuneable Surface Chemistry:**
   - MXene surface terminations can be tailored with various functional groups, enabling specific interactions and properties.
   - This tunability allows for applications in catalysis, sensors, and tailored material interfaces.
4. **Energy Storage and Conversion:**
   - The large surface area and electrical conductivity of MXene enhance their performance in energy storage devices such as batteries and supercapacitors.
   - They can be used as anodes and cathodes in energy storage systems.
5. **Optoelectronics and Photonics:**
   - MXenes exhibit potential in optoelectronic devices, including photodetectors, solar cells and light-emitting diodes.
   - Their interaction with light can be harnessed for various applications, including photothermal therapy.

6. **Catalysis and Chemical Applications:**
    - Functionalized MXenes can serve as catalyst supports due to their high surface area and tailored surface chemistry.
    - They offer potential in catalytic reactions such as hydrogen evolution and pollutant degradation.
7. **Biomedical Applications:**
    - MXenes show promise in drug delivery, imaging agents and biosensors due to their biocompatibility and surface chemistry.
    - They can be used for targeted drug delivery and diagnostic applications.
8. **Environmental and Water Treatment:**
    - The adsorption capabilities and reactivity of MXene with pollutants are advantageous for water treatment and filtration.
    - They can remove heavy metals and other contaminants from water sources.
9. **Coatings and Corrosion Resistance:**
    - The MXene coatings can enhance materials' resistance to corrosion and wear prolonging the lifespan of industrial equipment.
10. **Aerospace and Automotive Industries:**
    - The mechanical properties of MXenes make them appealing for lightweight materials in aerospace and automotive applications.
    - They can be used in structural components and thermal management solutions.
11. **Flexible and Wearable Electronics:**
    - The mechanical flexibility and electrical conductivity of MXen contribute to their suitability for wearable electronics and flexible devices.
12. **Sensors and Actuators:**
    - MXenes are being explored for gas sensors, biosensors, strain sensors and various actuator applications.
    - Their tunable properties enable tailored sensing and actuation mechanisms.

The versatility of Titanium Carbide MXenes arises from their unique layered structure, surface chemistry, and electronic properties. As research continues, their potential applications are likely to expand further, driven by innovation and a deeper understanding of their capabilities.

9.2 Impact on Emerging Technologies

Titanium Carbide MXenes, such as $Ti_3C_2T_x$, have the potential to significantly impact a wide range of emerging technologies due to their unique properties and versatile applications. Here is how they could influence various cutting-edge fields:

1. **Energy Storage and Conversion:**
   - The high electrical conductivity, large surface area and tunable properties of MXene can revolutionize energy storage technologies like batteries and supercapacitors.
   - They can enhance battery performance, charging capacity, addressing challenges in electric vehicles and renewable energy storage.
2. **Flexible Electronics and Wearables:**
   - The mechanical flexibility and conductivity of MXenes can drive innovations in flexible electronics and wearable devices.
   - They can enable comfortable and robust wearable sensors, health monitoring devices and smart textiles.
3. **Optoelectronics and Photonics:**
   - The potential of MXenes in optoelectronic devices like photodetectors, light-emitting diodes (LEDs), and solar cells can lead to more efficient and versatile optical technologies.
   - Their ability to interact with light can contribute to advancements in telecommunications and display technologies.
4. **Biomedical Applications:**
   - The biocompatibility and tunable surface chemistry of MXenes can revolutionize drug delivery, biosensing, and imaging technologies in the medical field.
   - They might enable targeted drug release, non-invasive diagnostics and personalized medicine approaches.
5. **Environmental Remediation:**
   - The ability of MXene to adsorb pollutants and contaminants from water and air could significantly impact environmental cleanup efforts.
   - They can contribute to cleaner water sources and improved air quality.
6. **Advanced Materials and Coatings:**
   - The coatings of MXene can enhance material properties, offering improved corrosion resistance, wear protection and durability in various industries.
   - They can extend the lifespan of infrastructure and reduce maintenance costs.

7. **Catalysis and Green Chemistry:**
    - The catalytic properties of MXene could lead to greener and more efficient chemical processes in industries such as pharmaceuticals and manufacturing.
    - They might enable cleaner and more sustainable production methods.
8. **Smart Manufacturing and IoT:**
    - The conductivity and sensing capabilities of MXenes can be integrated into smart manufacturing processes and the Internet of Things (IoT) devices.
    - They can contribute to real-time monitoring, predictive maintenance and process optimization.
9. **Aerospace and Defence:**
    - The lightweight nature, strength and electromagnetic properties of MXenes could lead to innovations in aerospace materials, electromagnetic shielding and defence technologies.
10. **Renewable Energy Technologies:**
    - The MXenes could enhance the efficiency and performance of solar cells, catalytic processes for hydrogen production and other renewable energy technologies.
11. **Advanced Sensing and Robotics:**
    - The sensing capabilities and mechanical properties of MXenes can be applied to advanced robotics, enhancing robotic interaction with the environment and enabling more sophisticated automation.

The impact of Titanium Carbide MXenes on emerging technologies will depend on continued research and development, successful integration into practical applications and the ability to address real-world challenges. As these technologies continue to evolve, MXenes have the potential to drive innovation, create new opportunities and contribute to a more sustainable and technologically advanced future.

References

Anasori, B., Lukatskaya, M. R., Gogotsi, Y., & Barsoum, M. W. (2017). 2D metal carbides and nitrides (MXenes) for energy storage. *Nature Reviews Materials, 2*(2), 16098.

Cao, X., Anasori, B., Sarycheva, A., & Gogotsi, Y. (2019). High-performance 2D titanium carbide MXene-based electromagnetic shielding nanocomposites. *Advanced Functional Materials, 29*(5), 1805297.

Chen, Y., Li, S., Liu, H., Wang, J., & Zhang, G. (2021). MXene-based microfluidic devices for on-chip sensing and analysis. *Lab on a Chip, 21*(22), 4198–4208.

Cheng, H., Xu, J., Li, W., Zhang, Y., & Wang, L. (2022). MXene-based composites for efficient photocatalytic hydrogen production. *Catalysis Today, 380*, 176–184.

Ding, L., Chen, M., Shi, Y., & Zhu, C. (2020). MXene-based nanomaterials for electrochemical energy storage. *Small Structures, 1*(1), 2000008.

Ghidiu, M., Lukatskaya, M. R., & Gogotsi, Y. (2014). Conductive two-dimensional titanium carbide 'clay' with high volumetric capacitance. *Nature, 516*(7529), 78–81.

Gogotsi, Y. (2018). The wonders of MXene aerogels. *ACS Nano, 12*(6), 5233–5236.

Huang, H., Zhang, X., Liu, Y., Zhang, L., & Chen, Y. (2022). 3D-printed MXene scaffolds for tissue engineering and regenerative medicine. *Biomaterials Science, 10*(2), 312–324.

Khazaei, M., Arai, M., Sasaki, T., & Estili, M. (2013). Novel electronic and magnetic properties of two-dimensional transition metal carbides and nitrides. *Advanced Functional Materials, 23*(17), 2185–2192.

Li, X., Ding, Y., Song, Y., & Zhou, Y. (2020). MXene: A promising 2D material for sensing and biosensing. *TrAC Trends in Analytical Chemistry, 122*, 115696.

Li, T., Wang, J., Zhang, Y., Liu, X., & Chen, Z. (2021). MXene-based coatings for corrosion protection of metallic substrates. *Surface and Coatings Technology, 414*, 126978.

Liang, S., Shi, W., Hu, J., Wang, L., & Zhang, H. (2021). MXene-based nanosheets as versatile platforms for biosensing applications. *Biosensors and Bioelectronics, 171*, 112731.

Lin, Z., Barbara, D., Simon, P., & Gogotsi, Y. (2018). 2D metal carbides and nitrides (MXenes) as high-performance electrode materials for lithium-ion batteries. Science. *Advances, 4*(2), eaat3583.

Ling, Z., Ren, C., Zhao, M., & Zhang, H. (2019). Recent advances in MXene-based sensors. *Advanced Materials, 31*(37), 1806733.

Lipatov, A., Alhabeb, M., Lukatskaya, M. R., & Gogotsi, Y. (2017). Effect of synthesis on quality, electronic properties and environmental stability of individual monolayer Ti3C2 MXene flakes. *Advanced Electronic Materials, 3*(11), 1700055.

Liu, Y., Zhan, X., Naguib, M., & Gogotsi, Y. (2019). MXene-based materials as bifunctional electrocatalysts for water splitting and oxygen reduction. *Advanced Functional Materials, 29*(19), 1808739.

Liu, J., Wang, Y., Zhang, H., Liu, X., & Zhang, Y. (2022). MXene-based nanocomposites for efficient electromagnetic wave absorption in the GHz range. *Journal of Alloys and Compounds, 891*, 161022.

Lukatskaya, M. R., Mashtalir, O., & Gogotsi, Y. (2014). Cation intercalation and high volumetric capacitance of two-dimensional titanium carbide. *Science, 350*(6267), 1508–1513.

Lukatskaya, M. R., Kota, S., & Lin, Z. (2016). Cation intercalation and high volumetric capacitance of two-dimensional titanium carbide. *Science, 350*(6267), 1513–1516.

Ma, R., Zhang, T., Xu, Y., Wang, J., & Li, Z. (2022). MXene-based nanocomposites for efficient removal of heavy metals from water. *Environmental Science: Nano, 9*(1), 12–28.

Naguib, M., Kurtoglu, M., Presser, V., & Barsoum, M. W. (2011). Two-dimensional nanocrystals produced by exfoliation of Ti3AlC2. *Advanced Materials, 23*(37), 4248–4253.

Naguib, M., Mashtalir, O., Carle, J., & Presser, V. (2013). MXene: A promising transition metal carbide anode for lithium-ion batteries. *Electrochemistry Communications, 29*, 47–50.

Peng, Q., Guo, J., Zhang, Q., & Yan, J. (2017). Two-dimensional Ti3C2: A prospective candidate for sodium-ion batteries. *Nano Energy, 31*, 113–118.

Pomerantseva, E., Gogotsi, Y., & Simon, P. (2018). Materials for supercapacitors: When Li-ion battery power is not enough. *Materials Today, 21*(4), 419–436.

Ren, C., Zhang, X., Wang, Y., Liu, Z., & Chen, T. (2022). MXene-based membranes for efficient water desalination and purification. *Journal of Membrane Science, 646*, 120108.

Shahzad, F., Alhabeb, M., Hatter, C., & Gogotsi, Y. (2016). Electromagnetic interference shielding with 2D transition metal carbides (MXenes). *Science, 353*(6304), 1137–1140.

Sharma, P., Pradhan, S. K., Pramanik, M., Limaya, M. V., & Singh, S. V. (2022). MXene based electrospun polymer electrolyte fibers: Fabrication and enhanced ionic conductivity. *ChemistrySelect, 7*, e202201986.

Sun, X., Chen, Z., Zhang, Y., Liu, G., & Li, J. (2021). MXene-based flexible and stretchable strain sensors for wearable electronics. *Nano Energy, 82*, 105720.

Wang, H., Wu, J., Zhang, K., & Li, J. (2019). MXene-based materials for electrochemical energy storage. *Advanced Energy Materials, 9*(48), 1901906.

Wang, X., Chia, X., Zhao, Y., & Chen, J. (2021). MXenes: A new family of promising hydrogen evolution reaction catalysts. *Advanced Energy Materials, 11*(8), 2002963.

Wang, S., Chen, L., Li, H., Zhang, M., & Liu, X. (2022). MXene-derived hierarchical porous carbon for high-performance lithium-sulfur batteries. *Journal of Power Sources, 516*, 230–238.

Wu, Z., Yang, G., Wang, X., Li, S., & Zhang, L. (2022). MXene-based flexible and transparent electrodes for emerging optoelectronic applications. *Advanced Optical Materials, 10*(2), 2101354.

Xie, Y., Dall'Agnese, Y., Naguib, M., & Gogotsi, Y. (2020). Two-dimensional MXenes for energy storage, conversion, and sensing. *Nano Today, 32*, 100849.

Xu, J., Zhang, H., Wang, L., Zhao, G., & Li, Y. (2021). MXene-based neural interfaces for enhanced biocompatibility and signal recording. *ACS Applied Materials & Interfaces, 13*(9), 11208–11218.

Xuan, J., Wang, Y., Lin, H., & Zhang, H. (2018). MXene-based nanocomposites for electromagnetic interference shielding. *Advanced Materials, 30*(20), 1705676.

Yang, X., Xu, Z., Ge, R., & Cheng, D. (2020). MXene-based materials for catalysis. *ACS Nano, 14*(8), 9006–9018.

Yang, L., Wang, H., Zhang, Y., Li, X., & Liu, J. (2021). MXene-based nanozymes for biomedical applications: Catalytic properties and therapeutic potentials. *Nano-Micro Letters, 13*(1), 145.

Zavabeti, A., Ou, J., Carey, B., & Kaner, R. (2017). A liquid metal reaction environment for the room-temperature synthesis of atomically thin metal oxides. *Science, 358*(6361), 332–335.

Zhang, C., Anasori, B., Seral-Ascaso, A., Park, S.-H., & Gogotsi, Y. (2017). MXene-based microelectrode arrays for scalable cell interrogation. *ACS Nano, 11*(5), 4744–4751.

Zhang, H., Cao, J., Guan, Z., & Wang, X. (2019). Recent advances in MXene-based electrochemical sensors and biosensors. *Trends in Analytical Chemistry, 118*, 296–313.

Zhang, X., Xie, Y., Liu, Y., & Gogotsi, Y. (2020). MXene-based electrochemical micro supercapacitors: A review. *Advanced Energy Materials, 10*(21), 1902995.

Zhang, Q., Wang, L., Li, Z., Liu, G., & Chen, H. (2022). MXene-based gas sensors for environmental monitoring and health applications. *Sensors and Actuators B: Chemical, 350*, 130972.

Zheng, Y., Wang, J., Zhang, H., Liu, X., & Chen, Z. (2021). MXene-based coatings for flame-retardant applications: Mechanisms and properties. *Journal of Hazardous Materials, 412*, 125170.

Zhou, J., Ren, W., Cheng, H., & Zhang, H. (2020). Recent advances in MXene-based materials for water purification. *Advanced Materials, 32*(39), 2001419.

Zhu, J., Yin, H., Guo, C., & Zhang, Y. (2017). MXene: A promising photocatalyst for water splitting. *Journal of Materials Chemistry A, 5*(47), 2445–2453.

Molybdenum Disulfide: A 2D Material

Neha Sharma, Prashant Kumar, and Manoj Kumar Khanna

1 Introduction

Molybdenum disulfide (Fig. 1) is a material that offers a enticing combination of properties. It can be utilized in a wide variety of fields including lubrication, electronics, catalysis, optics to name some. The atypical characteristics of its layered composition, together with its capacity to accommodate a variety of synthesis techniques, underlines its potential to revolutionize a variety of industries and drive the growth of technology. It is a compound composed of Molybdenum (Mo) and Sulphur (S) atoms. It has garnered noteworthy attention in the fields of nanotechnology, materials science, electronics, energy and other related disciplines. The incomparable characteristics along with diverse range of uses have propelled it to the forefront of scientific investigation and industrial advancement. The importance of MoS_2 arises from its layered configuration, wherein a solitary layer comprises a hexagonal pattern of molybdenum atoms positioned between two layers of Sulphur atoms. The arrangement of atoms in this system gives rise to a wide range of distinct features (Kobayashi & Yamauchi, 1995; Sebenik et al., 2005).

N. Sharma (✉)
Keshav Mahavidyalaya, University of Delhi, New Delhi, Delhi, India
e-mail: neha.sharma@keshav.du.ac.in

P. Kumar
Department of Physics, MLN College(M), University of Delhi, New Delhi, Delhi, India

M. K. Khanna
Department of Electronics, University of Delhi, South Campus, New Delhi, Delhi, India

S. A. Bansal et al. (eds.), *Emerging Applications of Novel Nanoparticles*, Lecture Notes in Nanoscale Science and Technology 37, https://doi.org/10.1007/978-3-031-57843-4_6

Fig. 1 Molybdenum disulfide crystal. (Courtesy: By John Chapman (Pyrope)—Own work, CC BY-SA 4.0, https://commons.wikimedia.org/w/index.php?curid=3926425)

Although it exhibits insulating properties in its bulk form, it undergoes a transition to a semiconductor state with a tunable band gap when reduced to a single-layer or few-layer configuration. The aforementioned behavior has generated considerable interest within the domains of electronics, optoelectronics, and nano-electromechanical systems. Despite possessing a multilayer structure, MoS_2 has considerable mechanical strength, rendering it suitable for applications that necessitate the combination of electronic functionality with mechanical integrity. The exceptional lubricating properties of the substance can be attributed to the relatively modest Van der Waals forces that exist between its layers. Solid lubricants are employed in situations characterised by elevated temperatures and pressures in order to mitigate friction and minimise wear. MoS_2 demonstrates remarkable catalytic efficacy in chemical processes, namely in Hydrogen evolution reactions (HER), which play a pivotal role in the generation of renewable energy. The catalytic properties of this substance encompass a broad spectrum of industrial processes and environmental applications. The optical properties of MoS_2, such as photoluminescence and absorption, are influenced by the number of layers present. Consequently, it finds utility in the fields of photonics, sensors, and various other applications pertaining to light.

2 Crystal Structure and Properties

2.1 Crystal Structure

The crystal structure of Molybdenum Disulfide plays a crucial role in determining its distinct properties and potential applications. MoS_2 belongs to the transition metal dichalcogenides (TMDs) class of materials, which are distinguished by their layered structure. In the crystal structure of MoS_2, molybdenum atoms (Mo) are sandwiched between two layers of sulfur atoms (S). Each MoS_2 layer has a hexagonal lattice with triangular molybdenum atoms, with each sulfur atom connected to three molybdenum atoms. This arrangement results in a flat hexagonal lattice plane. (Fig. 2).

Because of the weak van der Waals forces between the MoS_2 layers, they can readily slide over each other. This property contributes to MoS_2's strong lubricating qualities, making it a solid lubricant of choice in a variety of high-friction applications. Strong covalent connections hold the molybdenum and sulfur atoms together in a stable configuration within each layer. Sulfur atoms coordinate the molybdenum atoms octahedrally, and each sulfur atom is bound to three molybdenum atoms. Because of van der Waals forces, the inter-layer bonding is rather weak. This permits the layers to be easily separated, enabling techniques such as exfoliation, in which single or few-layer sheets of MoS_2 can be extracted from bulk crystals.

When studying single-layer MoS_2, also known as monolayer MoS_2, the crystal structure changes. Each molybdenum atom is connected to only two sulfur atoms in this example, resulting in a straight band gap. Due to its inherent semiconducting

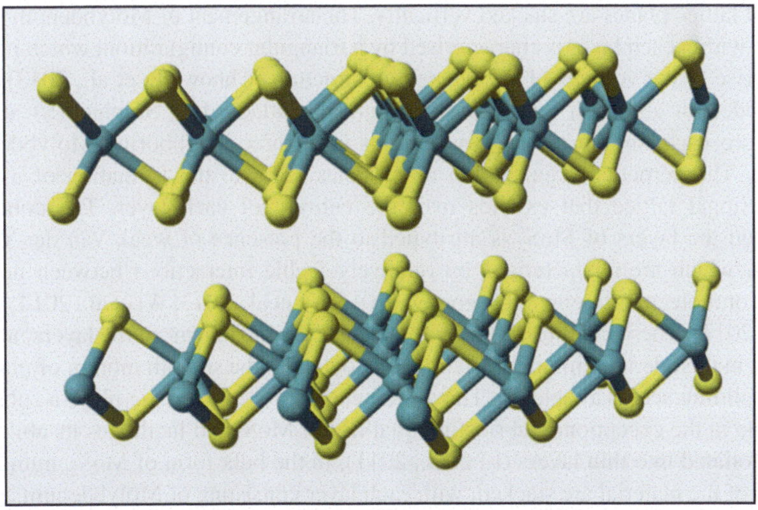

Fig. 2 Crystal structure of Molybdenum Disulfide. (Courtesy By Ben Mills—Own work, Public Domain, https://commons.wikimedia.org/w/index.php?curid=2976497)

properties, monolayer MoS_2 has emerged as a promising contender for several electrical and optoelectronic applications. In bulk, MoS_2 exhibits a stacked structure where multiple layers of MoS_2 are arranged atop one another, with the Sulphur atoms of each layer aligning directly over the molybdenum atoms of the layer immediately below. The arrangement of MoS_2 in a stacking structure has the potential to significantly influence its electrical and optical properties. The unique crystal structure of MoS_2, characterised by its layered configuration of Molybdenum and Sulphur atoms, gives rise to a diverse array of properties, encompassing semi conductivity, lubrication, and catalytic behaviour. The capacity for adaptability has been a driving force behind extensive research efforts aimed at exploring its potential uses across diverse industries and scientific domains.

2.2 Properties of MoS_2

2.2.1 Layered Structure

The compound MoS_2 is composed of hexagonal lattice layers in which Molybdenum atoms (Mo) are intercalated between Sulphur atoms (S). The present arrangement yields exceptional attributes and a wide range of applications. Molybdenum Disulfide (MoS_2), which belongs to the family of transition metal dichalcogenides (TMDs), is widely recognized for its remarkable characteristics attributable to its layered crystalline arrangement. The distinctive atomic configuration of MoS_2 contributes significantly to its wide range of uses and underlies its intriguing atomic and macroscopic properties (Li & Zhu, 2015; Ganatra & Zhang, 2014; Lembke et al., 2015; Huang et al., 2015). The MoS_2 material has a layered structure wherein hexagonal lattice planes are stacked vertically. The arrangement of Molybdenum (Mo) atoms within each layer is characterised by a triangular configuration, which in turn gives rise to the vertices of the hexagonal structure (Chhowalla et al., 2013). The Molybdenum atoms in question are chemically attached to Sulphur (S) atoms, therefore each Sulphur atom forms bonds with three neighboring Molybdenum atoms. The perpetual organization of particles leads to the formation of a two-dimensional lattice that extends over the entirety of each layer. The cohesion between the layers of MoS_2 is attributed to the presence of weak Van der Waals forces, which are characterised by relatively feeble interactions between neutral atoms or molecules (Geim & Grigorieva, 2013; Yu et al., 2013; Wi et al., 2014; Ding et al., 2012). These compressive forces facilitate the movement of the layers, allowing them to slide with minimal friction, analogous to the smooth motion of playing cards gliding across a tabletop. The aforementioned characteristic plays a substantial role in the exceptional lubricating qualities of MoS_2 and facilitates its ability to be exfoliated into thin layers (Li et al., 2011). In the bulk form of MoS_2, numerous layers of the material are stacked, with each layer consisting of Molybdenum atoms and Sulphur atoms. Notably, the Sulphur atoms in one layer are positioned precisely above the Molybdenum atoms in the layer immediately below. This stacking

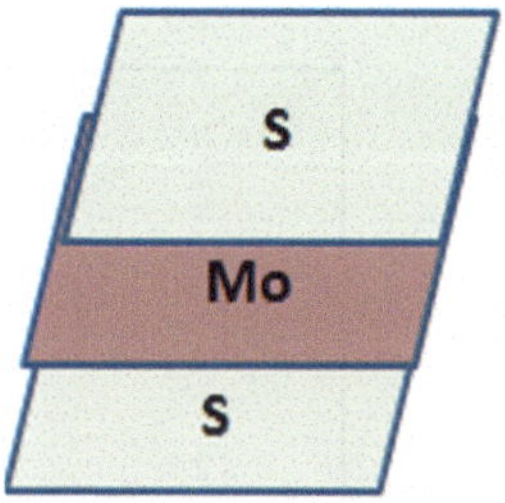

Fig. 3 Layered structure of MoS_2

configuration has an effect on the material's properties, particularly its electrical and optical performance. Significantly, the reduction of the structure of MoS_2 to a single layer, referred to as a monolayer, manifests distinctive characteristics. In the monolayer configuration, the bonding between Molybdenum atoms and Sulphur atoms is limited to a binary association, leading to the emergence of a direct band gap (Wu et al., 2014; Jariwala et al., 2014; Eda et al., 2011; Akinwande et al., 2014). The possession of this particular attribute confers semiconducting capabilities upon monolayer MoS_2, a feature of significant importance in the fields of electronics and optoelectronics. The phenomenon of quantum confinement effects arises when MoS_2 is stacked, leading to the modulation of its electrical properties based on the nano-scale dimensions of the material (Fig. 3). As the number of layers decreases, the material demonstrates distinct features that deviate from its bulk properties. The aforementioned observations have incited scholars to conduct inquiries into the prospective utilization of MoS_2 in quantum dots, nanoelectronics, and several other nanoscale applications. The study of behaviour of this kind holds significant importance in the field of electronics and optoelectronics.

2.2.2 Semiconductor Behavior

Molybdenum Disulfide (MoS_2), traditionally recognized as a solid lubricant, has garnered significant interest in contemporary times owing to its distinctive semiconductor properties. The presence of this characteristic, which arises from the atypical stratified composition of MoS_2, has expanded the realm of potential applications in the fields of electronics, optoelectronics, and other related domains (Yazyev & Kis, 2015). The present investigation aims to investigate the semiconductor properties of MoS_2 and the subsequent ramifications. The semiconductor behaviour of MoS_2 is attributed to its layered crystal structure.

Bulk MoS_2 exhibits the characteristics of an indirect band gap semiconductor (Fig. 4). As seen from figure, it indicates that the energy levels that are accessible to electrons in the valence band (VB) and the conduction band (CB) of the material are distinct from one another. No intermediate energy states are present (Frindt, 1966). The term "band gap energy" refers to the smallest amount of energy required by an electron to jump from the valence band to the conduction band. Indirect band gap is of importance as it plays an important part in the process of shaping the electrical characteristics of a material.

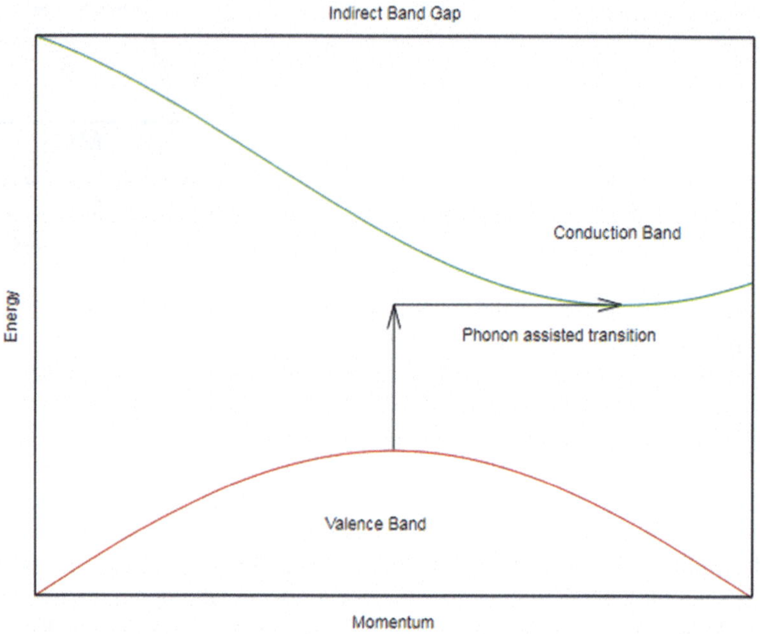

Fig. 4 Courtesy public domain, https://en.wikipedia.org/w/index.php?curid=14519297

Monolayer MoS_2 is achieve by initially reducing Molybdenum disulfide to a single layer. In this configuration, the atoms of Molybdenum have a pattern which is binding, they are only connected to two Sulphur atoms, as opposed to all three. Reason behind such a behavior is as there are only two Molybdenum atoms in the overall structure. As a result, a linear band gap is observed. It is widely accepted as a semiconductor due to the fact that monolayer MoS_2 has the feature of a straight band gap. It is exceptionally well-suited for a wide variety of applications in the domains of photonics, electronics, and optoelectronics. In electronics industry, many new opportunities have generated by making use of the monolayer MoS_2's semiconducting capabilities. It is a potential contender for use in transistors and logic circuits. This is possible as it is capable of regulating and modifying the passage of electrons. Since it is compatible with technology based on silicon, this widens the range of applications for which it can be used. Because MoS_2 has a straight band gap, it is a highly promising option for use in optoelectronic applications such as light-emitting devices, photodetectors, and solar cells. These applications require a material that has a gap between its conduction and valence bands. This material is able to effectively absorb visible light while also allowing it to pass through, which has sparked an interest in the investigation of novel technologies that have the potential to revolutionize the sectors of energy harvesting and data transit. The interesting tunability of MoS_2's behavior as a semiconductor is one of its defining characteristics (Novoselov et al., 2005). The band gap of MoS_2 can be altered by making adjustments to the number of layers or by applying strain to the material. Researchers are able to modify the electrical properties of MoS_2 and tune its characteristics to meet the requirements of specific applications. This creates a versatile foundation for the development and improvement of devices because of the material's adaptability.

2.2.3 Mechanical Strength

The intricate interplay between atomic configurations and mechanical forces inside the complex framework of MoS_2 gives rise to a multitude of unexpected phenomena. The combination of this particular attribute, coupled with the semiconducting properties and lubrication capabilities of MoS_2, has positioned the material as a prominent subject of investigation and advancement across various disciplines (Xiong & Cao, 2015). Researchers and engineers are captivated by its mechanical durability as they explore its complete spectrum of capabilities.

Molybdenum Disulfide, renowned for its semiconductor characteristics and lubrication properties, possesses remarkable mechanical strength that is sometimes overlooked. The innate characteristic of this feature has sparked the interest of scholars in various fields like nanotechnology, tribology, and materials engineering. The unique layered structure of MoS_2 is closely associated with its mechanical strength. The composition of each layer consists of Molybdenum (Mo) atoms that are intercalated between Sulphur (S) atoms, forming a hexagonal lattice structure (Mak et al., 2010). The cohesion between the layers is facilitated by the presence of weak Van der Waals contacts. Although each of these forces may be considered insignificant when examined individually, their combined impact on a larger scale is substantial, hence playing a crucial role in determining the mechanical stability of the material as a whole.

Despite its multilayer nature, MoS_2 exhibits remarkable rigidity and strength. Despite the relatively limited interlayer connections, they effectively impede the separation of layers, hence conferring an unexpected level of durability to the material. The aforementioned characteristic holds particular significance in nanoscale contexts, when materials are required to endure rigorous mechanical stresses (Mak et al., 2014). Due of its robust mechanical properties, Molybdenum disulfide exhibits exceptional suitability for implementation in nanomechanical and nanoelectromechanical systems (NEMS). The application of nanoscale systems necessitates the utilisation of materials that have the capacity to bear pressures, strains, and tensions without undergoing any kind of deformation or failing in any way (Deng et al., 2014). Because of its amazing resistance to mechanical stress, MoS_2 is an excellent choice for the manufacture of microscopic mechanical components that are distinguished by an exceptional level of precision and a long service life. Since it is characterized by a high level of mechanical strength, it is a favorable material in applications involving surface coatings and protective layers. Coatings is excellent For materials requiring increased wear resistance, MoS_2 is the best choice. These coatings also help in reducing friction and increasing the durability of mechanical components. This particular quality finds applications in aerospace, automotive etc.

2.2.4 Lubricating Properties (Fleisch & Bauer, 1988; Winer, 1967; Holinski & Gänsheimer, 1972; Dimigen et al., 1985; Zeming, 1982)

A complex relationship exists between Molybdenum disulfide atomic structure and the fundamental rules of physics. It is regarded as an exceptional lubricating substance. This property can be attributed to the fact that Molybdenum disulfide is a disulfide (Fleisch & Bauer, 1988; Winer, 1967). Many research labs have recommended Molybdenum disulfide as an extremely promising lubricant which has the potential to revolutionize industries involved in mechanical systems requiring smooth and efficient operation. Its extraordinary capacity to lower friction, to provide surface protection, and to operate efficiently even when it is subjected to adverse climatic conditions makes it a material of choice. Manufacturing units that are dependent on effective friction reduction and wear prevention have taken keen note of the brilliant lubricating characteristics of Molybdenum disulfide. The lubricating qualities that it possesses are the result of a complex interaction between the many atomic structures that make up (Holinski & Gänsheimer, 1972). The layered crystal structure of MoS_2 and the relatively weak Van der Waals forces that promote the cohesion between these layers are the mainly responsible for its lubricating qualities. It is s structure is made up of hexagonal layers, each of which contains molybdenum atoms (Mo) that are surrounded by Sulphur atoms (S). The hexagonal layers readily slide over one another due to relatively weak bonding forces that exist between these layers. When MoS_2 is utilized as a lubricant, result are reduction in both friction and wear.

Conventional lubricants like oils, are vulnerable to deterioration at high temperatures and pressures, which results in increased levels of friction and wear. In contrast, MoS_2 shows an exceptional property to keep its lubricating effectiveness even under such tough operating situations. As a result, it is a highly preferred choice for applications that place an emphasis on dependability and efficiency. One notable attribute of MoS_2 is its exceptional thermal stability at elevated temperatures (Dimigen et al., 1985). The efficacy of numerous lubricants tends to diminish and worsen with increasing temperatures. In contrast, MoS_2 exhibits stability and possesses lubricating properties, rendering it suitable for application in sectors such as aviation, manufacturing, and other domains that involve elevated temperatures. Molybdenum disulfide exhibits the ability to reduce friction and serve as a protective coating on surfaces, hence mitigating direct contact between metals and minimizing wear (Zeming, 1982). The longevity of mechanical components is enhanced and the frequency of maintenance is reduced through the formation of a protective layer resulting from the interaction between MoS_2 and the metal surface.

2.2.5 Catalytic Activity

Molybdenum Disulfide, renowned for its stratified configuration and semiconductor characteristics, unveils an additional facet of its adaptability: exceptional catalytic efficacy. The aforementioned attribute has generated a notable increase in the scholarly investigation of MoS_2's catalytic capabilities across a range of chemical processes and practical applications. This analysis (Pourabbas & Jamshidi, 2008) focuses on the catalytic activity of MoS_2 and its significant contributions to the advancement of sustainable energy, industrial processes, and related domains. The catalytic activity of MoS_2 is closely associated with its unique surface chemistry, which is dictated by the layered configuration of Molybdenum (Mo) and Sulphur (S) atoms. Exposure of specific regions on MoS_2 surface, provide active sites for chemical reactions involving Hydrogen, Sulphur, and a variety of other chemical species. Lack of coverage of surface is used for identification in such cases. In the field of hydrogen evolution reaction (HER) (Lee et al., 2021), which is an important step in the generation of eco- friendly energy., MoS_2 has garnered a substantial amount of recognition due to catalytic properties.

Molybdenum disulfide, as a catalyst, in the process of water splitting, enhances the creation of Hydrogen and Oxygen. This particular catalytic activity offers a potentially productive path for hydrogen fuel production, which is both renewable and kind to the environment. It is an engaging option to the usual noble metal catalysts that are typically used, such as platinum, due to its catalytic effectiveness and durability. The catalytic activity of Molybdenum disulfide highlights the intricate nature of the material by highlighting how the arrangement of the material's atoms has a substantial impact on the functioning of the material. The catalytic capability of MoS_2 marks an intersection between the discipline of materials science and its practical ramifications. As a result, it provides a promising chance for achieving a future that is both more sustainable and efficient.

2.2.6 Optical Properties

Its ability to absorb, transmit, and change light lays the path for the development of technologies which range from lighting to complicated systems for transmitting data. The optical brightness of MoS_2, ascertains the material's position as a candidate with immense potential in the disciplines of materials science and photonics (Novoselov et al., 2005). It has attracted substantial attention in the domains of photonics, optoelectronics, and other light-based applications due to its layered structure and unusual electrical capabilities. MoS_2's layered structure and optical properties are intertwined with one another.

When light energy interacts with layers of MoS_2, quantum confinement effects is observed resulting in noteworthy results (Wilson & Yoffe, 1969). The aforementioned phenomena reveal that when the physical dimensions of a material reach a scale comparable to the wavelength of light, it induces alterations in its electrical and optical characteristics. When it is exposed to light, Molybdenum disulfide exhibits a significant degree of light absorption and photoluminescence. The photons, which are incident, are absorbed, resulting in excitation of electrons from the Valence band to the Conduction band. Subsequently, the excited electrons undergo recombination with holes, which results in the emission of photons (Novoselov et al., 2004). The energy of such (emitted) photons is dependent upon the band gap of MoS_2, a property that can be modified by altering the number of layers or applying mechanical strain. By making changes in the number of layers, optical properties of MoS_2 show variations. The absorption and photoluminescence efficiency of single-layer MoS_2 is observed to be higher than the bulk MoS_2 due to direct band gap. This attribute has created a considerable attention in novel optoelectronic devices that use the layer-dependent optical properties of MoS_2. Valley polarization is a significant optical phenomenon observed in the material known as MoS_2. The electrons in MoS_2 exhibit distinct energy valleys, each characterised by its individual optical and spin characteristics. The aforementioned attribute holds significant implications for the fields of valleytronics and spintronics, which encompass the exploration of methods to alter these qualities in order to develop innovative electrical and optical devices (Geim, 2009; Katsnelson et al., 2006). The optical characteristics of MoS_2 have numerous applications like photodetectors, LEDs, modulators etc. Photodetectors utilizing MoS_2 material exhibit the capability to convert optical impulses into electrical signals, hence finding utility in several domains such as imaging, sensing, and communication. Molybdenum disulfide exhibits potential as a viable active material for utilization in light-emitting diodes (LEDs), offering both efficient light emission and the capacity to adjust its properties. Optical modulators based on MoS_2 are utilized to regulate the intensity of light in optical communication systems, hence facilitating the transmission of data.

2.2.7 Thermal Properties

Molybdenum Disulfide (MoS_2), well-known for its semiconducting and lubricating characteristics, has another intriguing aspect to its personality: thermal properties. The capacity of MoS_2 to conduct and manage heat is an important property that has applications ranging from electronics to thermal management. The thermal characteristics of Molybdenum Disulfide provide insight into the complicated ways in which its atomic structure interacts with thermal energy. MoS_2's ability to conduct and control heat paves the way for creative solutions to the difficulties of heat dissipation, energy conversion, and thermal stability, cementing its place as a remarkable material in the area of materials science and technology. While the thermal characteristics of MoS_2 have enormous potential, there are still obstacles in completely understanding its heat transport pathways and improving its thermal conductivity for specific applications (Xiaokun et al., 2016). Researchers are looking exploring ways to improve its thermal qualities and incorporate it into practical products. The thermal characteristics of MoS_2 are highly dependent on the layered structure of the material. The thermal conductivity of the material is affected not only by the weak Van der Waals forces that exist between the layers (Wang et al., 2012), but also by the interactions that take place between the atoms of Molybdenum (Mo) and Sulphur (S) that are present in each layer. The capacity of a substance to promote the movement of heat is referred to as its thermal conductivity, and MoS_2 demonstrates an unusually high thermal conductivity (Chhowalla et al., 2013). Strong atomic connections can be found inside the material's layers, which contributes to the material's relatively high in-plane thermal conductivity. Facilitating efficient heat transmission within a single layer. This particular attribute enables it to effectively conduct heat in two dimensions. Nevertheless, due to the limited interlayer bonding, the thermal conductivity in the out-of-plane direction is comparatively lower than the thermal conductivity in the in-plane direction (Jariwala et al., 2014).

In the realms of electronic engineering and thermal management, the thermal properties of MoS_2 have attracted a significant amount of attention. The importance of having efficient heat dissipation has significantly increased in recent years as a result of the ongoing miniaturization and performance improvements made to electronic devices (Mak et al., 2010). This is necessary in order to reduce the risk of the devices overheating. Because it has such high heat conductivity in the plane, it presents itself as a potential candidate for use in heat sinks and thermal interface materials. This would allow for more efficient heat dissipation and would ensure the dependability of electronic equipment. In the realms of electronic engineering and thermal management, the thermal properties of MoS_2 have attracted a significant amount of attention. Effective heat dissipation has taken on increased significance as a technique of preventing electronic devices from overheating as a direct result of the ongoing miniaturization and performance improvements brought about by technological innovation. as a result of its remarkable heat conducting capabilities within the plane (Radisavljevic et al., 2011). Molybdenum disulfide exhibits promising potential as a viable candidate for application in heat sinks and thermal interface materials, owing to its ability to facilitate effective heat dissipation and ensure the reliability of electronic devices (Lopez-Sanchez et al., 2013). The thermal properties of MoS_2 present opportunities for innovative applications at the nanoscale, where materials demonstrate unique characteristics. Its capacity to efficiently conduct heat on a nanoscale could be used for thermoelectric applications, in which heat is transformed into electricity, as well as for the construction of nanoscale heat switches and sensors (Yin et al., 2012; Kayyalha et al., 2015; Ganatra & Zhang, 2014).

2.2.8 Electrical Conductivity

The electrical conductivity of molybdenum disulfide adds another degree of complication to its multifarious character. Its capacity to conduct electricity, whether as a semiconductor or in other forms, opens up possibilities for advancement in electronics, sensors, and energy technology (Wang et al., 2014). The evolution of MoS_2 from its atomic structure to its conductive qualities shows the interdependence of materials research, fundamental physics, and the practical world of technological progress. Molybdenum Disulfide exhibits another noteworthy aspect aside from its mechanical strength and semiconducting nature. Its capacity to conduct electricity has sparked interest in electronics, sensors, and energy-related applications due to its layered structure and electrical capabilities,. Layered crystal structure accounts for it's electrical conductivity. Electrical characteristics of the material is influenced by the arrangement of atoms inside the layers, as well as the interactions between layers. As stated it has semiconductor-like properties in bulk form. This points to its capability to exhibit electrical conductivity under particular conditions. By application of an external electric field or by manipulating the number of layers, its conductive qualities can be synchronized Its conductive properties can also be altered by the application of an external electric field. MoS_2 in its monolayer configuration demonstrates the characteristics of a direct band gap semiconductor, in which the band gap plays an essential part in determining the conductivity characteristics of the material. Movement of charge carriers, electrons or holes, within the material results in the electrical conductivity of MoS_2. Defects present, impurities, and grain boundaries can have an impact on the mobility of charge carriers, thus influencing the total electrical conductivity.

The electrical properties of MoS_2 enable the manipulation of current flow by the utilisation of an external electric field, rendering it a compelling choice for prospective electronic devices (Tang, 2017). The utilization of MoS_2's electrical conductivity extends to the field of sensors as well. The capacity of this entity to engage with external stimuli, like as gases or molecules, can be harnessed for the purpose of sensing. Moreover, owing to its role in electrochemical reactions and adjustable conductive properties, it proves to be advantageous in the realm of energy storage and conversion apparatuses, such as batteries and super- capacitors.

2.2.9 Biocompatibility

The versatility and promise of Molybdenum Disulfide is exemplified by its transition from a solid lubricant to a biocompatible material, showcasing its applicability across various industries. When coupled with its distinct characteristics, the compatibility of this entity with biological systems heralds a novel epoch of innovative biomedical remedies, equipping researchers and medical practitioners with instruments to confront formidable obstacles and redefine the frontiers of medical science. Molybdenum Disulfide, renowned for its inherent semiconducting properties and impressive mechanical robustness, has demonstrated its utility in the field of biomedicine, extending its applicability beyond its conventional uses. The biocompatibility of MoS_2, referring to its capacity to interact with living organisms without inducing adverse effects, has rendered it a highly appealing candidate for numerous biological applications (Wang et al., 2015). The biocompatibility of the material is of utmost importance in order to facilitate its integration into various biomedical applications. The comprehension of MoS_2's interactions with biological systems, including cells and tissues, is of utmost importance in guaranteeing its safety and efficacy.

Multiple investigations have demonstrated that MoS_2 exhibits favorable biocompatibility with diverse cell types, including mammalian cells. The interaction between cells and MoS_2 is influenced by various factors, including its layered structure, surface chemistry, and nanoscale features. Due to its compatibility, this material exhibits potential applications in various fields such as cell culture, medicine delivery, and tissue engineering (Shah et al., 2015). Due to the fact that molybdenum disulfide is biocompatible, it is an excellent candidate for use in drug delivery systems.

When coupled with medicinal chemicals, functionalized MoS_2 nanoparticles have the ability to be selectively targeted towards certain cells or tissues. This can be accomplished by targeting specific cell types. Opportunities exist for personalized treatment and improved therapeutic outcomes thanks to the possibility of carrying medications to selected sites, regulating their release, and monitoring the entire process. Due to the fact that MoS_2 has desirable optical characteristics, it is an excellent candidate for use in bioimaging applications. The capacity of this substance to absorb visible light and then release that light enables the non-invasive visualization of cellular structures and tissue structures.

Researchers have showed that the utilization of fluorescent markers to identify MoS_2 enables the monitoring of cellular activity, as well as the acquisition of valuable insights into the course of disease. Because of its interaction with biomolecules, MoS_2 is a very intriguing prospect for use in biosensor applications. It is possible to modify the surface chemistry of the material in order to preferentially bind particular molecules. This opens the door to the identification and detection of biomarkers that are associated with various diseases. The use of sensors that make use of molybdenum disulfide has the potential to give diagnostic capabilities that are both quick and highly responsive.

2.2.10 Environmental Stability

The resistance of molybdenum disulfide to the effects of its surroundings is demonstrated by the compound's high level of environmental stability. It has proved to be a very reliable material in different industries due to its resistance to corrosive agents, high temperatures, and a wide variety of chemicals (Lee et al., 2019). Environmental stability of MoS_2 will continue to be an important component in laying the future road map for materials research and engineering. Its exceptional mechanical strength, semiconducting characteristics, and wide range of application makes it popular for industrial use, MoS_2 is highly attractive for use in a wide variety of industries and applications. to its outstanding resistance to deterioration and corrosion when exposed to a wide variety of conditions.

The atomic arrangement and chemical make-up of MoS_2 both have an impact on the substance's resistance to the effects of the environment. The resistance of molybdenum disulfide (MoS_2) to moisture, oxidation, and chemical agents can be attributed to its layered structure and the presence of strong chemical bonds between the Molybdenum (Mo) and Sulphur (S) atoms. The corrosion resistance exhibited by MoS_2 is quite commendable. In contrast to numerous other materials, the layers of MoS_2 function as a protective barrier that hinders direct interaction between metal atoms and corrosive chemicals, hence impeding the formation of oxides and other compounds through reactions with oxygen and moisture. Due to this characteristic, it finds application in environments where corrosion may compromise both performance and longevity.

The thermal stability of MoS_2 complements its corrosion resistance. It keeps structural integrity at high temperatures, making it appropriate for use in aircraft, manufacturing, and other high-temperature industries (Paredes et al., 2016). MoS_2's thermal stability ensures that its mechanical and functional qualities remain intact under extreme conditions. Because of its strong intralayer connections and van der Waals interactions between layers, MoS_2 is chemically inert. This feature renders MoS_2 less vulnerable to interactions with diverse chemicals, acids, and bases, making it appropriate for applications requiring exposure to a variety of substances.

2.2.11 Tunable Properties

Molybdenum Disulfide, a material with numerous outstanding features, has one quality that distinguishes it as a versatile chameleon: its adjustable properties. MoS_2's ability to vary its behaviors, from electrical properties to mechanical responses, has piqued the interest of researchers from a wide range of disciplines. MoS_2's ability to change and modify its properties demonstrates the breadth of materials research. Its tunable nature connects fundamental research to practical applications, allowing researchers and engineers to create materials and systems that address specific demands. The fact that it is a chameleon material highlights the power of atomic manipulation and demonstrates the boundless prospects for advances in a wide variety of fields (Samy et al., 2021). The layered structure of MoS_2 and the diverse behaviors that emerge when the number of layers is varied are the primary contributors to the material's adaptable properties. The behavior of monolayer MoS_2 is distinct from that of bulk MoS_2, which results in a variety of characteristics that can be fine-tuned in accordance with the application in question.

Changing the number of layers in a MoS_2 crystal allows one to play with the material's electrical characteristics. Bulk MoS_2 has an indirect band gap, whereas monolayer MoS_2 has a direct band gap. This band gap tunability influences its electrical conductivity and optical characteristics, making it suitable for applications ranging from electronics to optoelectronics. MoS_2's mechanical strength can also be modified by varying its thickness. While bulk MoS_2 has amazing mechanical qualities, due to its atomic arrangement, monolayer MoS_2 has even greater stiffness and strength. This tunability has significance for the development of lightweight, resilient materials for structural and nanoscale applications. The thermal conductivity of MoS_2 is affected by its layer-dependent behavior. Out-of-plane thermal conductivity diminishes with decreasing layer thickness, whereas in-plane thermal conductivity stays strong. Thermal management in electrical devices and energy-related applications can benefit from this variable thermal conductivity. Controlling the number of layers and altering the surface chemistry of MoS_2 surfaces can vary their reactivity. Because of its tunability, it is excellent for a wide range of chemical reactions, including hydrogen evolution and catalytic conversions (He & Que, 2016).

2.2.12 2D Material

With its atomic-thin structure and diverse features, two dimensional MoS_2 has enchanted the world of nanotechnology. Its quantum effects, semiconducting nature, and varied properties have sparked a surge of curiosity, propelling us toward a future in which this extraordinary material plays a vital role in changing electronics, photonics, and beyond. 2D Molybdenum Disulfide (MoS_2) has emerged as a preferred choice in the field of nanomaterials, capturing academics and engineers alike due to its unique features and numerous uses. It defies traditional expectations with its atomic thickness, this single-layer material, presents a wide spectrum of possibilities.

In two-dimensional MoS_2, a single layer of Molybdenum (Mo) atoms is sandwiched between two layers of Sulfur (S) atoms. Quantization of electron behavior due to spatial constraints occurs as a consequence of quantum confinement, which arises from the atomic-scale thickness. Distinct electrical and optical properties deviations are observed in bulk materials due to these interactions.

As mentioned earlier, its bulk form is characterized by an indirect band gap semiconductor property, whereas, the monolayer configuration of MoS_2 exhibits a direct band gap, This specific behavior leads to attention within the domains of electronics and optoelectronics. The ability to govern and manage electron mobility within a thin material holds great promise for the advancement of future transistors, sensors, and light-emitting devices. Its atomic-scale thickness attributes to the exceptional optical characteristics exhibited by 2D MoS_2. Its absorption spectrum spans the visible region due to which it exhibits a significant absorption and photoluminescence properties, This leads to its acceptance in photonics, photodetectors, and bioimaging applications. The capacity of the material to interact with light provides a wide variety of capabilities when applied in certain circumstances (Li & Zhu, 2015).

Despite its delicate look, two-dimensional (2D) MoS_2 has remarkable mechanical robustness and flexibility. Because of its toughness and atomic thinness, it is used in flexible electronics and lightweight materials. The capacity to produce mechanically resilient structures on a nanoscale opens the door to new device design and material engineering possibilities. The properties of 2D MoS_2 can also be tuned by varying the number of layers. The behavior of the material approaches that of its bulk form as the number of layers grows. This layer-dependent tunability provides a range of features, allowing for customized solutions for specific applications.

2.2.13 Exfoliation

Exfoliation of MoS_2 demonstrates our power to alter matter at the atomic level. This method has turned MoS_2 from a bulk material to a versatile platform with applications in a variety of fields. As scientists continue to unravel the mysteries of exfoliated MoS_2, the influence on science, technology, and industry is expected to be substantial and long-lasting. With Molybdenum Disulfide (MoS_2), the skill of delicately peeling away layers takes on a novel dimension in the field of materials research. This method of splitting MoS_2 into ultrathin layers has transformed the way we use its extraordinary features. Let's dive into MoS_2 exfoliation and see how this process has pushed the boundaries of study and technology.

The controlled separation of MoS_2's layered structure into individual atomic or few-layer thick sheets is referred to as exfoliation. The goal is to separate single or few-layer MoS_2 flakes from the bulk material while keeping their integrity and characteristics intact. There are many approaches for exfoliation, like, 1. Mechanical Exfoliation (Scotch Tape approach): Inspired by graphene isolation, this approach involves repeatedly attaching and removing adhesive tape off a MoS_2 crystal, resulting in the transfer of thin flakes onto a substrate (Ma et al., 2017). While easy, it is labor-intensive and yields only a little amount of flakes. 2. Liquid Exfoliation: This approach includes immersing bulk MoS_2 in a solvent and breaking apart the layers with mechanical or ultrasonic agitation. The resulting dispersion contains flakes of varied thickness that can be placed onto substrates for use in a variety of applications. Last but not the least Chemical Exfoliation: Intercalating compounds are used in chemical exfoliation to reduce the van der Waals interactions between MoS_2 layers. The addition of solvents or surfactants aids in layer separation, resulting in exfoliated MoS_2.

3 Methods of Synthesis

Molybdenum Disulfide exhibits significant promise across several domains, encompassing electronics and energy, among others. The synthesis of MoS_2, which possesses a distinctive layered structure and a wide range of features, is a prominent area of investigation within the field of materials science. Numerous techniques have been devised to generate MoS_2 in diverse configurations, tailored to specific applications. Every synthesis approach possesses its own set of advantages and disadvantages, and the selection of a particular method is contingent upon the desired characteristics and intended applications of the resulting MoS_2 material. The diverse range of synthesis techniques employed underscores the inherent versatility and potential of MoS_2 to catalyze significant advancements across multiple sectors. Some of the frequently used methods for synthesis are listed as below.

3.1 Chemical Vapor Deposition (CVD)

In the production of high-quality MoS_2 films, chemical vapor deposition (CVD) is a very popular and preferred choice. The procedure involves introduction of precursor molecules, namely Molybdenum Hexafluoride (MoF_6) and Hydrogen Sulphide (H_2S), into a reactor. The reactor at high temperatures (Hersam, 2015). MoS_2 layers are created when precursors engage in chemical interactions with a substrate. For controlling layer thickness, shape, and homogeneity, making it a viable technique

Fig. 5 Chemical vapor deposition set up. (https://commons.wikimedia.org/wiki/File:Chemical_Vapor_Deposition_Setup.jpg) (Aksy88, CC BY-SA 4.0 https://creativecommons.org/licenses/by-sa/4.0, via Wikimedia Commons)

for various applications in the domains of electronics and optoelectronics, CVD is best suited (Fig. 5).

3.2 *Hydrothermal Synthesis*

Chemical transformation resulting when Molybdenum salts and Sulphur sources are exposed to an aqueous solution under conditions of elevated temperature and pressure is known as hydrothermal synthesis. Nanoparticles or nanosheets comprising MoS_2 are synthesized through this technology (Zhao et al., 2019). Dimensions and shape of the synthesized MoS_2 can be controlled in hydrothermal synthesis. It is known for its inherent simplicity and adaptability (Fig. 6).

3.3 *Solvothermal Synthesis*

Solvothermal synthesis for conducting chemical reactions employs organic solvents as the medium. The utilisation of this methodology facilitates the manipulation of the particle dimensions, morphology, and crystal structure of the resultant MoS_2. The solvothermal synthesis method is employed to fabricate MoS_2 materials

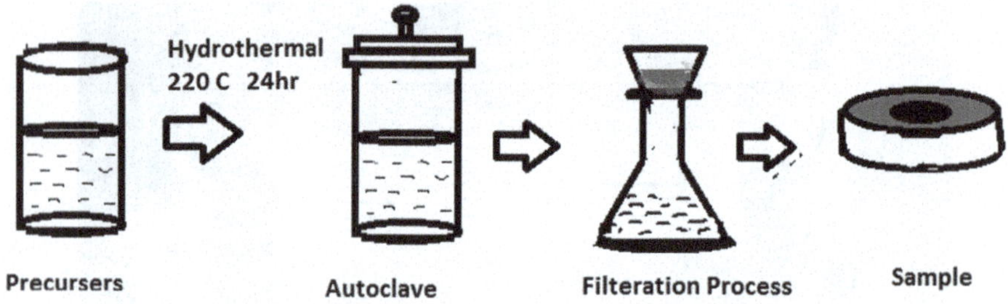

Fig. 6 Hydrothermal synthesis

Fig. 7 Solvothermal synthesis. (https://commons.wikimedia.org/wiki/File:Autoclave_01.svg) (Autoclave_01.JPG: Strikingdragonderivative work: Malyszkz, Public domain, via Wikimedia Commons)

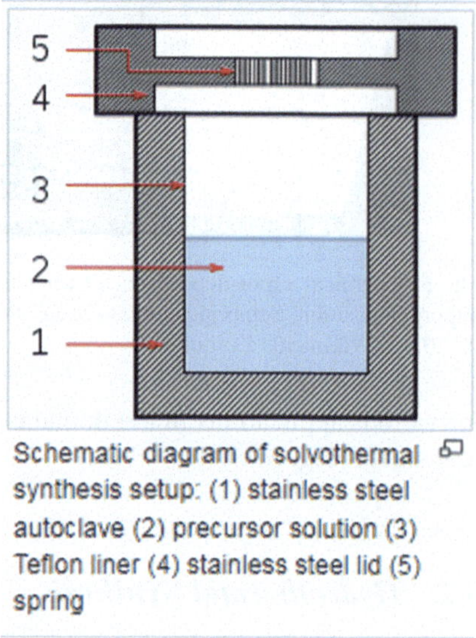

possessing tailored properties suitable for applications in energy storage, sensors, and catalysis (Del Alamo, 2011; Wang et al., 2023) (Fig. 7).

3.4 Mechanical Exfoliation (Scotch Tape Method)

The process of mechanical exfoliation, which draws inspiration from the isolation technique employed for Graphene, entails the removal of layers of MoS_2 from larger crystal structures through the utilization of sticky tape. This time-consuming approach yields high-quality, atomically thin flakes. It is utilized for fundamental research as well as the fabrication of devices with single or few layers of MoS_2, particularly in electronics and optoelectronics (Bhat, 2018).

3.5 *Liquid Phase Exfoliation*

Liquid phase exfoliation includes distributing bulk MoS_2 in a solvent and separating the layers using mechanical or ultrasonic agitation (Yoo et al., 2015). This technique produces MoS_2 nanosheets of different thicknesses suspended in solution. Liquid exfoliation is appropriate for use in inks, coatings, and composite materials.

3.6 *Electrochemical Synthesis*

Electrochemical synthesis is the process of reducing molybdenum precursors and sulfur sources in an electrochemical cell to generate MoS_2. Control over the morphology and properties of the generated material is possible using this technology (Hiraki, 1992). Catalysis, sensors, and energy storage devices all benefit from electrochemical synthesis (Fig. 8).

3.7 *Template-Assisted Synthesis*

MoS_2 is manufactured within a template structure, such as a porous substrate or polymer matrix, via template-assisted synthesis (Sha et al., 2017). Following synthesis, the template is eliminated, leaving MoS_2 with a distinct shape. This technique is used to make MoS_2 nanotubes, nanowires, and other complicated structures.

4 Applications

MoS_2's distinct features have led to its incorporation into a variety of industries: It has found application in variety of areas.

4.1 *Electronics and Optoelectronics Industry*

The adaptability of Molybdenum Disulfide has ushered in a new era of electronics and optoelectronics. Its atomic-scale operation, along with its unique electrical and optical features, is guiding the development of new gadgets that will define the future of technology (Hijazi & Moutaouakil, 2019). MoS_2-based devices have the potential to alter industries, improve communication, and enable new frontiers of discovery as they mature and find their place in real-world applications.

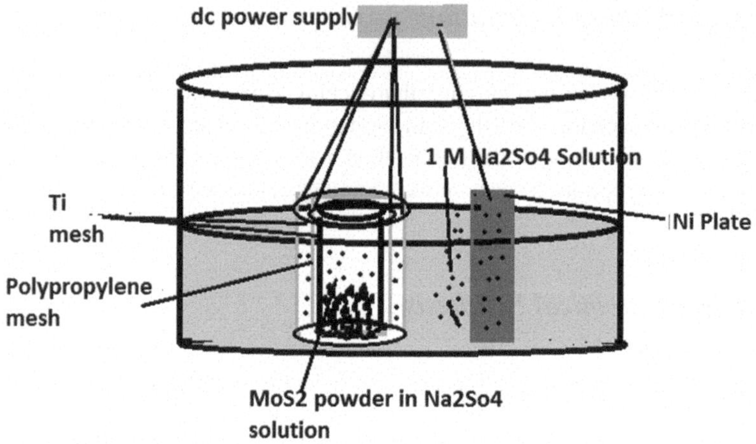

Fig. 8 Electrochemical exfoliation of MoS_2

MoS_2's semiconducting properties, particularly in monolayer form, make it an excellent choice for field-effect transistors (FETs). Low power consumption, rapid switching speeds, and compatibility with flexible substrates are all advantages of MoS_2-based FETs (Moutaouakil et al., 2010). They have the potential to replace or supplement conventional silicon-based transistors. The unusual electrical characteristics of MoS_2 allow for the development of logic gates and memory devices. Its compatibility with scalable fabrication techniques opens the door to the development of high-performance integrated circuits. Because of its mechanical flexibility and resilience, MoS_2 is well suited for flexible and wearable electronics. It can be incorporated into bending and stretchable technologies, allowing for advancements in healthcare, communication, and other fields. Because of its high light-matter interactions, MoS_2 can be used in photodetectors, which efficiently transform light signals into electrical signals. Because of its straight band gap in monolayer form, it is very effective at absorbing and detecting light across a wide spectrum.

MoS_2 is a potential material for LEDs due to its straight band gap and efficient light emission features. It emits light that can be tuned across a wide range of wavelengths when electrically activated. LEDs based on MoS_2 have the potential to be used in energy-efficient lighting and displays. MoS_2 absorbs light in the visible range, making it a viable candidate for solar cells. MoS_2 can be included into photovoltaic devices to improve light absorption and charge separation by forming heterojunctions with other materials. Because of its optical characteristics, MoS_2 is useful in the fabrication of photonic devices such as modulators and switches. Its capacity to control light at the nanoscale is critical for the development of efficient and compact optical communication systems. The atomically thin nature of MoS_2 and the quantum confinement effects in monolayers have piqued the interest of researchers in employing it for quantum devices. These devices could enable quantum information manipulation and storage, paving the path for quantum computing and communication technologies.

4.2 Catalysis and Energy Applications

In the fields of catalysis and energy applications, molybdenum disulfide is a substance that acts as a catalyst for change. Because of its catalytic qualities, sustainable hydrogen production, cleaner fuels, and a reduction in carbon dioxide emissions are being driven by it. Concurrently, its applications in the energy sector are opening the way for more effective methods of energy storage, conversion to solar power, and generation of thermoelectric power. It is at the forefront of the revolution towards sustainable energy, which is assisting us in making the shift towards a future that is greener and friendlier to the environment. It is notable for the versatility of its properties, has recently emerged as a potential game-changer in the applications of catalysis and energy (Moutaouakil, 2018). Because of its exceptional catalytic activity and distinctive electronic structure, new opportunities for the generation, storage, and transformation of renewable energy have become available as a result.

When it comes to catalyzing the Hydrogen Evolution Reaction (HER), an essential step in the generation of hydrogen, MoS_2 is in a league of its own. Because of its layered structure, it reveals a great number of catalytic sites, which makes the process of separating hydrogen and oxygen from water more easier. Catalysts based on MoS_2 have demonstrated remarkable efficacy in the generation of environmentally friendly and sustainable hydrogen fuel, hence presenting significant prospects for a hydrogen-powered future. In the context of the petrochemical sector, the utilization of catalysts derived from MoS_2 is crucial for the hydro desulfurization (HDS) process (Moutaouakil et al., 2012). The ability to effectively extract Sulphur compounds from crude oil and natural gas has led to the production of cleaner fuels with reduced Sulphur emissions. The catalytic efficiency of molybdenum disulfide (MoS_2) in the process of hydrodesulfurization (HDS) is of paramount importance in the reduction of pollutant emissions into the environment.

Molybdenum disulfide has gained prominence in several developing fields, including the electrochemical conversion of carbon dioxide into economically valuable carbon-based fuels and chemicals. Catalysts generated from molybdenum disulfide have demonstrated significant promise in accelerating the conversion of carbon dioxide into valuable commodities. This presents a viable solution to the pressing issue of reducing greenhouse gas emissions, while simultaneously promoting the advancement of sustainable resources. The use of MoS_2 as a catalyst demonstrates extensive versatility in various chemical processes, encompassing the transformation of biomass into biofuels and the production of numerous important organic molecules. The material's capacity to support these transitions presents prospects for the advancement of production techniques that prioritise environmental friendliness and sustainability.

Given its exceptional energy storage capabilities, this material emerges as a prominent candidate for utilization in batteries and supercapacitors. Because of its enormous capacity as well as its fast charge and discharge rates, it is an appealing option for use in energy storage solutions. Energy storage systems based on MoS_2

have the potential to completely disrupt the portable electronics and electric car industries. Because of its ability to absorb light and have a bandgap that can be tuned, MoS_2 is an excellent candidate for use in photovoltaic and solar cell technologies (Zhuo et al., 2023). Researchers hope that by integrating MoS_2 into these devices, they will be able to improve light absorption and charge separation, which would ultimately lead to more efficient conversion of solar energy.

The electrical conductivity of MoS_2 can be adjusted, making it a potential candidate for use in thermoelectric devices because of its thermal properties. It is able to transform waste heat into electricity, which not only improves energy efficiency in industrial processes but also provides a unique technique to harvesting the energy that may be derived from differences in temperature. MoS_2's function as a solid lubricant is essential to the reduction of mechanical friction and wear; nevertheless, its application does not directly involve the use of energy. By improving the operational efficacy of machines and cutting down on energy waste, this feature makes a contribution to the conservation of energy.

4.3 Lubrication and Tribology Applications

Molybdenum Disulfide may not be in the public eye very often, but behind the scenes, it is playing a crucial part in the reduction of friction, the prevention of wear, and the improvement of the efficiency and dependability of machinery used in a wide variety of industries. MoS_2's contribution to the field of lubrication and tribology continues to be important, enabling the smooth and dependable operation of innumerable mechanical systems despite the advancement of technology and our deeper understanding of tribology (Moutaouakil et al., 2009; El Moutaouakil et al., 2011; Cao et al., 2016; Goel et al., 2021; Gupta et al., 2020; Iwai et al., 2006). In the field of lubrication and tribology (the science of friction, wear, and lubrication), Molybdenum Disulfide may not be a household name, but it performs an essential and frequently overlooked function in a variety of applications. The excellent properties of MoS_2 as a solid lubricant have been empirically established, as it effectively mitigates friction, safeguards against wear, and enhances the operational efficiency of diverse machinery and mechanical systems.

Solid lubricants, such as Molybdenum disulfide, are employed in order to mitigate the deleterious effects of friction and wear that arise during the operation of mechanical components in relative motion. Solid lubricants, in contrast to their liquid counterparts like oils and greases, exhibit the advantageous characteristic of being stationary and not need frequent replenishment. The aforementioned characteristic renders solid lubricants particularly useful for utilization in specific applications. The chemical MoS_2 has a layered structure wherein Molybdenum atoms and Sulphur atoms are arranged in alternating layers. The layers possess the capability to undergo sliding motion with minimal opposition, hence facilitating their role as "slip planes."

The interlayer shear strength of MoS_2 is rather low, indicating that the layers can easily slide against one another with less applied stress. This characteristic reduces friction and prevents direct contact between metallic surfaces. Due to its chemical inertness, MoS_2 exhibits a lack of reactivity towards the majority of other compounds. Due to its versatile properties, it is well-suited for application in a wide range of contexts, including those involving the presence of corrosive substances or elevated temperatures. Despite its extremely thin atomic-scale thickness, it has a notable capacity to bear loads, suggesting its potential to endure significant pressures without experiencing breaking.

In order to lessen the amount of friction and wear that occurs in engines, transmissions, and other automotive components, MoS_2 is added as an addition to motor oils and greases. It lessens the amount of fuel needed to run the engine, increases overall efficiency, and lengthens the lifespan of essential engine parts. Cutting tools and other machine parts can have MoS_2 coatings added to them to reduce the amount of wear that occurs during the machining process. This boosts machining efficiency while also improving surface polish and extending the life of the tools. Lubricants for aircraft engines and landing gear often contain MoS_2 because of its usefulness in the aerospace sector (Ganatra & Zhang, 2014; Gao et al., 2015; Saha & Kruse, 2020; Wang & Mi, 2017). It is able to tolerate significant temperature swings as well as large loads, which guarantees that it will work dependably. MoS_2 is used in a variety of industrial applications, such as conveyor systems, bearings, and gearboxes, where it reduces the amount of friction that occurs and increases the amount of time that crucial components are able to perform effectively. The atomically thin layers of MoS_2 are being investigated for usage in nanoscale bearings and moving components in upcoming nanotechnology applications. These layers have the potential to provide precise lubrication in micro electronics.

4.4 Biomedical and Sensing

Molybdenum disulfide, which is more commonly linked with electronic components and lubricants, has insidiously made its way into the world of biomedical applications and sensing technology. Because of its one-of-a-kind characteristics, such as biocompatibility and sensitivity, it has proven to be an extremely significant asset in the process of revolutionizing healthcare and sensing solutions. Its nanoparticles are showing great promise as potential medication delivery vehicles. They provide controlled and targeted drug release because of their vast surface area and their capacity to encapsulate medicinal substances. Drug delivery systems based on MoS_2 can improve therapeutic efficacy while simultaneously reducing unwanted side effects.

Because of their powerful optical characteristics, MoS_2 nanoparticles are ideally suited for use in imaging applications. They are able to be functionalized with contrast agents for use in a variety of imaging modalities, such as magnetic resonance imaging (MRI) and fluorescence imaging, which helps in the early diagnosis and tracking of disease. Because of its photothermal characteristics, MoS_2 can take in light in the near-infrared spectrum and turn that energy into heat (Gao et al., 2016; Chen et al., 2018; Shree et al., 2021). Photothermal therapy makes use of this ability by delivering MoS_2 nanoparticles selectively to tumor areas and then irradiating them with light in order to destroy malignant cells while sparing healthy tissue.

Biosensors based on MoS_2 are developed with the purpose of identifying particular biomolecules, such as proteins, DNA, and even diseases like viruses. MoS_2 surfaces that have been functionalized in some way can grab target molecules, which results in quantifiable signals. The diagnosis and monitoring of diseases are two potential applications for these sensors. Because of their antibacterial qualities and high surface area, MoS_2 nanosheets are being investigated for use as wound dressings. This would be beneficial because it would speed up the process of tissue regeneration. They are also capable of being included into bandages for controlled drug release to expedite the healing process (Peng et al., 2020).

Even though MoS_2 has a lot of potential in biomedicine and sensing, there are still a lot of obstacles to overcome. Some of these issues include increasing sensitivity, maintaining stability, and increasing selectivity in sensing applications. Researchers are continuing their investigations into more sophisticated methods of functionalization and integration in an effort to find solutions to these issues.

4.5 Nanotechnology Applications

Molybdenum disulfide has made the transition from the macroscopic world to the world of nanotechnology in a manner that is both smooth and successful. Its one-of-a-kind characteristics and malleability at the nanoscale are driving innovation in a wide variety of fields, including electronics and sensing as well as materials engineering and catalysis. MoS_2 is a testament to our capacity to harness the amazing potential of materials on the atomic and molecular scale, which opens up new vistas in science and industry as the revolution in nanotechnology continues.

A single layer or a few layers of monolayer MoS_2 can function as a building block for nanoelectronic devices. Transistors, diodes, and interconnects that are built on MoS_2 have the potential to either replace or supplement components that are traditionally based on silicon. The capability to alter electronic properties at the nanoscale paves the way for the development of electronics that are smaller, quicker, and more efficient in their use of energy. Strong light-matter interactions at the nanoscale facilitated innovation in the field of nanophotonics brought about by MoS_2. It is employed in the construction of nanoscale light sources, detectors, and modulators, which enables the fabrication of compact and efficient photonic devices for communication and sensing. These devices can be used for both communication and sensing (Fan et al., 2015; Kumar et al., 2020; Kaur et al., 2018; Wu et al., 2019). Because of its high sensitivity to strain and outstanding mechanical properties, MoS_2 is an ideal material for use in NEMS applications. It has applications in areas such as healthcare robotics, precision instruments, and instrumentation, and it can be combined into very small sensors and actuators.

5 Conclusion

To summarize, molybdenum disulfide, also known as MoS_2, is a versatile substance that has found applications in a wide variety of subfields within the scientific and technological communities. Because of its layered structure, extraordinary qualities, and adaptability at both the macro and nanoscales, it has become a genuine chameleon, influencing innovation and progress in a wide variety of fields. In the realms of electronics and optoelectronics, where MoS_2 is positioned to redefine device miniaturization and efficiency, as well as in the realms of catalysis and energy applications, where it is playing a pivotal role in driving the transition towards cleaner and more sustainable energy sources, it has demonstrated that it possesses the potential to undergo radical transformations.

In the fields of biomedicine and sensing, MoS_2's biocompatibility and sensitivity are paving the way for advancements in areas such as environmental monitoring, drug delivery, and healthcare diagnostics. Nanotechnology has facilitated significant advancements in the domains of nanoelectronics, nanophotonics, and the synthesis of intricate nanomaterials.

Notwithstanding the considerable achievements, challenges persist. The researchers are encountering challenges pertaining to the scalability, stability, and optimization of the features of MoS_2 in order to cater to certain applications. As the exploration of the remarkable characteristics of this substance continues, it is anticipated that further findings and advancements will soon emerge. Molybdenum disulfide, characterised by its stratified arrangement and vast range of potential applications, serves as a testament to human inventiveness and our ability to use material properties for societal advancement. The development of MoS_2 was undertaken by the Materials Science and Engineering Department at the Massachusetts Institute of Technology. As the trajectory of study and investigation involving MoS_2 persists, it is reasonable to expect such paradigm-shifting findings that will profoundly impact our future, albeit with a limited comprehension of their implications at present. These advancements will have a significant influence.

References

Akinwande, D., Petrone, N., & Hone, J. (2014). Two-dimensional flexible nanoelectronics. *Nature Communications, 5*(5678), 1–12.

Bhat, N. (2018). Tunable steep slope MoS_2 transistor. In *Proceedings of the 2018 IEEE International Conference on Semiconductor Electronics (ICSE), Kuala Lumpur, Malaysia, 15–17 august 2018* (p. C1). IEEE.

Cao, Y., Gan, S., Geng, Z., Liu, J., Yang, Y., Bao, Q., & Chen, H. (2016). Optically tuned terahertz modulator based on annealed multilayer MoS_2. *Scientific Reports, 6*, 22899.

Chen, X., Shinde, S. M., Dhakal, K. P., Lee, S. W., Kim, H., Lee, Z., & Ahn, J.-H. (2018). Degradation behaviors and mechanisms of MoS_2 crystals relevant to bioabsorbable electronics. *NPG Asia Materials, 10*, 810–820.

Chhowalla, M., Shin, H. S., Eda, G., Li, L.-J., Loh, K. P., & Zhang, H. (2013). The chemistry of two-dimensional layered transition metal dichalcogenide nanosheets. *Nature Chemistry, 5*, 263–275.

Del Alamo, J. A. (2011). Nanometre-scale electronics with III-V compound semiconductors. *Nature, 479*, 317–323.

Deng, Y. X., et al. (2014). Black phosphorus-monolayer MoS_2 van der waals heterojunction p-n diode. *ACS Nano, 8*, 8292–8299.

Dimigen, H., Hübsch, H., Willich, P., & Reichelt, K. (1985). Stoichiometry and friction properties of sputtered MoS_x layers. *Thin Solid Films, 129*, 79–91.

Ding, S., Zhang, D., Chen, J. S., & Lou, X. W. (2012). Facile synthesis of hierarchical MoS_2 microspheres composed of few-layered nanosheets and theirlithium storage properties. *Nanoscale, 4*, 95–98.

Eda, G., Yamaguchi, H., Voiry, D., Fujita, T., Chen, M., & Chhowalla, M. (2011). Photoluminescence from chemically exfoliated MoS_2. *Nano Letters, 11*(511), 1–6.

El Moutaouakil, A., Suemitsu, T., Otsuji, T., Videlie, H., Boubanga-Tombet, S.-A., Coquillat, D., & Knap, W. (2011). Device loading effect on nonresonant detection of terahertz radiation in dual grating gate plasmon-resonant structure using InGaP/InGaAs/GaAs material systems. *Physica Status Solidi (C), 8*, 346–348.

Fan, X., Zheng, W. T., Kuo, J.-L., & Singh, D. J. (2015). Structural stability of single-layer MoS_2 under large strain. *Journal of Physics. Condensed Matter, 27*, 105401.

Fleisch, P. D., & Bauer, R. (1988). The aerospace CoChemical and structural effects on the lubrication properties of sputtered MoS_2 films. *Tribology Transactions, 31*(2), 239–250.

Frindt, R. F. (1966). Single crystals of MoS_2 several molecular layers thick. *Journal of Applied Physics, 37*, 1928–1929.

Ganatra, R. B., & Zhang, Q. (2014). Few-layer MoS_2: A promising layered semiconductor. *ACS Nano, 8*(5), 4074–4099.

Gao, M.-R., Chan, M. K. Y., & Sun, Y. (2015). Edge-terminated molybdenum disulfide with a 9.4-Å interlayer spacing for electrochemical hydrogen production. *Nature Communications, 6*, 7493.

Gao, X., Wang, X., Ouyang, X., & Wen, C. (2016). Flexible Superhydrophobic and Superoleophilic MoS_2 sponge for highly efficient oil-water separation. *Scientific Reports, 6*, 27207.

Geim, A. K. (2009). Graphene: Status and prospects. *Science, 324*(5934), 1530.

Geim, A. K., & Grigorieva, I. V. (2013). Van der Waals heterostructures. *Nature, 499*, 419–425.

Goel, N., Bera, J., Kumar, R., Sahu, S., & Kumara, M. (2021). MoS_2-PVP Nanocomposites decorated ZnO microsheets for efficient hydrogen detection. *IEEE Sensors Journal, 1*. https://doi.org/10.1109/JSEN.2021.3054038

Gupta, D., Chauhan, V., & Kumar, R. (2020). A comprehensive review on synthesis and applications of molybdenum disulfide (MoS_2) material: Past and recent developments. *Inorganic Chemistry Communications, 121*, 108200.

He, Z., & Que, W. (2016). Molybdenum disulfide Nanomaterials: Structures, properties, synthesis and recent Progress on hydrogen evolution reaction. *Applied Materials Today, 3*, 23–56.

Hersam, M. C. (2015). Emerging device applications for two-dimensional nanomaterial Heterostructures. In *Proceedings of the 2015 73rd annual device research conference (DRC), Columbus, OH, USA, 21–24 June 2015* (p. 209). IEEE.

Hijazi, A., & Moutaouakil, A. E. (2019). Graphene and MoS_2 structures for THz applications. In *Proceedings of the 2019 44th International Conference on Infrared, Millimeter, and Terahertz Waves (IRMMW-THz), Paris, France, 1–6 September 2019* (pp. 1–2). Paris.

Hiraki, A. (1992). Recent developments on metal-silicon interfaces. *Applied Surface Science, 56*(58), 370–381.
Holinski, R., & Gänsheimer, J. (1972). A study of the lubricating mechanism of molybdenum disulfide. *Wear, 19*, 329–342.
Huang, Y., Guo, J., Kang, Y., Ai, Y., & Li, C. M. (2015). Two dimensional atomically thin MoS_2 nanosheets and their sensing applications. *Nanoscale, 7*(46), 19358–19376.
Iwai, H., Kakushima, K., & Wong, H. (2006). Challenges for future semiconductor manufacturing. *International Journal of High Speed Electronics and Systems*, 43–81.
Jariwala, D., Sangwan, V. K., Lauhon, L. J., Marks, T. J., & Hersam, M. C. (2014). Emerging device applications for semiconducting two-dimensional transition metaldichalcogenides. *ACS Nano, 8*, 1102–1120.
Katsnelson, M. I., Novoselov, K. S., & Geim, A. K. (2006). Chiral tunnelling and the Klein paradox in graphene. *Nature Physics, 2*(9), 620.
Kaur, J., Singh, M., Dell'Aversana, C., Benedetti, R., Giardina, P., Rossi, M., Valadan, M., Vergara, A., Cutarelli, A., Montone, A. M. I., et al. (2018). Biological interactions of biocompatible and water-dispersed MoS_2 Nanosheets with bacteria and human cells. *Scientific Reports, 8*, 16386.
Kayyalha, M., Shi, L., & Chen, Y. P. (2015). Gate-tunable and thickness dependent electronic and thermoelectric transport in few-layer MoS_2. *arXiv:1505.05891*, preprint.
Kobayashi, K., & Yamauchi, J. (1995). Electronic structure and scanning-tunneling-microscopy image of molybdenum dichalcogenide surfaces. *Physical Review B, 51*(23), 17085–17095.
Kumar, P., Horwath, J. P., Foucher, A. C., Price, C. C., Acero, N., Shenoy, V. B., Stach, E. A., & Jariwala, D. (2020). Direct visualization of out-of-equilibrium structural transformations in atomically thin Chalcogenides. *npj 2D Materials and Applications, 4*, 16.
Lee, T.-W., Chen, C.-C., & Chen, C. (2019). Chemical stability and transformation of molybdenum disulfide Nanosheets in environmental media. *Environmental Science & Technology, 53*(11), 6282–6291.
Lee, S. J., et al. (2021). Morphology and catalytic performance of MoS_2 hydrothermally synthesized at various pH values. *Catalysts, 11*(10), 1229.
Lembke, D., Bertolazzi, S., & Kis, A. (2015). Single-layer MoS_2 electronics. *Accounts of Chemical Research, 48*(1), 100–110.
Li, X., & Zhu, H. (2015). Two-dimensional MoS_2: Properties, preparation, and applications. *Journal of Materiomics, 1*(1), 33–44.
Li, Y., Wang, H., Xie, L., Liang, Y., Hong, G., & Dai, H. (2011). MoS_2 nanoparticles grown on graphene: An advanced catalyst for the hydrogen evolution reaction. *Journal of the American Chemical Society, 133*, 7296–7299.
Lopez-Sanchez, O., Lembke, D., Kayci, M., Radenovic, A., & Kis, A. (2013). Ultrasensitive photodetectors based on monolayer MoS_2. *Nature Nanotechnology, 8*, 497–501.
Ma, H., Zhigang, S., & Ben, S. (2017). Understanding the exfoliation and dispersion of MoS_2 Nanosheets in pure water. *Journal of Colloid and Interface Science, 517*, 204–212.
Mak, K. F., et al. (2010). Atomically thin MoS_2: A new direct-gap semiconductor. *Physical Review Letters, 105*, 136805.
Mak, K. F., et al. (2014). The valley hall effect in MoS_2 transistors. *Science, 344*, 1489–1492.
Moutaouakil, A.E. (2018) Two-dimensional electronic materials for terahertz applications: Linking the physical properties with engineering expertise. In Proceedings of the 2018 6th International Renewable and Sustainable Energy Conference (IRSEC), Rabat, Morocco, 5(8):1–4.
Moutaouakil, A. E., Watanabe, T., Haibo, C., Komori, T., Nishimura, T., Suemitsu, T., & Otsuji, T. (2009). Spectral narrowing of terahertz emission from super-grating dual-gate Plasmon-resonant high-electron mobility transistors. *Journal of Physics Conference Series, 193*, 012068.
Moutaouakil, A. E., Komori, T., Horiike, K., Suemitsu, T., & Otsuji, T. (2010). Room temperature intense terahertz emission from a dual grating gate Plasmon-resonant emitter using InAlAs/InGaAs/InP material systems. *IEICE Transactions on Electronics, E93-C*, 1286–1289.
Moutaouakil, A. E., Suemitsu, T., Otsuji, T., Coquillat, D., & Knap, W. (2012). Nonresonant detection of terahertz radiation in high-electron-mobility transistor structure using InAlAs/InGaAs/

InP material systems at room temperature. *Journal of Nanoscience and Nanotechnology, 12*, 6737–6740.

Novoselov, K. S., Geim, A. K., Morozov, S. V., Jiang, D., Zhang, Y., Dubonos, S. V., Grigorieva, I. V., & Firsov, A. A. (2004). Electric field effect in atomically thin carbon films. *Science, 306*(5696), 666.

Novoselov, K. S., Jiang, D., Schedin, F., Booth, T. J., Khotkevich, V. V., Morozov, S. V., & Geim, A. K. (2005). Two-dimensional atomic crystals. *Proceedings of the National Academy of Sciences of the United States of America, 102*(30), 10451–10453.

Paredes, J. I., Munuera, J. M., Villar-Rodil, S., Guardia, L., Ayan-Varela, M., Pagan, A., Aznar-Cervantes, S. D., Cenis, J. L., Martinez-Alonso, A., & Tascon, J. M. D. (2016). Impact of covalent functionalization on the aqueous processability, catalytic activity, and biocompatibility of chemically exfoliated M nanosheets. *ACS Applied Materials & Interfaces, 8*(41), 27974–27986.

Peng, Z., Chen, X., Fan, Y., Srolovitz, D. J., & Lei, D. (2020). Strain engineering of 2D semiconductors and Graphene: From strain fields to band-structure tuning and photonic applications. *Light: Science & Applications, 9*, 190.

Pourabbas, B., & Jamshidi, B. (2008). Preparation of MoS_2 nanoparticles by a modified hydrothermal method and the photo-catalytic activity of MoS_2/TiO_2 hybrids in photo-oxidation of phenol. *Chemical Engineering Journal, 138*(1), 55–62.

Radisavljevic, B., Radenovic, A., Brivio, J., Giacometti, V., & Kis, A. (2011). Single-layer MoS_2 transistors. *Nature Nanotechnology, 6*, 147.

Saha, D., & Kruse, P. (2020). Editors' choice—Review—Conductive forms of MoS_2 and their applications in energy storage and conversion. *Journal of the Electrochemical Society, 167*, 126517.

Samy, O., Zeng, S., Birowosuto, M. D., & Moutaouakil, A. E. (2021). A review on MoS_2 properties, synthesis, sensing applications and challenges. *Crystals, 11*(4), 355.

Sebenik, R. F., et al. (2005). *Molybdenum and molybdenum compounds, Ullmann's encyclopedia of chemical technology*. Wiley-VCH.

Sha, J., Xu, W., Yuan, Z., Xu, B., & Chen, Y. (2017). Fabrication of liquid-gated molybdenum disulfide field-effect transistor. In *Proceedings of the 2017 IEEE 12th international conference on Nano/Micro Engineered and Molecular Systems (NEMS), Los Angeles, CA, USA, 9–12 April 2017* (pp. 788–791). IEEE.

Shah, P., Narayanan, T. N., Li, C. Z., & Alwarappan, S. (2015). Probing the biocompatibility of MoS_2 nanosheets by cytotoxicity assay and electrical impedance spectroscopy. *Nanotechnology, 26*(31), 315102.

Shree, S., Paradisanos, I., Marie, X., Robert, C., & Urbaszek, B. (2021). Guide to optical spectroscopy of layered semiconductors. *Nature Reviews Physics, 3*, 39–54.

Tang, W. (2017). *Electrical, electronic and optical properties of MoS_2 & WS_2*. Thesis, New Jersey Institute of Technology Digital Commons @ NJIT. https://core.ac.uk/download/pdf/232274113.pdf

Wang, Z., & Mi, B. (2017). Environmental applications of 2D molybdenum disulfide (MoS_2) Nanosheets. *Environmental Science & Technology, 51*, 8229–8244.

Wang, Q. H., Kalantar-Zadeh, K., Kis, A., Coleman, J. N., & Strano, M. S. (2012). Electronics and optoelectronics of two-dimensional transition metal dichalcogenides. *Nature Nanotechnology, 7*, 699–712.

Wang, S. Y., Ko, T. S., Huang, C. C., Lin, D. Y., & Huang, Y. S. (2014). Optical and electrical properties of MoS_2 and Fe-doped MoS_2. *Japanese Journal of Applied Physics, 53*, 04EH07.

Wang, S., Li, K., Chen, Y., Chen, H., Ma, M., Feng, J., Zhao, Q., & Shi, J. (2015). Biocompatible PEGylated MoS_2 nanosheets: Controllable bottom-up synthesis and highly efficient photothermal regression of tumor. *Biomaterials, 39*, 206–217.

Wang, C., Qianqian, W., Ding, Y., Zhang, X., Wang, W., Guo, X., Ni, Z., Lin, L., Cai, Z., Xiaofeng, G., Xiao, S., & Nan, H. (2023). High-responsivity and broadband MoS_2 photodetector using interfacial engineering. *ACS Applied Materials & Interfaces, 15*(39), 46236–46246.

Wi, S., Kim, H., Chen, M., Nam, H., Guo, L. J., Meyhofer, E., et al. (2014). Enhancement of photovoltaic response in multilayer MoS_2 induced by plasma doping. *ACS Nano, 8*, 5270–5281.
Wilson, J., & Yoffe, A. (1969). The transition metal dichalcogenides discussion and interpretation of the observed optical, electrical and structural properties. *Advances in Physics, 18*(73), 193.
Winer, W. O. (1967). Molybdenum disulfide as a lubricant: A review of the fundamental knowledge. *Wear, 10*, 422–452.
Wu, W., Wang, L., Li, Y., Zhang, F., Lin, L., Niu, S., et al. (2014). Piezoelectricity of single-atomic-layer MoS_2 for energy conversion and piezotronics. *Nature, 514*, 470–474.
Wu, B., Chen, L., Wu, X., Hou, H., Wang, Z., & Liu, S. (2019). Differential influence of molybdenum disulfide at the nanometer and Micron scales in the intestinal Metabolome and microbiome of mice. *Environmental Science. Nano, 6*, 1594–1606.
Xiaokun, G., Li, B., & Yang, R. (2016). Layer thickness-dependent phonon properties and thermal conductivity of MoS_2 journal of applied physics. *Journal of Applied Physics, 119*, 085106.
Xiong, S., & Cao, G. (2015). Molecular dynamics simulations of mechanical properties of monolayer MoS_2. *Nanotechnology, 26*, 185705.
Yazyev, O. V., & Kis, A. (2015). MoS_2 and semiconductors in the flatland. *Materials Today, 18*(1), 20–30.
Yin, Z., Li, H., Li, H., Jiang, L., Shi, Y., Sun, Y., Lu, G., Zhang, Q., Chen, X., & Zhang, H. (2012). Single-layer MoS_2 phototransistors. *ACS Nano, 6*, 74–80.
Yoo, G., Lee, S., Yoo, B., Han, C., Kim, S., & Oh, M. S. (2015). Electrical contact analysis of multilayer MoS_2 transistor with molybdenum source/drain electrodes. *IEEE Electron Device Letters, 36*, 1215–1218.
Yu, G. L., Jalil, R., Belle, B., Mayorov, A. S., Blake, P., Schedin, F., et al. (2013). Interaction phenomena in graphene seen through quantum capacitance. *Proceedings of the National Academy of Sciences (PNAS), 110*, 3282–3286.
Zeming, G. (1982). A discussion of the relationship between the structure of the transition metal Dichalcogenides and their lubrication performance. *A S L E Transactions, 25*, 207–212.
Zhao, G.-Y., Deng, H., Tyree, N., Guy, M., Lisfi, A., Peng, Q., Yan, J.-A., Wang, C., & Lan, Y. (2019). Recent Progress on irradiation-induced defect engineering of two-dimensional 2H-MoS_2 few layers. *Applied Sciences, 9*, 678.
Zhuo, Y., Kinloch, I. A., & Bissett, M. A. (2023). Simultaneous electrochemical exfoliation and functionalization of 2H-MoS_2 for Supercapacitor electrodes. *ACS Applied Nano Materials, 6*(19), 18062–18070.

Surface Functionalization of 2D MOs for Enhanced Biocompatibility and Biomedical Applications

Mohamed Abdelmonem, Emmellie Laura Albert, Ashreen Norman, Emma Ziezie Tarmizie, Samer H. Zyoud, and Che Azurahanim Che Abdullah

1 Introduction

In recent years, the fabrication of nanomaterials for diverse applications has emerged as a critical research domain. Significant advancements in nanomaterials production techniques have substantially enhanced the development of nanostructured materials. The term "nano" signifies small dimensions, and nanotechnology encompasses the interdisciplinary approach to the development, characterization, and synthesis of materials, with meticulous control over their dimensions and morphologies at the nanoscale (Kolahalam et al., 2019).

Nanomaterials have attracted interest owing to their substantial surface-to-volume ratios, leading to properties that are markedly superior to those of their bulk counterparts. These materials can be categorized based on their dimensional characteristics: zero-dimensional (0D) nanomaterials, wherein all three dimensions fall within the nanoscale range. Typically, these materials manifest as spherical or quasi-spherical particles with diameters under 100 nm (Liu et al., 2017a, b).

One-dimensional (1D) nanomaterials represent another category, characterized by their extension in two dimensions at the nanoscale. Examples of these include

M. Abdelmonem
Department of Physics, Faculty of Science, Universiti Putra Malaysia, Serdang, Selangor, Malaysia

E. L. Albert · A. Norman
Nanomaterial Synthesis and Characterization Laboratory, Institute of Nanoscience and Nanotechnology, Universiti Putra Malaysia, Serdang, Selangor, Malaysia

E. Z. Tarmizie
Centre of Foundation Studies for Agricultural Science, Universiti Putra Malaysia, Serdang, Selangor, Malaysia

S. A. Bansal et al. (eds.), *Emerging Applications of Novel Nanoparticles*, Lecture Notes in Nanoscale Science and Technology 37, https://doi.org/10.1007/978-3-031-57843-4_7

nanotubes, nanofibers, nanowires, nanorods, nanobelts, and nanofilaments. Notably, these nanomaterials are recognized for their extensive applications in the biomedical field, particularly in biosensing (Segal & Bussi, 2022), hyperthermia (Saliev, 2019), drug delivery systems (Pandey et al., 2021), and bioimaging (Haine & Niidome, 2017).

Two-dimensional (2D) nanomaterials are characterized by having two of their dimensions in the nanoscale range, leading to notable potential in diverse applications including sensing and drug delivery. These nanomaterials can manifest in various morphologies, including nanoplates, nanodisks, and nanosheets. The defining characteristic of their 2D architecture imparts distinctive physical and chemical properties. Among these is an outstanding quantum size effect, setting them apart from materials of zero- and one-dimensional nature (Wang et al., 2020).

The final category in nanomaterials is the three-dimensional (3D) class. This class encompasses materials composed of an aggregation of nanocrystals, each oriented differently. It can include dispersions of nanoparticles, conglomerations of nanowires and nanotubes, as well as multiple stacked nanolayers. The characteristics and prospective uses of 3D nanomaterials are influenced by a range of factors including their size, shape, dimensional attributes, and morphological features (Tiwari et al., 2012). These materials have demonstrated potential across diverse domains, notably in the field of catalysis. Bai et al., 2013), biomedicine (Wang et al., 2013), and energy (Xu et al., 2013).

The aim of this book chapter is to present an extensive review of the latest developments in the field of 2D nanomaterials, with a particular focus on their potential applications in biomedicine, especially concerning 2D MOs. The chapter primarily concentrates on the functionalization strategies employed to augment the biomedical efficacy of these materials, along with a thorough examination of the existing studies on their biocompatibility. While there are numerous scholarly reviews addressing the applicability of other dimensional nanomaterials in biomedicine, there is a discernible scarcity of research specifically on the biomedical applications of 2D MOs. This gap in the literature underscores the significance and necessity of this chapter.

S. H. Zyoud
Department of Mathematics and Science, Ajman University, Ajman, United Arab Emirates

Center of Medical and Bio-Allied Health Sciences Research, Ajman University, Ajman, United Arab Emirates

C. A. C. Abdullah (✉)
Department of Physics, Faculty of Science, Universiti Putra Malaysia, Serdang, Selangor, Malaysia

Nanomaterial Synthesis and Characterization Laboratory, Institute of Nanoscience and Nanotechnology, Universiti Putra Malaysia, Serdang, Selangor, Malaysia

UPM-MAKNA Cancer Research Laboratory, Institute of Bioscience, Universiti Putra Malaysia, Serdang, Selangor, Malaysia

Centre for Diagnostic Nuclear Imaging, Universiti Putra Malaysia, Serdang, Selangor, Malaysia
e-mail: azurahanim@upm.edu.my

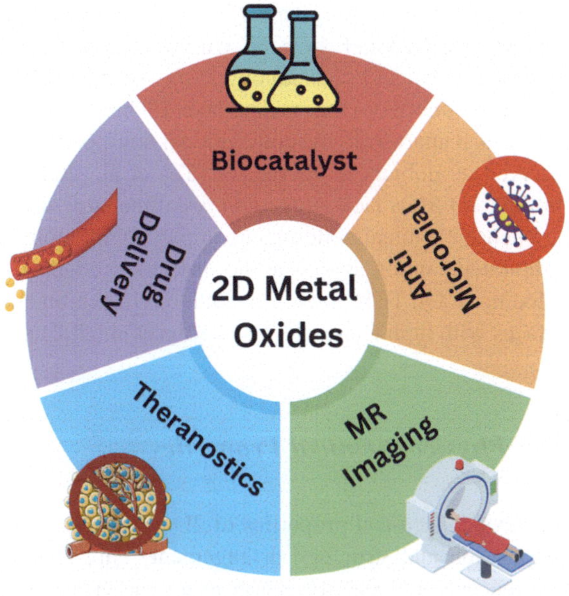

Fig. 1 Biomedical applications of 2D MOs

Recently, 2D MOs nanomaterials have attracted substantial attention owing to their unique attributes and advantages. Key among these is their high surface area-to-volume ratio, which provides a multitude of active sites for the efficient adsorption of molecules or ions and facilitates the enhanced transport of ionic species. Their exceptional physicochemical properties, coupled with high biocompatibility and low cytotoxicity, position 2D MOs as favorable candidates for various biological and biomedical applications. These include, but are not limited to, drug and gene delivery systems, therapeutic interventions, biosensing technologies, and bioimaging applications (Song et al., 2016). Figure 1 illustrates the diverse biomedical applications of 2D MOs, a factor that has catalyzed a surge in research activity within this domain. This increased investigative focus is aimed at fully exploring and understanding the capabilities and potential of these materials in biomedical contexts.

2 Properties of 2D Mos

2.1 Structural Properties

Nanomaterials composed of metal oxides in 2D are classified into two distinct groups based on their crystalline structures: layered and non-layered. The layered category of these nanomaterials is characterized by an assembly of numerous octahedral units, each comprising M-O (where M represents elements such as V, Mo, or Mn), organized in a stratified formation (Apte et al., 2020; Haque et al., 2018; Nagyné-Kovács et al., 2020; Ren et al., 2020).

Research has uncovered a particular crystal structure present in various oxides of elements like Ti, Mn, Fe, Co, Ni, Cu, Al, Gd, and Ge. This structure is distinguished by robust chemical bonds among atoms within the same plane and feeble van der Waals forces between the stacked layers. This distinct hexagonal coordination in the plane sets it apart from the conventional non-polarized crystal structures commonly observed at ambient temperatures (Yan et al., 2021). Conversely, the majority of ultrathin 2D MOs lack a layered crystalline structure. This absence is attributed to the intense chemical bonding encompassing all atoms, leading to a proliferation of unsaturated dangling bonds at the surfaces or edges. Consequently, these materials, including In_2O_3, TiO_2, SnO_2, ZnO, and WO_3, exhibit heightened activity and possess surfaces with high energy levels (Radha et al., 2021).

2.2 Physicochemical Properties

The physicochemical properties of 2D materials can vary based on their elemental composition and structural arrangement. This variability endows these materials with the potential for utilization in a wide array of applications (Dhenadhayalan et al., 2017). Additionally, 2D MOs display extensive absorption spectra and demonstrate significant fluorescence quenching capacities. These characteristics stem from their spectra overlapping with the excitation and emission spectra of fluorescent nanomaterials. Consequently, this attribute renders them as potential candidates for application in fluorescence sensing, particularly in techniques reliant on Förster Resonance Energy Transfer (FRET). Furthermore, the expansive specific surface areas characteristic of 2D MOs offer optimal substrates for substantial drug-loading capacities, facilitating their employment in diverse drug delivery and cancer treatment endeavors. Moreover, the robust oxidative and catalytic properties of 2D MOs allow for their rapid decomposition in the presence of reducing agents like glutathione (GSH), which in turn triggers the release of the encapsulated drugs. 2D MOs are also applicable in identifying reducing substances within biological specimens. Additionally, the physicochemical properties and electronic band structures of these 2D MOs can be tailored through ion incorporation or surface modification. This adaptability offers a pathway to engineer these materials with specific optical and electrical properties, aligning them with desired functionalities (Liu et al., 2009).

2.3 Magnetic Properties

Another frequently observed characteristic of metal oxides is their magnetic properties. These properties in nanomaterials are influenced by factors such as their size, structural composition, volume, surface area, and the presence of step atoms (Chen et al., 2018). In the realm of biomedical applications, encompassing bioimaging, drug delivery, magnetic hyperthermia, and biosensor technologies, there is a preference for magnetic nanomaterials exhibiting anisotropic characteristics due to their ferromagnetic behavior. Contrastingly, isotropic nanostructures fabricated from magnetic metals such as Fe, Co, and Ni tend to exhibit superparamagnetic properties at ambient temperatures, a phenomenon largely attributable to their diminutive size (Maldonado-Camargo et al., 2017).

Certain phases of iron oxide nanoparticles (IONPs) such as magnetite or maghemite, cannot exist in a 2D form. However, they can be deposited onto the surface of select 2D MOs, such as MnO_2 (Pandey et al., 2019), In_2O_3 (Raha & Ahmaruzzaman, 2021), or ZnO (Długosz et al., 2021), to facilitate functionalization. Additionally, a phase of iron oxide known as hematine can be synthesized from hematite, possessing magnetic properties similar to those of 3D iron oxide phases (Puthirath Balan et al., 2018).

IONPs exhibit distinctive characteristics, including the capability to produce T1 or T2 contrasts in Magnetic Resonance Imaging (MRI). This attribute renders them a potent and non-toxic contrast agent for in vivo molecular imaging and cell tracking applications (Zhang et al., 2020). Furthermore, these nanoparticles are capable of generating heat when exposed to an alternating magnetic field. This property is instrumental in facilitating hyperthermia therapy for cancer treatment, where locally elevated temperatures are used therapeutically (Zhang & Song, 2016). Moreover, the magnetic characteristics of IONPs enable their attraction to designated locations or retention within a specified volume by magnetic forces, rendering them advantageous in applications that require force generation. Such attributes distinctly set IONPs apart in terms of functionality compared to other nanomaterials.

2.4 Stability

The stability of 2D MOs is a crucial aspect that must be considered when characterizing and utilizing these synthesized 2D metal nanomaterials. Numerous 2D metal nanostructures have been successfully fabricated by controlling growth kinetics. However, their elevated surface energy and the existence of unsaturated surface atoms contribute to the thermodynamic instability of these 2D structures. When provided with sufficient energy, these metal nanostructures are prone to morph into isotropic shapes, reducing their surface-to-volume ratio in an effort to attain stability. Consequently, despite their inherently unstable nature, 2D MOs are often regarded as having enhanced stability (Liz-Marzán & Grzelczak, 2017). In addition, the surface modification of 2D MOs can significantly improve their stability and biocompatibility in a microenvironment as will be discussed in the section on functionalization.

3 Synthesis of 2D-Metal Oxides

Two primary approaches exist for synthesizing 2D MOs: the top-down and bottom-up methods. The top-down method is typically applied to layered 2D MOs and entails fragmenting three-dimensional crystals into mono- and few-layered 2D MOs through various energy inputs. In contrast, the bottom-up approach involves the accumulation of atoms or molecules on a substrate, resulting in the formation of ultrathin 2D MOs. Figure 2 highlights the key synthesis techniques utilized in generating 2D MOs.

3.1 Top-Down Approach

The top-down approach serves as a direct and scalable technique for the exfoliation of 2D MOs, effectively maintaining their high crystallinity. Nevertheless, this methodology faces constraints in terms of controlling the thickness and lateral dimensions of the resultant nanosheets. Moreover, the choice of the layered host crystal significantly narrows its applicability. Frequently used top-down techniques for the exfoliation process encompass methods such as exfoliation assisted by mechanical force, exfoliation via soft chemical methods, and exfoliation facilitated by electrochemical means (Xie et al., 2022).

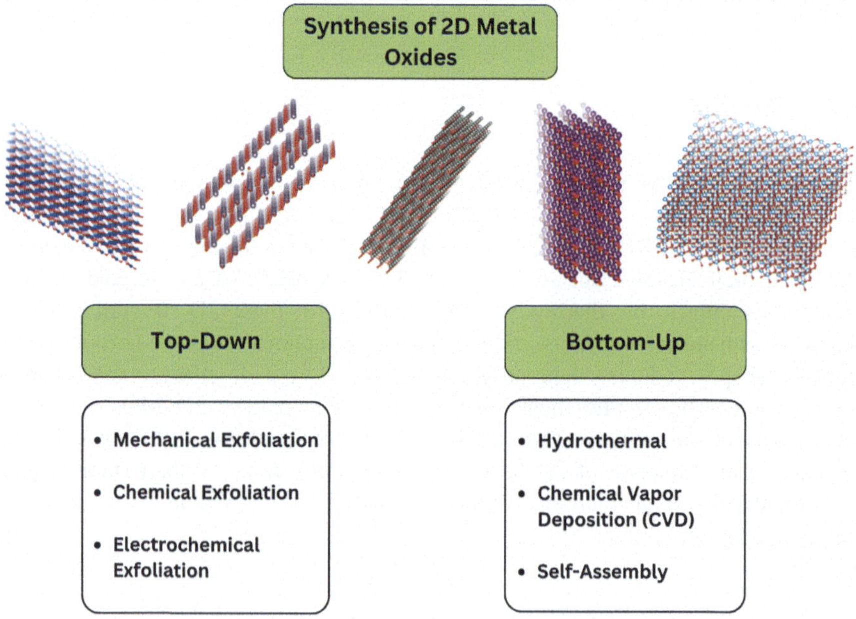

Fig. 2 Main synthesis techniques used for the fabrication of 2D MOs

3.2 Bottom-Up Approach

The bottom-up approach, often referred to as the gathering-up method, entails the synthesis of 2D nanomaterials from atomic or molecular precursors, initially present as nanoparticles or nanocrystals. These precursors participate in chemical reactions, leading to an increase in their size. Subsequently, they undergo a process of self-assembly into more intricate substances or are deposited onto specific substrates (Iqbal et al., 2012; Xiao et al., 2018). The bottom-up approach is favored owing to its capability to precisely manipulate features like the size, shape, composition, sequential stacking, and chemical stability of 2D nanomaterials. A variety of synthesis techniques are utilized to prepare these materials using the bottom-up approach. These include the hydrothermal or solvo-thermal method, chemical vapor deposition (CVD), and self-assembly synthesis (Han et al., 2018; Jiang et al., 2019). Table 1 delineates the principal fabrication methodologies utilized in the synthesis of 2D MOs.

4 Functionalization of 2D-Metal Oxides

2D MOs exhibit several beneficial properties, they can be utilized in various biomedical applications. Moreover, the latter 2D materials exhibit remarkable characteristics, including a substantial surface area and active-edge sites, which make them a promising candidate for utilization in advanced technologies. Nevertheless, specific properties of 2D MOs severely impede their exclusive usage for specific applications. However, owing to their instability in intricate physiological conditions, 2D MOs are prone to rapid aggregation. Furthermore, these materials are not inherently biocompatible and must be functionalized before being integrated into biological systems (Crowley et al., 2018). Therefore, it is imperative to implement surface modifications using chemical and physical methods to enhance the dispersal, biostability, and biocompatibility of 2D materials. Morevoer, the creation of oxygen vacancies and the implementation of doping techniques can result in distinctive electronic functionalities and catalytic characteristics in 2D Mos materials (Liu et al., 2016). Table 2 is a compilation of various studies that have utilized different materials for the purpose of functionalizing the surface of 2D MOs in many biomedical applications. Figure 3 depicts various methods of functionalizing 2D MOs.

4.1 Chemical Functionalization

There are various methods for surface modification; however, chemical functionalization, which involves either covalent or non-covalent bonding, is the most prevalent and efficacious approach. Organic components, typically in the form of polymers or smaller

Table 1 Techniques for the synthesis of 2D Mos

Synthesis method	Mechanism	2D materials	Advantages	Limitations	References
Top-down approach					
Mechanical exfoliation	Mechanical exfoliation through the process of pressing a polished metal surface onto a specified substrate	TiO_2 MnO Fe_2O_3 CoO Ni_2O_3 Cu_2O Al_2O_3 GeO_2 Gd_2O_3	Highly reproducible production of high-quality metal oxides	Low yield Poor thickness and size control Constrained to the production of 2Dlayered metal oxides	Chen et al. (2018)
Chemical exfoliation	Appropriate chemical force is required for ion intercalation, which weakens interlayer interaction, resulting in exfoliated dispersion formation	MoO_3 $LiCoO_2$	When combined with shear force reduce the reaction time tremendously	High cost Low efficiency	Xie et al. (2022)
Electrochemical exfoliation	This technique utilizes electrical current to expedite the migration of external molecules or ions into the bulk material. This process results in the enlargement of the spacing between the layers	ZnO MnO_2	No chemical waste (Eco-friendly) Short processing time High yield	This method is applicable for the preparation of only a limited number of metal oxides	Xie et al. (2022)
Bottom-up approach					
Hydrothermal method	Combining precursor materials, then transferring them into an autoclave vessel, followed by heating under elevated temperature and pressure conditions to initiate the reaction and facilitate the growth of nanostructures, utilizing solvents as catalysts	TiO_2@ MoS_2 CdS@ MoS_2 CeO_2 In_2O_3	High yield Size and shape can be controlled High quality crystals Low cost instrument	High temperature and pressure	Rana and Hussain (2022)

(continued)

Table 1 (continued)

Synthesis method	Mechanism	2D materials	Advantages	Limitations	References
Chemical vapor deposition (CVD)	The process involves decomposing volatile precursors and depositing them onto a suitable substrate. This is typically done by heating and depositing onto a substrate that is cooler	MoO_3-MoS_2 TiO_2 MoO_2	High quality ultrathin 2D MOs High crystallinity	The dimensions of nanosheets are generally too expansive to be suitable for biomedical applications	Jia et al. (2020)
Self-assembly	This method orchestrates the assembly of unstructured elements into organized patterns through specific interactions that occur autonomously, without the need for external human manipulation or stimuli. The interactions employed in this process include van der Waals forces, hydrophobic interactions, electrostatic forces, and a range of non-covalent interactions	CuO Eu_2O_3 TiO_2 ZnO Co_3O_4 Fe_2O_3 MnO_2	Production of highly uniform 2D MOs structures are suitable for use in biosensing applications	Time consuming and complicated Difficult to be used for industrial applications	Jia et al. (2020)

organic molecules, are often employed and are either covalently or non-covalently attached to MO for their utilization in biosensing and therapeutic domains. In such scenarios, Polyethylene glycol (PEG) is notably prevalent as a polymer choice for surface modification of 2D MO nanosheets, with the objective of improving their physiological stability (Gong et al., 2018; Song et al., 2016).

Research demonstrated that the process of PEGylation could be accomplished by simply immersing MnO_2 nanosheets in amino-PEG (PEG5000-NH_2). The surface-modified 2D MnO_2 nanosheets exhibited significantly enhanced dispersibility and stability in saline solutions relative to the unmodified versions. This improvement is ascribed to the formation of Mn-N coordinative bonds between PEG5000-NH2 and the nanosheets. Such findings carry significant potential for advancing the

Table 2 Previous studies of functionalized 2D MOs in biomedical applications

2D Metal oxide/ metal oxide based materials	Coating material	Biomedical application	References
TiO_2@ WS_2	Chitosan	The application of chitosan coating to 2D nanocomposites enhances their antibacterial and anticancer properties, as well as improves their photocatalytic activity	Kasinathan et al. (2022)
BN-TiO_2	PEDOT (poly 3,4-ethylene dioxythiophene)	The application of a PEDOT coating on 316 L stainless steel (SS) implants has demonstrated enhancements in both in vitro corrosion resistance and biocompatibility. Consequently, this nanocomposite has potential utilization within the orthopedic field	Kumar et al. (2022)
ZnO	TiO_2	ZnO@TiO_2, characterized by a diminished band gap energy, holds potential for application in the fields of photocatalytic reactions and photothermal therapy	Rajendran et al. (2018)
$TiO_{2@}$Au	Mitochondria-targeted triphenylphosphine (TPP) and AS1411 aptamer	The integration of mitochondria-targeting triphenylphosphonium (TPP) and AS1411 aptamer into the nanocomposite significantly enhanced its sonosensitization capability. This improvement led to an increased quantum yield of reactive oxygen species (ROS), thereby elevating the precision and effectiveness of sonodynamic therapy (SDT)	Cao et al. (2019)
ZnO	DOX-FA-PEG	This study's results reveal that Zinc Oxide, when functionalized with Folic Acid and coated with polyethylene glycol, displays thermally responsive effects, pH-sensitive behavior, and prolonged drug release properties. These characteristics establish it as a highly suitable material for drug delivery systems, particularly in the context of breast cancer treatment	Vimala et al. (2017)
GO	Chitosan oligosaccharide (CO) and γ-polyglutamic acid (γ-PGA) and doxorubicin (DOX)	A nanoparticle-assisted drug delivery system with high biocompatibility and responsiveness to stimuli has been developed for anti-cancer treatment. The findings suggest that GO-CO-γ-PGA-DOX triggers apoptosis through the intrinsic mitochondrial apoptotic pathway	Vimala et al. (2017)

(continued)

Table 2 (continued)

2D Metal oxide/ metal oxide based materials	Coating material	Biomedical application	References
MnO_2	Graphitic-phase C_3N_4 (G-C_3N_4)	G-C_3N_4–MnO_2 nanocomposite has been established to possess membrane permeability and it serves as a biosensor for in vivo monitoring and visualization of variations in intracellular GSH levels.	Zhang et al. (2014)
MnO_2	Palladium (Pd)/ Doxorubicin (DOX)	The efficacy of the nanocomposite in inhibiting MDA-MB-231 cells and tumor growth in murine models was substantiated by administering Pd@MnO_2/DOX. This finding corroborates the advantageous therapeutic effects attained by integrating photothermal therapy and chemotherapy using this compound	Sun et al. (2019)

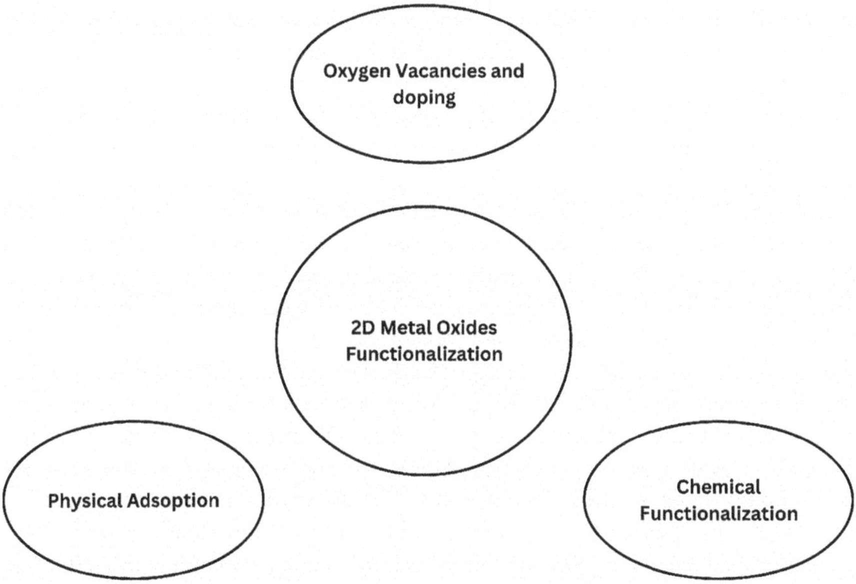

Fig. 3 Types of 2D MOs Functionalization

development of biomedical applications that leverage PEGylated MO nanosheets (Chen et al., 2014).

PEG molecules have been utilized for modifying the surface of 2D MO nanosheets, including MoOx and ultrathin $TeO_2/(NH_4)xWO_3$ nanoribbons, with the aim of augmenting their stability and biocompatibility in physiological environments. Additionally, silica has been used as a coating material for surface modification through chemical bonding. A study illustrated this concept by attaching silica-coated upconversion nanoprobes (UCSs) to MnO_2 nanosheets. This attachment was achieved through intermolecular hydrogen or coordination bonds formed between Mn atoms and O atoms in the silica shell. This method facilitated the creation of a dual-functional platform, suitable for upconversion luminescence (UCL) imaging and synergistic therapeutic applications (Fan et al., 2015a, b). In different research, the stabilization of MnO_2 nanosheets was achieved by employing sodium dodecyl sulphate (SDS) and lauric acid. This approach utilized the hydrophobic effect in conjunction with the coordination interactions between the carboxyl groups and the unsaturated Mn atoms (Liu et al., 2017a, b).

4.2 Physical Adsorption

An alternative strategy for altering the surface of 2D MOs is via physical adsorption. This method is particularly advantageous when it is essential to preserve the native properties of the nanoparticles, or in scenarios where the detachment of functional groups is integral to the diagnostic or delivery process. Physical adsorption entails the non-covalent binding of selected probes or drug carriers to the expansive basal plane of the nanoparticles, thereby optimizing loading capacity and interaction with the external environment. This type of functionalization is primarily driven by hydrophobic interactions, electrostatic forces, and van der Waals forces (Yang et al., 2018).

As example of physical adsorption functionalization, there is a study conducted by Deng et al. have synthesized assemblies of up conversion nanoparticles (UCNPs) and MnO_2 nanosheets via an in-situ reduction process involving $KMnO_4$ on the surface of UCNPs. The dominant binding force between the two components is attributed to electrostatic interactions (Deng et al., 2011).

In a research made by Fan et al., 2015a, b. They developed an advanced gene-silencing therapy utilizing a DNAzyme-MnO_2 nanocomposite. This system incorporates chlorin e6-labelled DNAzymes (Ce_6-DNAzyme) adsorbed on MnO_2 nanosheets through physisorption. The MnO_2 nanosheets act as a protective carrier against endogenous nuclease degradation of the DNAzyme. Intracellular reduction of the MnO_2 nanosheets by glutathione (GSH) initiates the release of Mn^{2+} ions, enhancing the efficacy of the 10–23 DNAzyme in gene silencing. Crucially, this nanosystem also enables real-time monitoring of its delivery effectiveness via activatable magnetic resonance and the restoration of fluorescent signals (Fan et al., 2015a, b).

4.3 Oxygen Vacancies and Doping Functionalization

The electronic characteristics and catalytic potential of 2D materials, including MOs, are significantly influenced by factors such as defects, the oxidation state of cations, and doping. Therefore, the introduction of oxygen vacancies and the incorporation of dopants are crucial methods for altering the surface of 2D MOs. These techniques facilitate the modification of the electronic band structure and the Fermi level, culminating in the functionalization of the material's surface (Grisolia et al., 2016; Kalantar-zadeh et al., 2016).

Molybdenum oxides display a variety of notable properties stemming from the existence of diverse oxygen vacancies. Fully stoichiometric MoO_3 is typically viewed as catalytically inert. In contrast, sub-stoichiometric MoO_3, characterized by an increased presence of oxygen vacancies, can generate a substantial number of active sites, consequently amplifying its catalytic efficacy (Haque et al., 2019; Zhang et al., 2018). Molybdenum oxides, characterized by varying levels of oxygen vacancies, demonstrate a variable band gap, which can be adjusted within a range of 2.8–3.6 eV (Haque et al., 2019; Khademi et al., 2009).

The stoichiometric alteration of molybdenum oxides can be achieved via chemical or electrochemical methods, which include the extraction of oxygen through chemical reduction processes employing appropriate reducing agents (Bai et al., 2018; Ou et al., 2018; Wang et al., 2012). Intentional doping with various ions and molecules represents a straightforward yet efficacious strategy for modulating the nature and magnitude of carriers in 2D MOs. This method induces substantial alterations in their physicochemical characteristics (Xu et al., 2017).

The doping of 2D MOs can be achieved via electrochemical or wet-chemical approaches, including processes like intercalation or the alteration of surface properties using charged or polar entities. Due to the extensive surface area inherent in 2D oxides, they possess a high capacity for accommodating doping agents. This characteristic renders these techniques highly efficient in altering the material's properties (Stamenkovic et al., 2006).

Additionally, the enhanced mechanical flexibility of certain 2D oxides allows for the incorporation of size-mismatched dopants (Nilius & Freund, 2015). The existence of free charge carriers in dopants can lead to a phenomenon known as ultra-doping, where electrons are transferred to the plane of the 2D oxide. This process can alter the plasmon resonance wavelengths of MO sheets into the visible-near-infrared (vis-NIR) spectrum, rendering them appropriate for applications in plasmonic biosensing (Zhang et al., 2018).

In research carried out by Ou et al., it was found that reversible doping of MoO_3 nano disks with H+ ions leads to alterations in their plasmon resonance characteristics. These H_+ ions are positioned in either the interstitial or intralayer locations of MoO_3, resulting in the formation of intercalated compounds ($HxMoO_3$). This process concurrently introduces delocalized electrons, inducing plasmon resonance within the visible-near-infrared (vis-NIR) spectrum.

The majority of 2D MOs exhibit low absorption in the near-infrared (NIR) biological window, which restricts their use in 2D MO-based photothermal therapy. Nonetheless, the introduction of dopants and the creation of oxygen vacancies in 2D MOs can elevate their free charge carrier concentration to a high level, ranging between 1020 and 1021 cm^{-3}. This increase leads to the emergence of plasmonic absorption peaks within the NIR spectrum. Such advancements open new avenues for more effective photothermal therapy using 2D MOs (Yu et al., 2017).

5 Biocompatibility Studies of 2D-Functionalized Metal Oxides

The biocompatibility of materials, a paramount characteristic for those that interface with living cells and biological substances, ensures the absence of adverse effects. Evaluating the biocompatibility of 2D MOs both in vitro and in vivo, along with their cellular response, is imperative for assessing their potential in clinical applications. These assessments are critical for ascertaining the capacity of 2D MOs to safely engage with biological systems.

Numerous studies have explored the toxicity of 2D MOs, with factors such as the crystal facets (Kalantar-zadeh et al., 2015). 2D MOs, such as V_2O_5 and MoO_3, have been documented to exhibit mild genotoxic effects (Altamirano-Lozano et al., 1999). Particularly, 2DMoO_3 nanosheets have demonstrated the ability to initiate the caspase pathway in MCF-7 breast cancer cell lines, consequently inducing apoptosis (Anh Tran et al., 2014). Comparable results regarding toxicity have been reported in other research.

In vivo toxicity assessments of 2DTiO_2 nanosheets in mouse models, following intraperitoneal injection, revealed low toxicity levels. However, extended exposure to these TiO_2 nanosheets caused particle accumulation, which led to minor abnormalities. On the other hand, several studies have indicated that 2D MOs are biocompatible and exhibit minimal or no cytotoxic effects on normal cells. Table 3 provides a detailed exploration of previous biocompatibility studies conducted on in-vitro cell lines using 2D MOs.

6 Biomedical Applications of 2D-Metal Oxides

6.1 Drug Delivery Systems

Nanoscience has gained significant prominence in the development of effective drug delivery systems. The field of nanotechnology is being extensively explored by researchers to create nano drugs capable of surpassing the challenges faced by traditional drugs, including issues like limited solubility and suboptimal

Table 3 Previous biocompatibility studies of 2D MOs to be used potentially in biomedical applications

2D MOs/ metal oxides-based nanomaterials	Toxicity test approach	Toxicity test assay	Findings	References
TiZrPdSiNb@ ZnO	In-vitro Saos-2 cell lines	Alamar blue	A composite material comprising TiZrPdSiNb alloy and ZnO nanosheets, synthesized via hydrothermal growth, is characterized by its non-cytotoxic nature and lack of inflammatory response induction. This material has been shown to promote a 1.5-fold increase in the proliferation of Saos-2 cells after a seven-day culture period. Additionally, the release of cytokines such as TNF-α, IL-6, and IL-1β is observed to be minimal or non-existent following 24 h of culture	Careta et al. (2021)
SP-modified 2D MnO_2	In vitro 4 T1 cell lines	CCK-8	Research findings indicated that 2D MnO_2 nanosheets, when altered with Soybean phospholipid (SP), exhibited negligible cytotoxic impacts on 4 T1 cells, even at a considerably high concentration of 600 mg mL^{-1}	Liu et al. (2018)
FA-MnO_2/ZnPc	In vitro/In vivo HeLa cells	CCK-8	Investigators explored the use of FA-MnO_2 nanosheets as carriers for Zinc Phthalocyanine (ZnPc) in photodynamic therapy, at concentrations reaching approximately 112 μg/mL. The incorporation of ZnPc onto the nanosheets did not compromise the viability of HeLa cells. Moreover, when the complex was administered to mice at a dosage of 0.5 mg ZnPc/kg, there was no evident systemic toxicity observed in major organs. These results imply that FA-MnO_2 nanosheets could serve as a safe and efficient delivery system for ZnPc in photodynamic therapy applications	Kim et al. (2017)

(continued)

Table 3 (continued)

2D MOs/ metal oxides-based nanomaterials	Toxicity test approach	Toxicity test assay	Findings	References
ZnO	In vitro A431, HepG2 and Vero cell lines	MTT	In the study, all ZnO samples synthesized and administered at a concentration of 120 μg/mL demonstrated a marked reduction in the viability of nonmelanoma skin cancer cells (A431) in a dose-dependent manner, while exhibiting negligible effects on normal cells (Vero). Furthermore, at a concentration of 150 μg/mL, the ZnO dispersion showed minimal impact on the cell viability of HepG2 cells	Jevapatarakul et al. (2020)
ZnO/SiO_2 and ZnO/SiO_2/RGO	MCF-7, KMBC/71, and 3 T3 Cell lines	MTT	The findings of the research demonstrated that concentrations above 31 μg/mL for ZnO/SiO_2 and 15 μg/mL for ZnO/SiO_2/RGO led to significant cytotoxicity in 3 T3-fibroblasts. Furthermore, the study highlighted that the cytotoxic influence of these nanocomposites on normal fibroblast cells was comparatively less marked than their impact on breast cancer cells	Chireh et al. (2022)
ZnO	U87, HeLa, and HEK cell lines	MTT	The investigation evaluated the efficacy and toxicity of the complexes in human U87 gliomas, HeLa, and HEK cells. The outcomes revealed that these complexes were more effective in suppressing the growth of U87 and HeLa cancer cells, while exhibiting relatively lower toxicity in normal human HEK cells	Wahab et al. (2011)

bioavailability. Nanoparticles have emerged as a key instrument in augmenting the biocompatibility of pharmaceuticals. They play a crucial role in enhancing the stability of drugs as they traverse through the bloodstream and aid in the efficient absorption by cells (Zeng et al., 2018).

The expanding interest in nanoparticles has led to an increased focus on the use of 2D nanomaterials for drug delivery applications. Owing to their distinctive characteristics, including expansive surface area, significant aspect ratio, and customizable properties, these 2D nanomaterials present as a highly promising avenue for the development of more effective drug delivery systems.

Various 2D MOs have been recognized for their efficacy as carriers in drug delivery systems. A notable example is MnO_2 nanosheets, which are reducible to Mn^{2+} by glutathione. This reduction is advantageous in MnO_2-based drug delivery systems, particularly because the microenvironment of cancer cells exhibits elevated levels of glutathione, which catalyzes the conversion process, enhancing the timely release of drugs directly within the tumor site. On a multifaceted drug delivery system, MnO_2 nanosheets were employed as carriers within Fe_3O_4@SiO_2/$NaYF_4$:Yb, Er (MSU) hybrid structures. The MSU's magnetic core and $NaYF_4$:Yb, Er shell functioned as tools for magnetic targeting and fluorescence bioimaging, respectively. Introducing glutathione led to the reduction of MnO_2 nanosheets, activating the previously quenched upconversion emission of MSU. Congo red, serving as a model drug, was used to examine drug binding and release, demonstrating its affinity to MnO_2 nanosheets via electrostatic interactions. The synthesized drug delivery system exhibited a drug loading capacity of 30 mg/g. The aforementioned study was focused on reviewing recent developments in the application of metal oxide-based drug delivery systems, with a particular emphasis on the distinctive features and benefits of MnO_2 nanosheets in this area. Controlled release of a model drug can be facilitated by the introduction of glutathione. When the drug is associated with the MnO_2 nanosheets, it remains unreleased for a duration of up to 5 days. However, introducing up to 5 mM of glutathione leads to an 80% discharge of the drug within an hour. This swift release of the model drug is attributed to the redox reaction occurring between the MnO_2 nanosheets and glutathione (Liu et al., 2018).

In addition, a study has developed an innovative photoactivated platinum (Pt) prodrug delivery system, termed MnO_2@Pt-SP, designed to address cisplatin-resistant non-small cell lung cancer (NSCLC). This system utilizes MnO_2 nanosheets to augment cisplatin's therapeutic effectiveness by facilitating its access to the nucleus and diminishing drug expulsion through the suppression of multidrug resistance-associated protein (MRP1). The photo-induced hyperthermia generated by the nanosheets further contributes to the inhibition of platinum drug efflux and lowers the consumption of intracellular GSH. This reduction minimizes the formation of GSH-Pt complexes and bolsters cellular uptake, leading to increased accumulation of Pt in cells. The combined chemo-photothermal therapy markedly enhances the therapeutic efficiency in treating cisplatin-resistant NSCLC, both in vitro and in vivo. Furthermore, the MnO_2@Pt-SP nanosheets can serve as on-demand activatable MRI contrast agents, providing real-time monitoring of redox reactions during GSH consumption and Pt drug release (Gao et al., 2020).

Recent research has shown that 2D MnO_2 nanosheets possess an abundance of anchoring sites for anticancer drugs, facilitating the effective dispensation of these loaded therapeutics. The study elucidates that drug release is attributed to the gradual disintegration of MnO_2 nanosheets within the slightly acidic microenvironments typical of tumor tissues. This disintegration, instigated by pH fluctuations, can hasten the release of encapsulated anticancer agents. Such a mechanism enables the concurrent release of drugs in response to specific pH conditions, potentially aiding in surmounting multidrug resistance observed in cancer cells (Chen et al., 2014).

6.2 Biosensing

The unique properties of 2D MOs present significant prospects for merging biological recognition mechanisms with electronic signal transduction. This integration paves the way for the creation of novel bioelectronic devices endowed with distinctive functionalities (Solanki et al., 2011). Recent research has documented the successful creation of a Prostate Specific Antigen (PSA) sensor, employing MoO_3 nanosheets within a 2D nanosheet structure. The study reveals that MnO_2 nanosheets, due to their exceptional sensitivity and specificity, are highly effective for PSA detection through a sensing mechanism. This efficacy is ascribed to the process of fluorescence quenching and its subsequent recovery in the aptamer, which is contingent on the target molecule's presence or absence. Furthermore, the research emphasizes the innovative application of MnO_2 nanosheets as a synthetic oxidase in colorimetric assays, thereby markedly improving the swift and sensitive detection of glutathione. By emulating the activity of an oxidase enzyme, these MnO_2 nanosheets contribute to a more effective detection methodology for glutathione (Liu et al., 2017a, b).

Tin oxide represents an additional instance of a 2D metal oxide (2D MO) material appropriate for biosensor applications. Recent research indicates that hybrid constructions like the indium tin oxide (ITO)–WS2 based surface plasmon resonance (SPR) biosensor demonstrate considerably improved efficacy relative to traditional metal-based SPR sensors. The enhanced performance of the biosensor is largely due to the advantageous characteristics of ITO, a composite of In_2O_3 and SnO_2. ITO's properties, including a tunable dielectric constant, minimal capacitance, and high optical transmittance, contribute significantly to its efficacy. Notably, the transmittance of ITO films reaches up to 95% in the visible light spectrum and 80% in the infrared range. This high transmittance markedly increases the sensitivity of the functional materials in both the modulator and SPR applications. The study further posits that by fine-tuning the thickness of Ag, ITO, and WS_2 layers, the sensitivity of the biosensor can be further amplified. Overall, the combination of ITO and WS_2 holds great potential for advancing biosensor technology (Han et al., 2019).

6.3 Bioimaging

MnO_2 nanosheets, characterized by their unique planar nanoarchitecture, represent a prevalent form of metal oxide that is well-suited for drug loading in chemotherapeutic applications. Moreover, the inherent capacity of MnO_2 to disassemble in reaction to various internal and external stimuli renders it beneficial for MRI within the tumor microenvironment. This advantage is primarily due to the paramagnetic properties of Mn^{2+} ions (Yang et al., 2018).

Consequently, the distinctive characteristics of 2DMnO_2 nanosheets have garnered significant interest in the realm of medical imaging. These nanosheets have demonstrated potential as efficient contrast agents for MRI, both in their unmodified state and when subjected to functionalization (Chen et al., 2014). MnO_2 displays sensitivity to both internal and external stimuli, such as changes in pH and redox state, rendering it useful for developing MRI platforms that can respond to fluctuations in pH or H_2O_2 levels, this was according to a study which reported the efficacy of coating of MnO_2 onto nanomaterials, with a specific example of Cu_2–xSe@MnO_2 demonstrating a three-fold increase in T_1-MRI contrast in a tumor-reducing environment (Fu et al., 2018).

Moreover, a study investigated the potential of ZnO nanosheets for imaging cultured cells due to their strong fluorescent properties. The study involved treating model leukemia K562 cells with the ZnO nanosheets for 20 min before imaging, with the differential interference contrast picture used for contrast. Results indicated that the yellow-orange light emission from the ZnO nanostructures was clearly visible around or inside the cells under UV excitation, indicating successful attachment onto or penetration into the cells. This discovery indicates that the synthesized ZnO nanosheets present a feasible alternative for application as labels in bioimaging, offering prospective uses in the realm of biomedicine (Jiang et al., 2011).

7 Challenges and Conclusion

The biomedical sector has increasingly recognized the potential of 2D MOs due to their distinct attributes, rendering them suitable for diverse applications including biosensing, bioimaging, and drug delivery. These materials offer numerous benefits, such as expansive surface area, straightforward synthesis, robust stability, magnetic properties, and the capacity for functionalization via several methods including chemical and physical adsorption, creation of oxygen vacancies, and doping. These characteristics establish 2D MOs as a promising candidate for various biological applications. Nonetheless, prior to their integration into biomedical applications, it is essential to thoroughly assess their biosafety and biocompatibility. This evaluation encompasses the study of interactions between 2D MOs and biological systems, encompassing both cellular and tissue levels. The lack of comprehensive characterization of 2D nanomaterials can result in uncertainties regarding their cellular interactions, posing challenges for biomedical use. Consequently, it is imperative to conduct detailed physicochemical characterizations of these materials, accompanied by exhaustive toxicological assessments utilizing diverse in vitro cell lines and in vivo studies. Such thorough evaluations are vital to enhance our comprehension of the safety and effectiveness of these materials, thereby facilitating their broader application in the biomedical domain.

References

Altamirano-Lozano, M., Valverde, M., Alvarez-Barrera, L., Molina, B., & Rojas, E. (1999). Genotoxic studies of vanadium pentoxide (V2O5) in male mice. II. Effects in several mouse tissues. *Teratogenesis, Carcinogenesis, and Mutagenesis, 19*(4), 243–255. https://doi.org/10.1002/(SICI)1520-6866(1999)19:4<243::AID-TCM1>3.0.CO;2-J

Anh Tran, T., Krishnamoorthy, K., Song, Y. W., Cho, S. K., & Kim, S. J. (2014). Toxicity of nano molybdenum trioxide toward invasive breast cancer cells. *ACS Applied Materials & Interfaces, 6*(4), 2980–2986. https://doi.org/10.1021/am405586d

Apte, A., Mozaffari, K., Samghabadi, F. S., Hachtel, J. A., Chang, L., Susarla, S., Idrobo, J. C., Moore, D. C., Glavin, N. R., Litvinov, D., Sharma, P., Puthirath, A. B., & Ajayan, P. M. (2020). 2D electrets of ultrathin MoO_2 with apparent piezoelectricity. *Advanced Materials, 32*(24), 2000006. https://doi.org/10.1002/adma.202000006

Bai, H., Li, X., Hu, C., Zhang, X., Li, J., Yan, Y., & Xi, G. (2013). Large-scale, three–dimensional, free–standing and mesoporous metal oxide networks for high–performance photocatalysis. *Scientific Reports, 3*(1), 2204. https://doi.org/10.1038/srep02204

Bai, S., Zhang, N., Gao, C., & Xiong, Y. (2018). Defect engineering in photocatalytic materials. *Nano Energy, 53*, 296–336. https://doi.org/10.1016/j.nanoen.2018.08.058

Cao, Y., Wu, T., Dai, W., Dong, H., & Zhang, X. (2019). TiO_2 nanosheets with the Au nanocrystal-decorated edge for mitochondria-targeting enhanced Sonodynamic therapy. *Chemistry of Materials, 31*(21), 9105–9114. https://doi.org/10.1021/acs.chemmater.9b03430

Careta, O., Fornell, J., Pellicer, E., Ibañez, E., Blanquer, A., Esteve, J., Sort, J., Murillo, G., & Nogués, C. (2021). ZnO nanosheet-coated TiZrPdSiNb alloy as a piezoelectric hybrid material for self-stimulating orthopedic implants. *Biomedicine, 9*(4), 352. https://doi.org/10.3390/biomedicines9040352

Chen, Y., Ye, D., Wu, M., Chen, H., Zhang, L., Shi, J., & Wang, L. (2014). Break-up of two-dimensional MnO_2 nanosheets promotes ultrasensitive pH-triggered theranostics of cancer. *Advanced Materials, 26*(41), 7019–7026. https://doi.org/10.1002/adma.201402572

Chen, Y., Fan, Z., Zhang, Z., Niu, W., Li, C., Yang, N., Chen, B., & Zhang, H. (2018). 2DMetal nanomaterials: Synthesis, properties, and applications. *Chemical Reviews, 118*(13), 6409–6455. https://doi.org/10.1021/acs.chemrev.7b00727

Chireh, M., Karam, Z. M., Naseri, M., Jafarinejad-Farsangi, S., & Ghaedamini, H. (2022). Synthesis, characterization and cytotoxicity study of graphene/doped ZnO/SiO2 nanocomposites. *Applied Physics A, 128*(4), 307. https://doi.org/10.1007/s00339-022-05367-6

Crowley, K., Ye, G., He, R., Abbasi, K., & Gao, X. P. A. (2018). α-MoO_3 as a conductive 2D oxide: Tunable n-type electrical transport via oxygen vacancy and fluorine doping. *ACS Applied Nano Materials, 1*(11), 6407–6413. https://doi.org/10.1021/acsanm.8b01600

Deng, R., Xie, X., Vendrell, M., Chang, Y.-T., & Liu, X. (2011). Intracellular glutathione detection using MnO_2 -Nanosheet-modified upconversion nanoparticles. *Journal of the American Chemical Society, 133*(50), 20168–20171. https://doi.org/10.1021/ja2100774

Dhenadhayalan, N., Yadav, K., Sriram, M. I., Lee, H.-L., & Lin, K.-C. (2017). Ultra-sensitive DNA sensing of a prostate-specific antigen based on 2D nanosheets in live cells. *Nanoscale, 9*(33), 12087–12095. https://doi.org/10.1039/c7nr03431h

Długosz, O., Szostak, K., Krupiński, M., & Banach, M. (2021). Synthesis of Fe3O4/ZnO nanoparticles and their application for the photodegradation of anionic and cationic dyes. *International journal of Environmental Science and Technology, 18*(3), 561–574. https://doi.org/10.1007/s13762-020-02852-4

Fan, W., Bu, W., Shen, B., He, Q., Cui, Z., Liu, Y., Zheng, X., Zhao, K., & Shi, J. (2015a). Intelligent MnO_2 nanosheets anchored with upconversion nanoprobes for concurrent pH-/H_2 O_2 -responsive UCL imaging and oxygen-elevated synergetic therapy. *Advanced Materials, 27*(28), 4155–4161. https://doi.org/10.1002/adma.201405141

Fan, H., Zhao, Z., Yan, G., Zhang, X., Yang, C., Meng, H., Chen, Z., Liu, H., & Tan, W. (2015b). A smart DNAzyme-MnO_2 Nanosystem for efficient gene silencing. *Angewandte Chemie, 127*(16), 4883–4887. https://doi.org/10.1002/ange.201411417

Fu, D., Ding, X., Wu, J., Li, C., Wang, Q., & Jiang, J. (2018). Cationic polyelectrolyte mediated synthesis of MnO_2 -based core-shell structures as activatable MRI theranostic platform for tumor cell ablation. *Particle & Particle Systems Characterization, 35*(7), 1800078. https://doi.org/10.1002/ppsc.201800078

Gao, F., Yang, X., Luo, X., Xue, X., Qian, C., & Sun, M. (2020). Photoactivated nanosheets accelerate nucleus access of Cisplatin for drug-resistant cancer therapy. *Advanced Functional Materials, 30*(49), 2001546. https://doi.org/10.1002/adfm.202001546

Gong, F., Chen, J., Han, X., Zhao, J., Wang, M., Feng, L., Li, Y., Liu, Z., & Cheng, L. (2018). Core–shell TaOx@MnO_2 nanoparticles as a nano-radiosensitizer for effective cancer radiotherapy. *Journal of Materials Chemistry B, 6*(15), 2250–2257. https://doi.org/10.1039/C8TB00070K

Grisolia, M. N., Varignon, J., Sanchez-Santolino, G., Arora, A., Valencia, S., Varela, M., Abrudan, R., Weschke, E., Schierle, E., Rault, J. E., Rueff, J.-P., Barthélémy, A., Santamaria, J., & Bibes, M. (2016). Hybridization-controlled charge transfer and induced magnetism at correlated oxide interfaces. *Nature Physics, 12*(5), 484–492. https://doi.org/10.1038/nphys3627

Haine, A. T., & Niidome, T. (2017). Gold nanorods as nanodevices for bioimaging, photothermal therapeutics, and drug delivery. *Chemical and Pharmaceutical Bulletin, 65*(7), 625–628.

Han, Z. J., Murdock, A. T., Seo, D. H., & Bendavid, A. (2018). Recent progress in plasma-assisted synthesis and modification of 2D materials. *2D Materials, 5*(3), 032002. https://doi.org/10.1088/2053-1583/aabb81

Han, L., Zhao, X., Huang, T., Ding, H., & Wu, C. (2019). Comprehensive study of phase-sensitive SPR sensor based on metal–ITO hybrid multilayer. *Plasmonics, 14*(6), 1743–1750. https://doi.org/10.1007/s11468-019-00968-z

Haque, F., Daeneke, T., Kalantar-zadeh, K., & Ou, J. Z. (2018). Two-dimensional transition metal oxide and chalcogenide-based photocatalysts. *Nano-Micro Letters, 10*(2), 23. https://doi.org/10.1007/s40820-017-0176-y

Haque, F., Zavabeti, A., Zhang, B. Y., Datta, R. S., Yin, Y., Yi, Z., Wang, Y., Mahmood, N., Pillai, N., Syed, N., Khan, H., Jannat, A., Wang, N., Medhekar, N., Kalantar-zadeh, K., & Ou, J. Z. (2019). Ordered intracrystalline pores in planar molybdenum oxide for enhanced alkaline hydrogen evolution. *Journal of Materials Chemistry A, 7*(1), 257–268. https://doi.org/10.1039/C8TA08330D

Iqbal, P., Preece, J. A., & Mendes, P. M. (2012). Nanotechnology: The "top-down" and "bottom-up" approaches. In *Supramolecular chemistry*. Wiley. https://doi.org/10.1002/9780470661345.smc195

Jevapatarakul, D., T-Thienprasert, J., Payungporn, S., Chavalit, T., Khamwut, A., & T-Thienprasert, N. P. (2020). Utilization of Cratoxylum formosum crude extract for synthesis of ZnO nanosheets: Characterization, biological activities and effects on gene expression of nonmelanoma skin cancer cell. *Biomedicine & Pharmacotherapy, 130*, 110552. https://doi.org/10.1016/j.biopha.2020.110552

Jia, Y., Yi, X., Li, Z., Zhang, L., Yu, B., Zhang, J., Wang, X., & Jia, X. (2020). Recent advance in biosensing applications based on two-dimensional transition metal oxide nanomaterials. *Talanta, 219*, 121308. https://doi.org/10.1016/j.talanta.2020.121308

Jiang, H., Wang, H., & Wang, X. (2011). Facile and mild preparation of fluorescent ZnO nanosheets and their bioimaging applications. *Applied Surface Science, 257*(15), 6991–6995. https://doi.org/10.1016/j.apsusc.2011.03.053

Jiang, J., Li, N., Zou, J., Zhou, X., Eda, G., Zhang, Q., Zhang, H., Li, L.-J., Zhai, T., & Wee, A. T. S. (2019). Synergistic additive-mediated CVD growth and chemical modification of 2D materials. *Chemical Society Reviews, 48*(17), 4639–4654. https://doi.org/10.1039/C9CS00348G

Kalantar-zadeh, K., Ou, J. Z., Daeneke, T., Strano, M. S., Pumera, M., & Gras, S. L. (2015). Two-dimensional transition metal dichalcogenides in biosystems. *Advanced Functional Materials, 25*(32), 5086–5099. https://doi.org/10.1002/adfm.201500891

Kalantar-zadeh, K., Ou, J. Z., Daeneke, T., Mitchell, A., Sasaki, T., & Fuhrer, M. S. (2016). Two dimensional and layered transition metal oxides. *Applied Materials Today, 5*, 73–89. https://doi.org/10.1016/j.apmt.2016.09.012

Kasinathan, K., Samayanan, S., Marimuthu, K., & Yim, J.-H. (2022). Facile synthesis of 2d chitosan based nanocomposite for in vitro biomedical and photocatalytic applications. *SSRN Electronic Journal*. https://doi.org/10.2139/ssrn.4211155

Khademi, A., Azimirad, R., Zavarian, A. A., & Moshfegh, A. Z. (2009). Growth and field emission study of molybdenum oxide nanostars. *The Journal of Physical Chemistry C, 113*(44), 19298–19304. https://doi.org/10.1021/jp9056237

Kim, S., Ahn, S. M., Lee, J.-S., Kim, T. S., & Min, D.-H. (2017). Functional manganese dioxide nanosheet for targeted photodynamic therapy and bioimaging *in vitro* and *in vivo*. *2D Materials, 4*(2), 025069. https://doi.org/10.1088/2053-1583/aa652f

Kolahalam, L. A., Kasi Viswanath, I. V., Diwakar, B. S., Govindh, B., Reddy, V., & Murthy, Y. L. N. (2019). Review on nanomaterials: Synthesis and applications. *Materials Today: Proceedings, 18*, 2182–2190. https://doi.org/10.1016/j.matpr.2019.07.371

Kumar, A. M., Khan, A., Hussein, M. A., Khan, M. Y., Dafalla, H., Suresh, B., & Ramakrishna, S. (2022). Hybrid nanocomposite coatings from PEDOT and BN-TiO2 nanosheets: Enhanced invitro corrosion resistance, wettability and biocompatibility for biomedical applications. *Progress in Organic Coatings, 170*, 106946. https://doi.org/10.1016/j.porgcoat.2022.106946

Liu, G., Wang, L., Sun, C., Chen, Z., Yan, X., Cheng, L., Cheng, H.-M., Lu, G. Q., & (Max). (2009). Nitrogen-doped titania nanosheets towards visible light response. *Chemical Communications, 11*, 1383. https://doi.org/10.1039/b820483g

Liu, K., Zhang, W., Lei, F., Liang, L., Gu, B., Sun, Y., Ye, B., Ni, W., & Xie, Y. (2016). Nitrogen-doping induced oxygen divacancies in freestanding molybdenum trioxide single-layers boosting electrocatalytic hydrogen evolution. *Nano Energy, 30*, 810–817. https://doi.org/10.1016/j.nanoen.2016.09.015

Liu, S., Jin, M., Chen, Y., Gao, H., Shi, X., Cheng, W., Ren, L., & Wang, Y. (2017a). High internal phase emulsions stabilised by supramolecular cellulose nanocrystals and their application as cell-adhesive macroporous hydrogel monoliths. *Journal of Materials Chemistry B, 5*(14), 2671–2678. https://doi.org/10.1039/C7TB00145B

Liu, J., Meng, L., Fei, Z., Dyson, P. J., Jing, X., & Liu, X. (2017b). MnO2 nanosheets as an artificial enzyme to mimic oxidase for rapid and sensitive detection of glutathione. *Biosensors and Bioelectronics, 90*, 69–74. https://doi.org/10.1016/j.bios.2016.11.046

Liu, Z., Zhang, S., Lin, H., Zhao, M., Yao, H., Zhang, L., Peng, W., & Chen, Y. (2018). Theranostic 2D ultrathin MnO2 nanosheets with fast responsibility to endogenous tumor microenvironment and exogenous NIR irradiation. *Biomaterials, 155*, 54–63. https://doi.org/10.1016/j.biomaterials.2017.11.015

Liz-Marzán, L. M., & Grzelczak, M. (2017). Growing anisotropic crystals at the nanoscale. *Science, 356*(6343), 1120–1121. https://doi.org/10.1126/science.aam8774

Maldonado-Camargo, L., Unni, M., & Rinaldi, C. (2017). *Magnetic characterization of iron oxide nanoparticles for biomedical applications* (pp. 47–71). https://doi.org/10.1007/978-1-4939-6840-4_4

Nagyné-Kovács, T., Studnicka, L., Lukács, I. E., László, K., Pasierb, P., Szilágyi, I. M., & Pokol, G. (2020). Hydrothermal synthesis and gas sensing of monoclinic MoO3 Nanosheets. *Nanomaterials, 10*(5), 891. https://doi.org/10.3390/nano10050891

Nilius, N., & Freund, H.-J. (2015). Activating nonreducible oxides via doping. *Accounts of Chemical Research, 48*(5), 1532–1539. https://doi.org/10.1021/acs.accounts.5b00018

Ou, G., Xu, Y., Wen, B., Lin, R., Ge, B., Tang, Y., Liang, Y., Yang, C., Huang, K., Zu, D., Yu, R., Chen, W., Li, J., Wu, H., Liu, L.-M., & Li, Y. (2018). Tuning defects in oxides at room temperature by lithium reduction. *Nature Communications, 9*(1), 1302. https://doi.org/10.1038/s41467-018-03765-0

Pandey, A. M., Agalave, S. G., Vinod, C. P., & Gnanaprakasam, B. (2019). $MnO_2@Fe_3O_4$ magnetic nanoparticles as efficient and recyclable heterogeneous catalyst for Benzylic $sp^3C{-}H$ oxidation. *Chemistry – An Asian Journal, 14*(19), 3414–3423. https://doi.org/10.1002/asia.201900810

Pandey, R. P., Mukherjee, R., Priyadarshini, A., Gupta, A., Vibhuti, A., Leal, E., ... & Lyu, X. (2021). Potential of nanoparticles encapsulated drugs for possible inhibition of the antimicrobial resistance development. *Biomedicine & Pharmacotherapy, 141*, 111943.

Puthirath Balan, A., Radhakrishnan, S., Woellner, C. F., Sinha, S. K., Deng, L., de los Reyes, C., Rao, B. M., Paulose, M., Neupane, R., Apte, A., Kochat, V., Vajtai, R., Harutyunyan, A. R., Chu, C.-W., Costin, G., Galvao, D. S., Martí, A. A., van Aken, P. A., Varghese, O. K., … Ajayan, P. M. (2018). Exfoliation of a non-van der Waals material from iron ore hematite. *Nature Nanotechnology, 13*(7), 602–609. https://doi.org/10.1038/s41565-018-0134-y

Radha, S. K., Crowley, K., Holler, B. A., Gao, X. P. A., Lambrecht, W. R. L., Volkova, H., Berger, M.-H., Pentzer, E., Pachuta, K. G., & Sehirlioglu, A. (2021). Ultrathin 2D-oxides: A perspective on fabrication, structure, defect, transport, electron, and phonon properties. *Journal of Applied Physics, 129*(22). https://doi.org/10.1063/5.0051093

Raha, S., & Ahmaruzzaman, M. (2021). Novel magnetically retrievable In2O3/MoS2/Fe3O4 nanocomposite materials for enhanced photocatalytic performance. *Scientific Reports, 11*(1), 6379. https://doi.org/10.1038/s41598-021-85532-8

Rajendran, K., Gajendiran, M., Kim, S., Kim, K., & Balasubramanian, S. (2018). Synthesis and characterization of biocompatible zinc oxide nanorod doped-titanium dioxide nanosheet. *Journal of Industrial and Engineering Chemistry, 57*, 387–395. https://doi.org/10.1016/j.jiec.2017.08.047

Rana, A. K., & Hussain, A. A. (2022). *An introduction to the wonder 2D nanomaterials: Synthetic approaches and fundamental properties* (pp. 1–24). https://doi.org/10.1007/978-981-16-8538-5_1

Ren, B., Wang, Y., & Ou, J. Z. (2020). Engineering two-dimensional metal oxides *via* surface functionalization for biological applications. *Journal of Materials Chemistry B, 8*(6), 1108–1127. https://doi.org/10.1039/C9TB02423A

Saliev, T. (2019). The advances in biomedical applications of carbon nanotubes. *C, 5*(2), 29.

Segal, E., & Bussi, Y. (2022). Semiconducting silicon nanowires and nanowire composites for biosensing and therapy. In *Semiconducting silicon nanowires for biomedical applications* (pp. 363–378). Woodhead Publishing.

Solanki, P. R., Kaushik, A., Agrawal, V. V., & Malhotra, B. D. (2011). Nanostructured metal oxide-based biosensors. *NPG Asia Materials, 3*(1), 17–24. https://doi.org/10.1038/asiamat.2010.137

Song, G., Hao, J., Liang, C., Liu, T., Gao, M., Cheng, L., Hu, J., & Liu, Z. (2016). Degradable molybdenum oxide Nanosheets with rapid clearance and efficient tumor homing capabilities as a therapeutic nanoplatform. *Angewandte Chemie International Edition, 55*(6), 2122–2126. https://doi.org/10.1002/anie.201510597

Stamenkovic, V., Mun, B. S., Mayrhofer, K. J. J., Ross, P. N., Markovic, N. M., Rossmeisl, J., Greeley, J., & Nørskov, J. K. (2006). Changing the activity of electrocatalysts for oxygen reduction by tuning the surface electronic structure. *Angewandte Chemie International Edition, 45*(18), 2897–2901. https://doi.org/10.1002/anie.200504386

Sun, C., Liu, Y., Zhou, R., Yao, L., Wang, R., Zang, W., & Meng, W. (2019). Nano Pd-decorated manganese dioxide nanosheets for effective photothermal chemotherapy. *ACS Applied Bio Materials, 2*(11), 4747–4755. https://doi.org/10.1021/acsabm.9b00490

Tiwari, J. N., Tiwari, R. N., & Kim, K. S. (2012). Zero-dimensional, one-dimensional, two-dimensional and three-dimensional nanostructured materials for advanced electrochemical energy devices. *Progress in Materials Science, 57*(4), 724–803. https://doi.org/10.1016/j.pmatsci.2011.08.003

Vimala, K., Shanthi, K., Sundarraj, S., & Kannan, S. (2017). Synergistic effect of chemo-photothermal for breast cancer therapy using folic acid (FA) modified zinc oxide nanosheet. *Journal of Colloid and Interface Science, 488*, 92–108. https://doi.org/10.1016/j.jcis.2016.10.067

Wahab, R., Kaushik, N. K., Verma, A. K., Mishra, A., Hwang, I. H., Yang, Y.-B., Shin, H.-S., & Kim, Y.-S. (2011). Fabrication and growth mechanism of ZnO nanostructures and their cytotoxic effect on human brain tumor U87, cervical cancer HeLa, and normal HEK cells. *JBIC Journal of Biological Inorganic Chemistry, 16*(3), 431–442. https://doi.org/10.1007/s00775-010-0740-0

Wang, G., Ling, Y., Wang, H., Yang, X., Wang, C., Zhang, J. Z., & Li, Y. (2012). Hydrogen-treated WO3 nanoflakes show enhanced photostability. *Energy & Environmental Science, 5*(3), 6180. https://doi.org/10.1039/c2ee03158b

Wang, Y., Zhao, Q., Hu, Y., Sun, L., Bai, L., Jiang, T., & Wang, S. (2013). Ordered nanoporous silica as carriers for improved delivery of water insoluble drugs: A comparative study between three dimensional and two dimensional macroporous silica. *International Journal of Nanomedicine, 4015*. https://doi.org/10.2147/IJN.S52605

Wang, S., Yang, X., Zhou, L., Li, J., & Chen, H. (2020). 2D nanostructures beyond graphene: Preparation, biocompatibility and biodegradation behaviors. *Journal of Materials Chemistry B, 8*(15), 2974–2989. https://doi.org/10.1039/C9TB02845E

Xiao, X., Wang, H., Urbankowski, P., & Gogotsi, Y. (2018). Topochemical synthesis of 2D materials. *Chemical Society Reviews, 47*(23), 8744–8765. https://doi.org/10.1039/C8CS00649K

Xie, H., Li, Z., Cheng, L., Haidry, A. A., Tao, J., Xu, Y., Xu, K., & Ou, J. Z. (2022). Recent advances in the fabrication of 2D metal oxides. *IScience, 25*(1), 103598. https://doi.org/10.1016/j.isci.2021.103598

Xu, G., Ding, B., Shen, L., Nie, P., Han, J., & Zhang, X. (2013). Sulfur embedded in metal organic framework-derived hierarchically porous carbon nanoplates for high performance lithium–sulfur battery. *Journal of Materials Chemistry A, 1*(14), 4490. https://doi.org/10.1039/c3ta00004d

Xu, K., Wang, Y., Zhao, Y., & Chai, Y. (2017). Modulation doping of transition metal dichalcogenide/oxide heterostructures. *Journal of Materials Chemistry C, 5*(2), 376–381. https://doi.org/10.1039/C6TC04640A

Yan, P., Ji, L., Liu, X., Guan, Q., Guo, J., Shen, Y., Zhang, H., Wei, W., Cui, X., & Xu, Q. (2021). 2D amorphous-MoO3−x@Ti3C2-MXene non-van der Waals heterostructures as anode materials for lithium-ion batteries. *Nano Energy, 86*, 106139. https://doi.org/10.1016/j.nanoen.2021.106139

Yang, B., Chen, Y., & Shi, J. (2018). Material chemistry of two-dimensional inorganic nanosheets in cancer theranostics. *Chem, 4*(6), 1284–1313. https://doi.org/10.1016/j.chempr.2018.02.012

Yu, N., Hu, Y., Wang, X., Liu, G., Wang, Z., Liu, Z., Tian, Q., Zhu, M., Shi, X., & Chen, Z. (2017). Dynamically tuning near-infrared-induced photothermal performances of TiO_2 nanocrystals by Nb doping for imaging-guided photothermal therapy of tumors. *Nanoscale, 9*(26), 9148–9159. https://doi.org/10.1039/C7NR02180A

Zeng, X., Luo, M., Liu, G., Wang, X., Tao, W., Lin, Y., Ji, X., Nie, L., & Mei, L. (2018). Polydopamine-modified black phosphorous nanocapsule with enhanced stability and Photothermal performance for tumor multimodal treatments. *Advanced Science, 5*(10), 1800510. https://doi.org/10.1002/advs.201800510

Zhang, Z.-Q., & Song, S.-C. (2016). Thermosensitive/superparamagnetic iron oxide nanoparticle-loaded nanocapsule hydrogels for multiple cancer hyperthermia. *Biomaterials, 106*, 13–23. https://doi.org/10.1016/j.biomaterials.2016.08.015

Zhang, X.-L., Zheng, C., Guo, S.-S., Li, J., Yang, H.-H., & Chen, G. (2014). Turn-on fluorescence sensor for intracellular imaging of glutathione using g-C_3N_4 Nanosheet–MnO_2 Sandwich Nanocomposite. *Analytical Chemistry, 86*(7), 3426–3434. https://doi.org/10.1021/ac500336f

Zhang, B. Y., Zavabeti, A., Chrimes, A. F., Haque, F., O'Dell, L. A., Khan, H., Syed, N., Datta, R., Wang, Y., Chesman, A. S. R., Daeneke, T., Kalantar-zadeh, K., & Ou, J. Z. (2018). Degenerately hydrogen doped molybdenum oxide nanodisks for ultrasensitive plasmonic biosensing. *Advanced Functional Materials, 28*(11), 1706006. https://doi.org/10.1002/adfm.201706006

Zhang, L., Tong, S., Zhang, Q., & Bao, G. (2020). Lipid-encapsulated Fe_3O_4 nanoparticles for multimodal magnetic resonance/fluorescence imaging. *ACS Applied Nano Materials, 3*(7), 6785–6797. https://doi.org/10.1021/acsanm.0c01193

Application of a Novel Nanotherapeutic Strategy in Ayurvedic Treatment

Sumanta Bhattacharya

1 Introduction

Herbal medicines played an important role in the healthcare services of the ancient world. Human civilization has survived for a long time due to the immersive medicinal properties of plant extracts. However, the emergence of allopathy reduced its use in the twentieth century A.D. due to the faster rate of healing of the chemically synthesised allopathic drug molecules. However, it induces side effects and reduces the body"s self-immunisation capability due to its target-specific activities (Singh et al., 2021). The increasing rate of environmental degradation and climate change increases the number of cases of acute diseases and serious genetic disorders. The modern lifestyle also has a serious impact on the mental health of humans. The existing medicinal resources are unable to resolve the issues permanently. The modern nuclear lifestyle is increasing the number of cases of stress and depression among people of all ages, from children to the elderly. Moreover, the changing nature of climate change, environmental degradation, and altered food habits increase the outbreak of serious health issues. The recent outbreak of COVID-19 exhibits the threat of the changing nature of infectious diseases. The inclination towards processed foods increases the risk of both overweight and malnutrition (Beslay et al., 2020). The increase in the therapeutic efficiency of medicinal products is necessary for the survival of the present as well as future generations under these changing circumstances.

Nanotechnology will play an important role in increasing the therapeutic efficiency of herbal extracts and expanding their application capacity in the

S. Bhattacharya (✉)
Maulana Abul Kalam Azad University of Technology, Kolkata, West Bengal, India

S. A. Bansal et al. (eds.), *Emerging Applications of Novel Nanoparticles*, Lecture Notes in Nanoscale Science and Technology 37, https://doi.org/10.1007/978-3-031-57843-4_8

human body. The bioactive components are unable to react effectively with the cellular components due to their lower solubility in biological fluids. Nanoparticles have improved absorption capacity and exhibit more chemical reactivity than macrocompounds (Valikarimwala et al., 2021). Moreover, in ancient times, ashes, or bhasma, were used as an important medicinal component, which is basically a nanomaterial (Yadav & Chaudhary, 2019). The nanoadsorbants lead to the effective adsorption of drug molecules, exhibiting more therapeutic activities due to an increase in bioavailability. The lower in vivo efficacy of phytomedicines is due to their poor permeability, low systematic availability, instability, extensive first-pass metabolism, and low aqueous solubility. The interaction of herbal drugs with nanoparticles increases their bioavailability in the human body and improves their potential for delivery of drug molecules to the target. Nanoparticles are widely used as carriers of drug molecules because of their increased chemical activity and ability to cross tissue barriers. Modern nanotechnology enables the development of different types of clinically active ashes. Nanomaterials improve the bioavailability and cellular uptake capacity of herbal drugs. It also helps to cross the blood-brain barrier effectively and deliver the bioactive components to the intracellular region. Nanocarriers like liposomes, polymeric nanoparticles, quantum dots, dendrimers, carbon nanotubes, etc. can effectively carry herbal bioactive components like flavonoids, terpenoids, tannins, etc. due to improved bioavailability and cellular uptake capacity. It improves the therapeutic action of herbal drugs. The use of nanomaterials increases the chemical reactivity of herbal drugs with the affected cells. It increases the therapeutic efficiency of acute diseases like cancer. It also helps in gene therapy, which leads to the effective treatment of serious genetic disorders. The nano-based neurophytomedicines help to cure serious neurological disorders effectively. The nanovaccines exhibit improved efficiency in the treatment of viral infections (Azharuddin et al., 2022). The importance of nanophytomedicines also increases in the pharmaceutical industry due to their increasing capacity to improve therapeutic efficiency (Hesari et al., 2021). The immunological characteristics of the nanoparticles derived from plant extracts have potential applications in the pharmaceutical industry. The use of biologically synthesised nanomaterials increases the therapeutic activity of existing pharmaceutical products. The use of nanotextiles in the preparation of smart bandages increases the rate of recovery of wounds and fractures. The use of nanobiomarkers and nanobiosensors facilitates the effective monitoring of drug molecules. The market share of nanomedicines is increasing exponentially every year due to their increased therapeutic efficiency. The integration of nanotechnology also increases the capacity for extracting bioactive components from plant-based waste products, which also encourages the process of waste management. The nanovaccines exhibit increased efficiency due to the changing nature of infectious diseases.

2 Application of Nanotherapy in Herbal Medicine

A nanotherapeutic strategy improves the pharmacokinetic activities of the drug molecules by increasing their bioavailability and biodistribution. Figure 1 depicts the application of nanotherapeutic strategies in different sectors of medical science viz. improvement of the bioavailability of drugs, ensure effective drug delivery etc. Nanotechnology increases the therapeutic action of the drug molecules by improving the drug metabolism, tissue retention, and cellular uptake capacity of the bioactive components present in the drug molecules through precision drug delivery. The coating of nanoparticles modifies the pharmacology of drug molecules by ensuring effective absorption of drugs in intracellular components. The drug molecules are even capable of interacting with the protein molecules, DNA, mRNA, and siRNA molecules that facilitate effective treatment of cancer and genetic diseases. It enables remediation of the root cause of health issues. It reduces the possibility of side effects by optimising drug release, target-oriented distribution, and excretion ability (Hunt et al., 2022). Nanoflares are specially engineered nanomaterials in the field of cancer treatment. It effectively detects the presence of cancer cells in the blood stream. It is able to bind to the genetic target in cancer cells and generate light when

Fig. 1 Application of nanotherapeutic strategy in medical science

the target genetic sequence is found (Chenab et al., 2019). Nanotherapeutic strategies increase the efficiency of cancer treatment through improved drug stability and solubility, enhanced concentration of drugs at the target site, therapeutic payloads into the blood stream via targeted drug delivery with minimum systemic toxicities, prolonged half-lives of drugs in blood plasma, specific targeting of tumours with improved absorption, the ability to encapsulate a range of drugs, etc. The nanocarriers facilitate the effective dissemination of drug molecules and bioactive agents in tumour microenvironments. It also helps in stem cell therapy by enabling active delivery of cancer stem cells to the desired target. Nanotherapy also improves cancer therapy strategies through non-coding RNA-based targeting, exosome-based targeting strategies, self-assembling pro-drug-based strategies, integrin targeting, etc. (Bukhari, 2022). Biomimetic and bioengineered nanoparticles derived from viruses, bacteria, and fungal pathogens are used for the preparation of nanovaccines that proved to be more effective than chemical-based vaccines as a preventive measure against infectious diseases due to increased drug solubility, prolonged circulation time, the capability of overcoming biological carriers, enhanced bioavailability, targeting infection sites, and modulating the drug release profile. The self-assembling characteristics of nanoparticles exhibit improved immunological characteristics in the body. The self-assembling nanoparticles bearing antigens are being used as novel vaccines against viral diseases. Nanomaterials derived from bacterial extracellular vesicles can act as potential sources of vaccines due to their immunogenic characteristics, improved cellular uptake capacity, self-adjuvant capacity, and capability to present exogenous antigens (Yang et al., 2019). Different novel nanocarriers enable them to cross the blood-cerebrospinal fluid barrier effectively, which helps in the treatment of neurodegenerative diseases (Harilal et al., 2019). The self-healing characteristics of the nanomaterials enable wound healing at a faster rate. It accelerates the normal wound healing process by enabling cell type specificity and regulating molecules as well as the pathophysiology of chronic wounds (Hamdan et al., 2017).

The bioactive components present in the herbs are not able to cross the blood-brain barrier due to the size mismatch. Nanocarriers like liposomes, polymeric nanoparticles, quantum dots, dendrimers, carbon nanotubes, etc. can effectively carry herbal bioactive components like flavonoids, terpenoids, tannins, etc. due to improved bioavailability and cellular uptake capacity. It improves the therapeutic action of herbal drugs as shown in Fig. 2 (Bonifaclo et al., 2014). The nanomaterials derived from the herbal extracts obtained from curcumin, quercetin, resveratrol, piperine, etc. increase the bioavailability of the drugs for the treatment of neurodegenerative diseases. Polymeric nanoparticles, nanocapsules, nanospheres, polymeric nanogels, nanosuspensions, carbon nanotubes, nanofibers, polymeric nanomiscelles, polymeric nanoliposomes, and eposomes are some nanomaterials used for the treatment of neurodegenerative diseases (Moradi et al., 2020). Bhasma, or the ashes of metals, act as a potential nanocarrier. Ashes are biologically derived organometallic or organo-mineralic oxides where the metals are converted into nanoparticles during incineration. The toxicity of the metals can be removed by converting the zerovalent metals to metallic ions through oxidation. The absorption

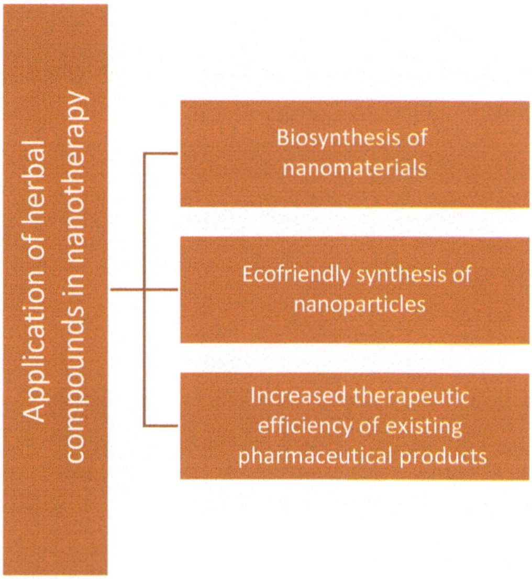

Fig. 2 Application of herbal compounds in nanotherapy

characteristics of the bhasma are higher than those of normal metallic drugs, which improves the bioavailability, bioassimilability, and bioadaptability of the bioactive components. It also increases the drug metabolism characteristics at the intracellular level (Valikarimwala et al., 2021). Figure 2 depicts applications of herbal compounds in nanotherapy.

The nanomaterials act as biomarkers and biosensors that facilitate the detailed analysis of treatment procedures and the pharmacokinetic study of drug molecules. The different sectors for the applications of nanobiosensors is shown in Fig. 2. The combination of nanomaterials, microfluidics, automatic samples, and transduction devices in a single chip acts as an effective biosensor to detect diseases and monitor therapeutic actions. It is a cost-effective, tiny, and user-friendly alternative to electrical and chemical sensors. Its capability for targeted drug delivery and improved pharmacokinetic ability reduce the possibility of side effects from the drugs. The nanobiomarkers effectively capture the chemical, biochemical, optical, and electronic signals of the desired target instead of interacting with them, which reduces the possibility of side effects (Jain, 2009). The nanobiosensors enable noninvasive transportation of clinically active biomarkers in several biofluids such as sweat, tears, and saliva without complex manipulation and treatment like chemical biosensors. The antibody fragment-based nanosensors facilitate the detection of antibodies through antigen-antibody interaction (Yun et al., 2009). Figure 3 depicts different applications of nanobiosensors.

Nanobots, the advanced robots developed with nanotechnology and robotics, act as miniature surgeons when inserted into the human body. It effectively repairs the intracellular structures, resulting in the healing of a particular disease within the human body as shown in Fig. 4. The DNA-based nanobots are also prepared for the eradication of genetic diseases by modifying the target genome sequence.

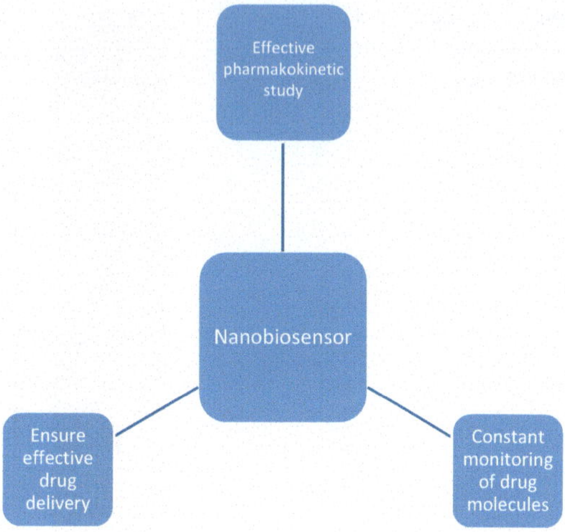

Fig. 3 Application of nanobiosensors

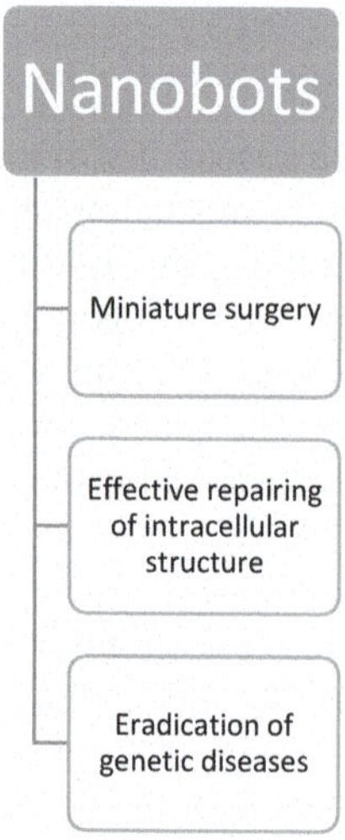

Fig. 4 Application of nanobots

Nanomaterials also act as effective carriers of genetic material in the body (Anjum et al., 2021).

Nanophytomedicine is an emerging branch of medical science that deals with the application of nanotechnology in herbal medicine, phytotherapy, or phytopharmacology as explained in Fig. 5. Nanotechnology plays an important role in the effective targeted and non-targeted delivery of drug molecules in medical science. The nanoadsorbants lead to the effective adsorption of drug molecules, exhibiting more therapeutic activities due to an increase in bioavailability. The lower in vivo efficacy of phytomedicines is due to their poor permeability, low systematic availability, instability, extensive first-pass metabolism, and low aqueous solubility. The interaction of herbal drugs with nanoparticles increases their bioavailability in the human

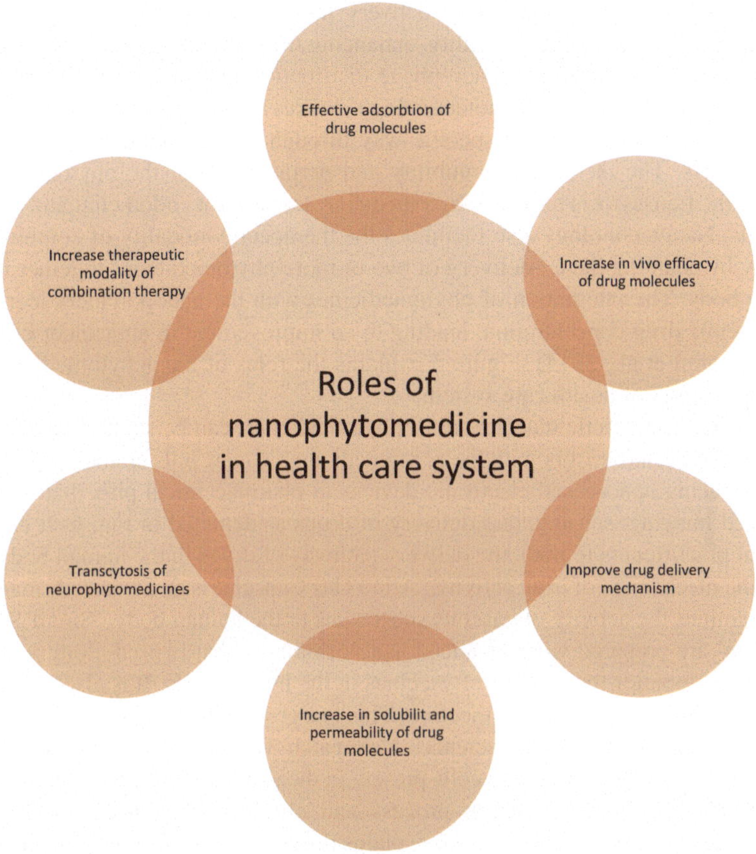

Fig. 5 Roles of nanophytomedicine in health care system

body and improves their potential for delivery of drug molecules to the target. Nanoparticles are widely used as carriers of drug molecules because of their increased chemical activity and ability to cross tissue barriers. Specially engineered nanomaterials are prepared that are able to carry drug molecules, proteins, plasmids, antibodies, oligonucleotides, fluorophores, ligands, polymers, radioisotopes, tissue-engineered products, etc. to the human body and affect desired cells. Dendrimers, liposomes, polymeric nanoparticles, polymeric miscells, carbon nanotubes, mesoporous silicon, quantum dots, nanocrystals, niosomes, nanospheres, phytosomes, etc. are some nanomaterials that participate in drug delivery mechanisms. The aqueous solubility and permeability of the blood-brain barrier of the phytomedicines can be improved by reducing the size of the phytomedicine drug molecules to the nanoscale. The flavonoids and lignins present in the Cucscuta chinensis plant exhibit improved bioavailability and adsorption capability in the human body while being reduced to nanosize by the nanosuspension method. The rate of diffusion of the suspension of nanophytomedicines is greater than that of phytomedicines into the gastrointestinal medium. The drug delivery mechanism of herbal drugs can be increased by increasing the solubility, enhancing the stability, reducing the toxicity effects, improving the tissue macrophage distribution mechanism, and enhancing pharmacological activities. Nanotechnology ensures the targeted delivery of phytomedicines in a cell- or tissue-specific way through the development of effective drug carriers. The increase in solubility and permeability of the phytomedicines leads to the transcytosis of neurophytomedicine across tight endothelial and epithelial cells. Nanotechnology also facilitates the therapeutic modality of combination therapy by enabling the co-delivery of two or more phytomedicines together in the human body. The integration of phytomedicines with the nanopolymers increases the systemic drug concentration, leading to an improvement in anticancer efficacy (Gunasekaran et al., 2014). Figure 5 explains the roles of nanophytomedicines in different sectors of health care system.

The pharmacokinetic mechanism of phytomedicines can be studied through the integration of nanosensors with the phytomedicines. Smart pills are being prepared by introducing nano-based electronic devices to pharmaceutical pills that perform advanced imaging, sensing, and delivery of drugs as depicted in Fig. 6. It enables medical practitioners to track the delivery pathway of drugs in the human body and study the mechanism of drug activity. 'Atmo Gas Capsule' is an advanced smart pill that examines the activity of gaseous substances in the human body. 'Smart Sensor Capsules' are prepared with advanced nanotechnology that is used along with the vaccines to monitor the activity of vaccines in the human body. It is also able to be consumed orally instead of by injection. This is a groundbreaking innovation in the field of research in medical science. Nanopatch vaccines use nanoparticles to deposit the vaccines on immune cells present in the skin and lower the risk of infection. It is an easier vaccination process and eliminates the need for vaccine refrigeration (Idrees, 2015). Figure 6 explains different roles of smart pills in medical science.

Fig. 6 Roles of smart pills in medical science

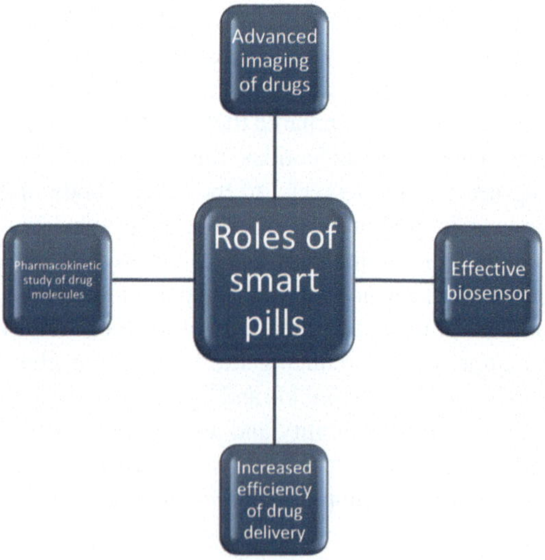

3 Novel Nanotherapeutic Implementation in Herbal Industry

Herbal drugs are gaining more importance than allopathic drugs due to their more effective therapeutic outcomes and low side effects. The application of nanotechnology will enhance the efficiency of herbal drugs by improving the bioavailability of the bioactive components in the intracellular environment. Nanotechnology can be used in every stage of the synthesis of herbal drugs. In the plantation stage, the use of nanofertilizer will facilitate the healthy growth of herbal plants. The application of nanobiotechnology enables the increased production of desired secondary metabolites in herbs. The nanocarriers ensure the targeted as well as non-targeted delivery of biologically incompatible secondary metabolites in the affected cell without producing any side effects. The nanobiomarkers and nanobiosensors facilitate the monitoring of the therapeutic action of the drug. The blending of the nanobiomarker and nanosensor with the herbal extracts facilitates the effective transportation of bioactive components in the targeted cell. Table 1 shows the functions of different types of nanobiosensors. The nanomaterials can also be prepared from herbal extracts, which can be used as a potential drug delivery route. The integration of nanobiosensors and nanocarriers facilitates the sensor-based smart delivery of herbal drugs in the affected cellular environment (Sarangi & Padhi, 2018).

The nanoadsorbants lead to the effective adsorption of drug molecules, exhibiting more therapeutic activities due to an increase in bioavailability. The lower in vivo efficacy of phytomedicines is due to their poor permeability, low systematic availability, instability, extensive first-pass metabolism, and low aqueous solubility. The interaction of herbal drugs with nanoparticles increases their bioavailability in the human body and improves their potential for delivery of drug molecules to the target. Nanoparticles are widely used as carriers of drug molecules because of their increased chemical activity and ability to cross tissue barriers. Specially engineered nanomaterials are prepared that are able to carry drug molecules, proteins, plasmids, antibodies, oligonucleotides, fluorophores, ligands, polymers, radioisotopes, tissue-engineered products, etc. to the human body and affect desired cells. Dendrimers, liposomes, polymeric nanoparticles, polymeric miscells, carbon nanotubes, mesoporous silicon, quantum dots, nanocrystals, nanoisomes, nanospheres, phytosomes,

Table 1 Function of some nanobiosensors

Nanobiosensors	Functions
Enzyme based sensor	Monitoring the activity of different enzymes like glucose, lactate, alcohol, etc.
Tissue based sensor	Detection of minute changes in the receptors when cells bind with on the transducer surface
Immunosensor	Detection of specific biomarker in a biological component
DNA biosensor	Detection of the complimentary sequence by DNA–DNA hybridization
Thermal biosensor	Monitor food metabolism process, enzyme-based physiological activities, fermentation systems

etc. are some nanomaterials that participate in drug delivery mechanisms (Patra et al., 2018). The aqueous solubility and permeability of the blood-brain barrier of the phytomedicines can be improved by reducing the size of the phytomedicine drug molecules to the nanoscale. The flavonoids and lignins present in the Cucscuta chinensis plant exhibit improved bioavailability and adsorption capability in the human body while being reduced to nanosize by the nanosuspension method. The rate of diffusion of the suspension of nanophytomedicines is greater than that of phytomedicines into the gastrointestinal medium. The drug delivery mechanism of herbal drugs can be increased by increasing the solubility, enhancing the stability, reducing the toxicity effects, improving the tissue macrophage distribution mechanism, and enhancing pharmacological activities. Nanotechnology ensures the targeted delivery of phytomedicines in a cell- or tissue-specific way through the development of effective drug carriers. The name of some suitable nanocarriers for some important plant based bioactive components are provided in Table 2. The increase in solubility and permeability of the phytomedicines leads to the transcytosis of the drug molecules across tight endothelial and epithelial cells. Nanotechnology also facilitates the therapeutic modality of combination therapy by enabling the co-delivery of two or more phytomedicines together in the human body. The integration of phytomedicines with the nanopolymers increases the systemic drug concentration, leading to an improvement in anticancer efficacy (Gunasekaran et al., 2014).

The pharmacokinetic mechanism of phytomedicines can be studied through the integration of nanosensors with the phytomedicines. Smart pills are being prepared by introducing nano-based electronic devices to pharmaceutical pills that perform advanced imaging, sensing, and delivery of drugs. It enables medical practitioners to track the delivery pathway of drugs in the human body and study the mechanism of drug activity. 'Atmo Gas Capsule' is an advanced smart pill that examines the activity of gaseous substances in the human body (Netscribes, 2020). 'Smart Sensor Capsules' are prepared with advanced nanotechnology that is used along with the vaccines to monitor the activity of vaccines in the human body (Sutton, 2015). It is

Table 2 Some plant based bioactive components and their corresponding nanocarriers

Name of bioactive components	Source	Suitable nanocarrier
Curcuminoids	Curcuma longa	Curcuminoids solid lipid nanoparticles
Glycyrrhizin acid	Glycyrrhiza glabra	Glycyrrhizin acid loaded nanoparticles
Flavonoids	Camellia sinensis	Nanoparticles of cuscuta chinesis
Taxel	Taxus brevifolia	Taxel loaded nanoparticles
Artemisinin	Artemisia annua	Artemisinin nanocapscules
Camptothecin	Camptotheca acuminata	CPT encapsulated nanoparticles
Berberine	Hydrastis canadensis	Berberine-loaded nanoparticles

also able to be consumed orally instead of by injection. This is a groundbreaking innovation in the field of research in medical science. Nanopatch vaccines use nanoparticles to deposit the vaccines on immune cells present in the skin and lower the risk of infection. It is an easier vaccination process and eliminates the need for vaccine refrigeration (Crichton & Kendall, 2017). Figure 7 explains the use of different nanotherapeutic strategies in enhancing the therapeutic efficiency of herbal medicines.

The integration of herbal bioactive components with nanocapsules increases the in vivo efficacy of the herbal drugs in health. Nanophytochemicals play a significant role in the treatment of neurodegenerative diseases. The crossing of the blood-brain barrier and the blood-cerebrospinal fluid barrier becomes easier with the application of nanocarriers. The phytokinetic study of the neurophytomedicines can be improved through the application of nanosensors, which will facilitate the detailed in vivo study of the mechanism of the neurophytomedicines (Kim et al., 2021). The plant extracts can act as potential chemotherapeutic agents that can be effectively delivered by nanocarriers in the affected cells. In the case of gene therapy, the nanomaterials derived from the herbal extracts facilitate the transfer of clinically engineered

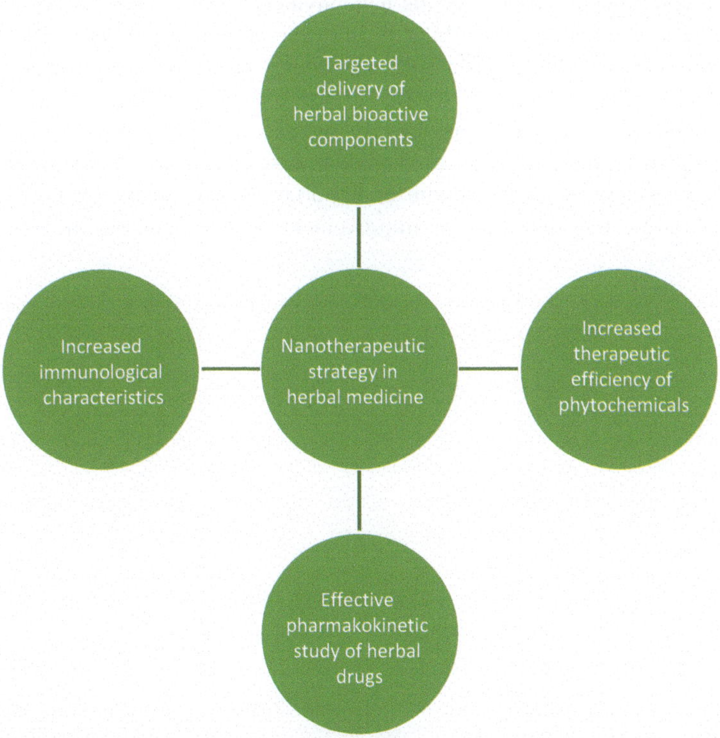

Fig. 7 Nanotherapeutic strategy of herbal medicines

genetic materials. The presence of nanosensors facilitates the easy identification of the affected genes, which mitigates the risk of side effects in gene therapy (Herranz et al., 2011).

The neurohormesis characteristics of several phytochemicals facilitate the preparation of optimum doses for the effective treatment of several neurodegenerative diseases like Alzheimer"s, Parkinson"s, epilepsy, etc. Different roles of nano neurophytomedicines in the treatment of neurodegenerative diseases are explained in Fig. 8. The hormesis effects of the apparently toxic phytochemicals increase immune responses and mitigate the risk of neurological diseases. The application of nanotechnology reinforces the neurohormesis effects of the phytochemicals. The hermetic responses of several phytochemicals like resveratrol, catechin, sulforaphane, hypericin, allicin, etc. induce adaptive stress response signalling pathways, resulting in the amelioration of the resistive potential towards injury and diseases. The hormesis characteristics of the phytochemicals increase the hemodynamic characteristics of the body, leading to the effective management of mental and physical stresses caused by ageing. The neuron cells are more susceptible to damage as they age due to the rapid rate of DNA damage. Neurohormesis enables the repair of damaged DNA, resulting in a decrease in the rate of cell death by radical scavenging and increased antioxidant activities of the nanophytochemicals (Salinaro et al., 2018). Mitohormesis is the process of exhibiting hormesis activities by interacting with the mitochondria, which play a central role in bioenergy production and nutrient metabolism (Ristow & Schmeisser, 2014). Berberine, an alkaloid extracted from Captidis rhizoma and Hydrastis canadensis, directly interacts with mitochondria through the electron transport chain and reduces the oxidative stresses generated in the neuronal cells. The consumption of epicatechin, an important bioactive component belonging to the flavonol group, leads to the improvement of the cognitive function of the

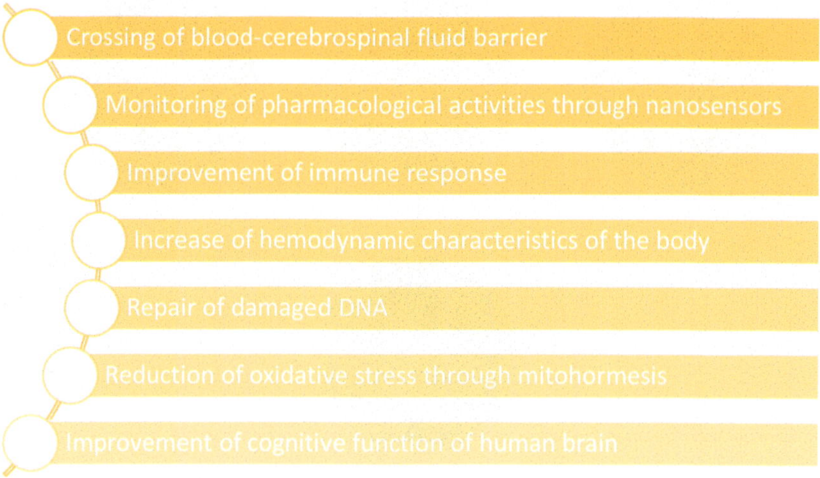

Fig. 8 Roles of nano neurophytomedicines in healthcare sector

human brain by modifying neuronal spine concentration, hippocampal angiogenesis, and memory function (Ngowi et al., 2021).

4 Statistical Analysis of the Use of Nanotechnology in Medical Science

The application of nanotechnology is increasing at a rapid pace from 2005 to 2020 as a drug delivery mechanism, biomarkers, and detection of pathogens in parasitology (Hikal et al., 2021). According to the forecast of Research and Markets, the application of nanotechnology in the medicinal market is expected to grow by 15.3% by 2026 (Research and Markets, 2022). Nanotechnology is mostly used as a drug carrier, in vivo diagnostics, and implants rather than directly as drug molecules. 56%, 16%, and 15% of the use of nanotechnology in medical science is as an effective drug carrier, in vivo diagnostics, and implants, whereas only 4% uses it as a drug directly (Jayaraj et al., 2013). The global market value of nanomedicine has followed an increasing trend from 2006 to 2021 as depicted in Fig. 9 (Business Research Company, 2023).

As per the data released by the Department of Commerce, Government of India, in 2023, the export value of herbal medicines is increasing rapidly, especially after the COVID-19 period. It indicates the increasing inclination of the world towards herbal medicines as depicted in Fig. 10 (The Economic Times, 2023).

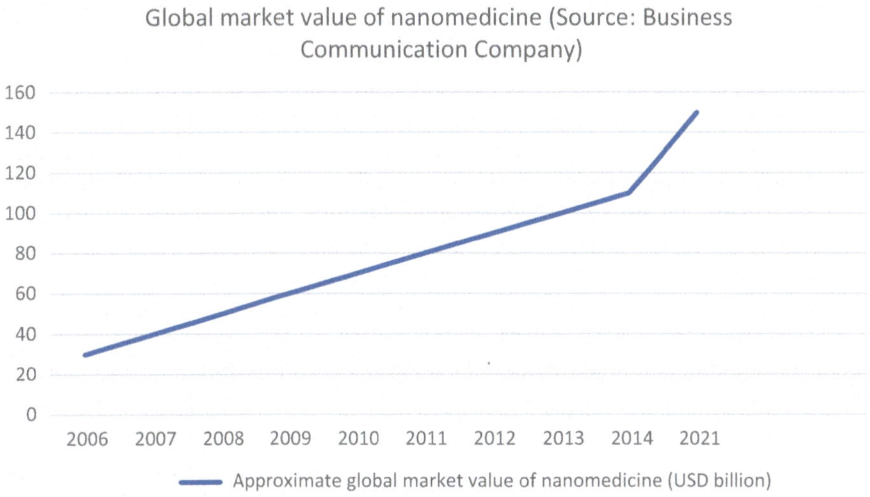

Fig. 9 Global market value of nanomedicine in last decade

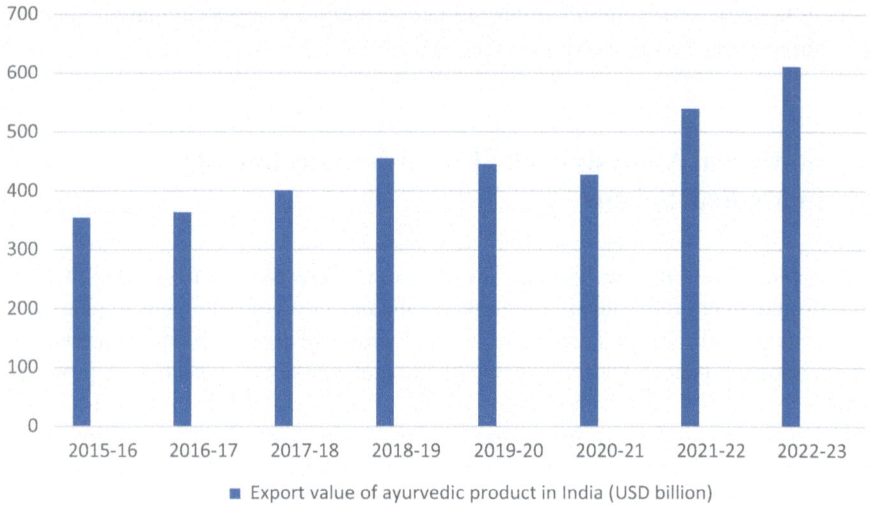

Fig. 10 Export value of ayurvedic product in India

5 Technological Requirement for the Use of Nanotherapy in Herbal Medicine

The use of cutting-edge technology is required to assure the creation of successful herbal nanomedicines. Figure 11 explains different technological advancements in the use of nanotherapeutic strategy in herbal medicine. Effective nanocarrier production requires cutting-edge technological innovation. A cost-effective method for synthesising nanomaterials will require the development of technologies such as vapour phase techniques, atomic or molecular gas condensation, arc discharge generation, laser ablation processes, plasma processes, chemical vapour deposition, sol-gel techniques, solvothermal methods, sonochemical methods, mechanical attrition, nanolithography, template fabrication, etc. To ensure the successful growth of the herbal plants, the use of agricultural biotechnology, microbial ecology, and biofertilizer microbiology is necessary. The use of nanobiotechnology and nanobioremediation ensures that herbal plants continue to grow and develop even in the presence of biotic and abiotic stress. Modern organic farming techniques aid in increasing the fertility of the soil. The soil bacteria"s metagenomic study aids in the production of efficient biofertilizer, which raises the soil fertility rate. The cultivation of herbs is improved by the use of current technology, such as robots, the internet of things, machine learning, data analytics, precision agriculture, drones, and minichromosome technology. The use of sophisticated nanobiosensors makes it possible to analyse the pharmacokinetics of herbal medicines in the body. The creation of smart pills that make it easier to monitor how a pill interacts with body parts is made possible by nanotechnology (Ansari et al., 2012).

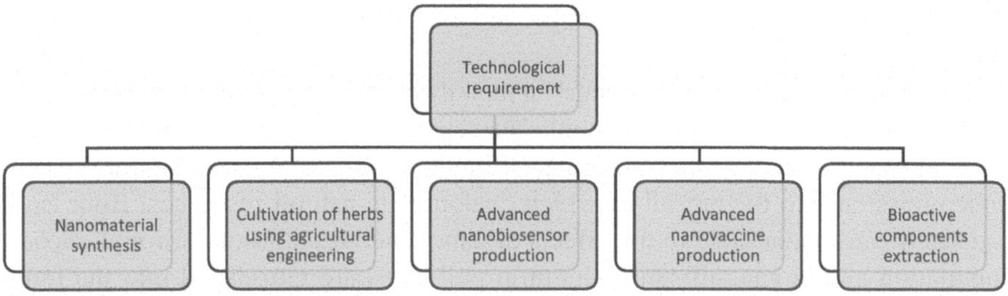

Fig. 11 Technological requirement for the use of nanotherapy in herbal medicine

It is important to make smart nanocarriers that can handle big changes in the pH and temperature of body fluids and that can also hide hydrophobic herbal medicine ingredients. At the targeted site, it increases the bioavailability of herbal bioactive compounds. Since these vaccines are more biocompatible than standard immunisations, the creation of nanovaccines using efficient bioactive materials also requires technological improvement (Kenchegowda et al., 2022).

Technology innovation is also required for the optimal extraction of bioactive components from herbal medicines. The successful extraction of bioactive compounds from the plant source would be facilitated by current qualitative and quantitative analytical methods such as superficial fluid chromatography, Fourier transform infrared spectroscopy, near infrared spectroscopy, gas chromatography, mass spectrometry, etc. (Noore et al., 2021).

6 Environmental Consideration of the Use of Nanotherapy in Herbal Medicine

The internal physiological activities of the herbal plants are what make them effective as a therapy. Secondary metabolites, which are formed when plants employ environmental nutrients and water for metabolism, are used as bioactive components in the manufacture of herbal medicines. Plant species" ability to translocate nutrients and metabolise waste products is disrupted as a result of the growing rate of environmental degradation, which accumulates toxicants in soil and groundwater reserves. The plant"s rate of producing bioactive secondary metabolites slows as a result. It will decrease the availability of essential ingredients used to make herbal remedies. Nanotechnology has the potential to enhance the drug delivery process and boost bioavailability. Overharvesting occurs as a result of an increase in the demand for medicinal plants from the pharmaceutical industry. It contributes to issues including extinction, habitat loss, deforestation, global warming, and invading species competition. Herb biodiversity is declining because to human interference with forests and soil. Multiple studies demonstrate that over 15,000 different types of therapeutic herbs are in danger of being lost forever (García-Quintero & Palencia, 2021).

The sustainability of the nanomaterial manufacturing process in the natural environment must be taken into account. The chemical synthesis of nanomaterials results in the production of wastewater effluent, solid wastes, sewage sludge, and toxic emission into the air, all of which contaminate natural resources. Both land and sea life are vulnerable to the effects of nanomaterial pollution. Through bioaccumulation and biomagnification, harmful pollutants are introduced into the food chain. Ecotoxicity results from an excess of nanomaterial contamination in vegetation and soil. To reduce environmental damage, it is advised that nanomaterials be synthesised in a sustainable manner using biological resources (Gupta et al., 2023).

Herbal medicine production results in pharmaceutical waste that is harmful to the environment. Figure 12 different strategies of environmental friendly synthesis of nanomedicines. The pharmaceutical industry generates a wide variety of byproducts, including unused chemicals, waste water, sample pharmaceuticals, partially filled vials and syringes, contaminated textile items, empty receptacles, medicine distribution devices, and so on. To keep pharmaceutical waste from polluting the environment, a long-term plan for its disposal is necessary (Rajan et al., 2017).

Large-scale medicinal plant cultivation necessitates adherence to organic farming practises. Because of their interference with plant physiochemistry, chemical fertilisers, pesticides, fungicides, herbicides, etc. lower the medicinal efficacy of plant species. Soil nutrient accumulation capability can be enhanced through the use of biofertilizers based on microorganisms and other organic fertiliser to promote plant health. The immunological features of plant species are enhanced through the use of biopesticides and herbicides, which enhances the defence mechanism against disease. Additionally, it stimulates the plant"s increased production of beneficial secondary metabolites (Jiang et al., 2022). Figure 13 depicts different environmen-

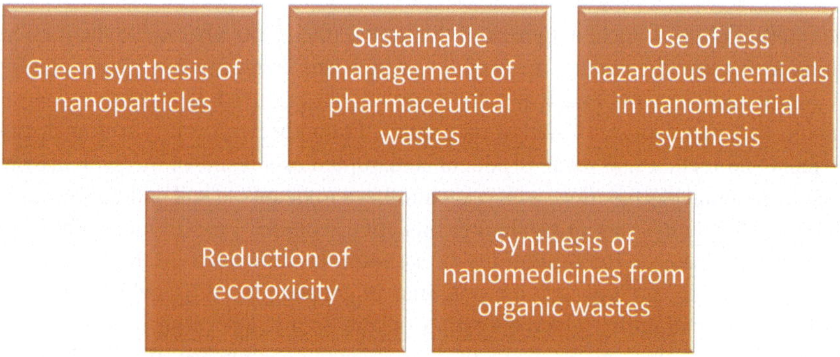

Fig. 12 Environment friendly synthesis of nanomedicines

tal considerations of nanotherapeutic strategy in the field of herbal medicine.

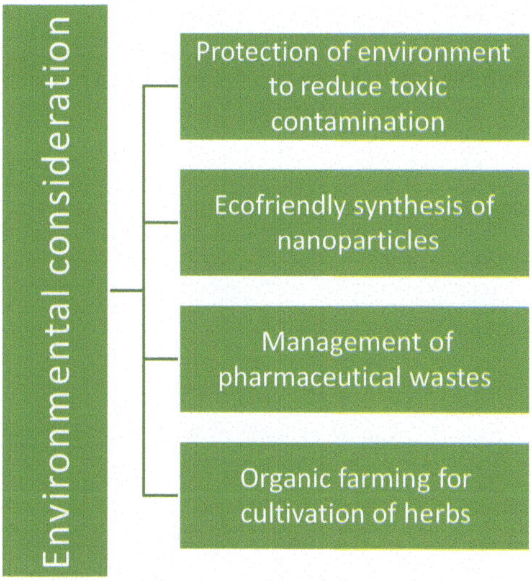

Fig. 13 Environmental consideration of the nanotherapeutic strategy in herbal medicines

7 Economic Consideration of the Use of Application of Nanotherapeutic Strategy in Herbal Medicines

This investment is required in order to begin industrial manufacturing of nanomaterials and nanocarriers of a high quality. Different economic strategies for the successful implementation of nanotherapeutic strategies have been explained in Fig. 14. In order to reduce the negative effects that this process has on the earth, giving the environmentally friendly synthesis process the priority it deserves is essential. Agricultural, fruit, food, vegetable, biomedical, and other types of waste need to be given financial priority in order for them to be utilised in the synthesis of highly equipped nanomaterials. This can only be accomplished through increased funding. For bionanosensor synthesis to become economically and ecologically feasible, additional financing is required.

The investment of capital into the commercial production of medicinal plants is absolutely necessary. The cultivation of organic herbs requires an increase in investment in order to improve soil quality and the yields of therapeutic herbs. Increased investments in agricultural technology, such as precision farming, intelligent pest control, geoinformatics, robotics, drones, and the internet of things, could be beneficial to sales of herbal medicines. It is critical to encourage the use of agricultural biotechnology in order to hasten the rate at which plants manufacture bioactive components (IBEF, 2020).

The manufacturing of herbal medicines is dependent on the effective extraction of the bioactive components contained in the plant species. In order to extract something, one must make use of a wide variety of scientific and analytical processes, such as hydrodistillation, maceration, and the Soxhlet extraction method. These days, the extraction process involves the use of pricey high-tech analytical tools such as cutting-edge spectroscopic and chromatographic techniques. Because of the invention of this data analytics tool, the effectiveness of these analytical procedures

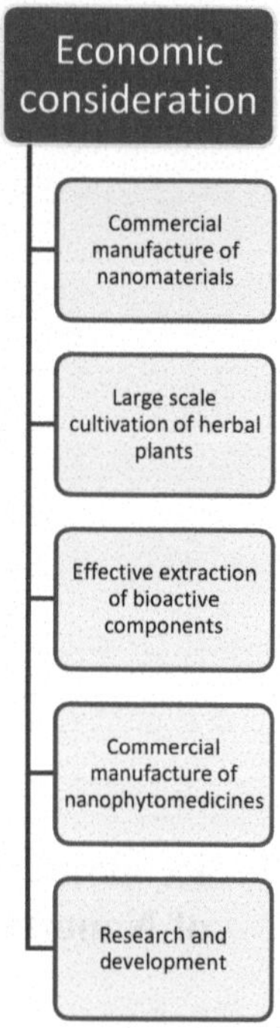

Fig. 14 Economic consideration of the implementation of nanotherapeutic strategies in herbal medicine

has been significantly improved. Optimum performance of the extraction process requires financial investment in the development and application of innovative extraction techniques (Kala et al., 2006).

In order to successfully manufacture high-quality nanophytomedicines, it is necessary to adhere to appropriate production practises. In order to complete the process, various methods such as homogenization, sonication, milling, emulsification, and the use of organic solvents are utilised. It will require financial resources to develop a long-term strategy for the incorporation of nanotechnology into herbal remedies. The development of a successful industrial sector for nanophytomedicine will require significant financial commitment. Because of the growing interest in herbal treatments, this industry should be profitable in the near future (Nouri et al., 2020).

More research needs to be done in this area so that we may develop a way to synthesise potent herbal medicinal substances using nanotechnology that is more environmentally friendly and less expensive. It is necessary to conduct additional research into the possibilities that nanotechnology holds beyond the realm of drug carriers. It is impossible to make progress in nanophytomedicine without financial support for research in allied sectors such as nanotechnology, nanobiotechnology, plant biotechnology, ayurveda, alternative medicine, pharmaceutical science, and so on. Increased financing for innovative research makes both the advancement of medicine and the development of treatment strategies that are more efficient use of resources (Tiwary et al., 2023).

8 Policy and Regulation Required for the Effective Implementation of Nanophytomedicines

There should be an emphasis in health policy on the use of herbal medicines and ayurvedic practises. Figure 15 depicts different policies and regulations required for effective implementation of nanophytomedicines. It is necessary to find ayurvedic research institutions and treatment centres. It is important to educate the public about the many ways in which herbal therapies can help them (Demeke et al., 2022).

The agricultural policy should prioritise the growing of medicinal plants. Herbal plant cultivation for commercial purposes should be encouraged. To cultivate efficient herbal plants in the face of biotic and abiotic stressors, the policy must incorporate the use of contemporary technology. Innumerable Indian tribes possess traditional knowledge of uncommon therapeutic plants. For this knowledge to be

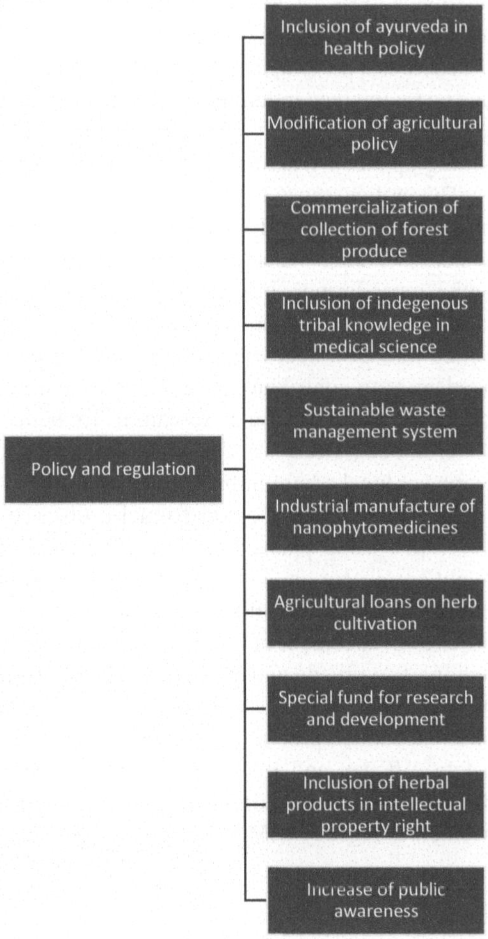

Fig. 15 Policy and regulation required for the effective implementation of nanophytomedicines

incorporated into standard treatment protocols, the tribes must be given the tools to create long-term tribal policy (Marcelino et al., 2023).

It will improve the ayurvedic healing methods. There are healing properties in a variety of forest products. Forest products with medicinal value should be collected and developed for commercial use. The poor and vulnerable who rely on harvesting forest goods will also gain from this.

Herbal plant cultivation and commercial forest product collection would boost rural economies. The policy for advancing rural areas should incorporate such measures. Furthermore, the raw materials for the synthesis of nanomaterials come from a wide variety of biowastes, including those generated from agriculture and livestock. For the synthesis of nanomaterials from agricultural and livestock wastes, rural development policies can foster the growth of village enterprises.

The medical education policy should prioritise ayurvedic training. Incorporating cutting-edge innovations such as nanotechnology, nanosensors, and data analytics will provide tomorrow"s medical professionals with cutting-edge tools. Reforming pharmaceutical science and technology curricula to emphasise natural treatments and nanotechnology is also necessary.

There is a pressing need to build a garbage collection and processing system that can be maintained indefinitely. For the synthesis of nanomaterials, biological wastes are crucial eco-friendly raw materials. In addition, waste management is crucial for setting up an environment-friendly system for synthesising nanomaterials.

The policy must be implemented so that nanophytomedicine can be mass-produced in industry. To build a sustainable herbal pharmaceutical sector on a global scale, researchers, academics, and business leaders must work together. In order to entice more farmers into growing medicinal herbs, rural banks need to create agricultural financing schemes tailored to this industry. Technology advancement in agriculture also necessitates monetary support. Attracting additional scholars and academics to contribute to the fields of ayurveda and nanomedicine requires dedicated financing for such research. To better support alternative medicine, the government"s R&D strategy needs to be adjusted. To improve tribal economies and empower local communities, intellectual property protections should be extended to traditional medicines made from locally sourced herbs. If any medications are developed using plants found in tribal areas and patented, the indigenous people ought to receive an economic share. It is important to raise awareness about the efficacy of ayurvedic medications among the general public (PIB, 2023).

The advancement of herbs" medicinal roles can be advanced through the combination of Ayurveda and Western medicine. Herbal medications based on nanotechnology must be subjected to the same stringent regulatory examination as all pharmaceutical products. Pharmaceutical companies must provide data on the safety, efficacy, and quality of their pharmaceuticals to regulatory bodies, including the Food and Drug Administration (FDA) in the United States and the European Medicines Agency (EMA) in Europe (Demeke et al., 2022).

9 Health and Safety Consideration of the Use of Nanothrepeutic Techniques in Herbal Medicine

In order for nanotechnology-manufactured herbal medications to be successfully implemented in medical science and treatment, strict regulation of their safety and efficacy is essential. Figure 16 depicts different health and safety considerations that needs to be considered to ensure the safe and successful use of nanothrepeutic techniques in the field of herbal medicine.

Before nanophytomedicines to be used on a broad scale in clinical practise, they need to undergo rigorous preclinical investigations. Laboratory research, animal models, and in vitro testing on disease models are all part of the medication approval process prior to human testing (Tiwary et al., 2023).

The drug"s clinical trials are crucial as well. To fully comprehend how medications work, active monitoring of their performance using biosensors and smart tablets is required. Research into the amount of adverse effects and other forms of toxicity in unaffected human body parts is also included. The efficacy of nanophytomedicines in vivo has been confirmed by their performance in clinical trials.

The pharmacovigilance of pharmaceutical products is necessary after the market approval of nanophytomedicines in order to evaluate the safety and harmful effects of the drugs on the market. To ensure proper monitoring and use, it is important for healthcare professionals, patients, medical retailers, laboratory technicians, etc. to monitor and report any bad effects or toxicity of the drugs.

Nanophytomedicine production should adhere to current good manufacturing practises. To prevent environmental degradation, it is imperative that the manufacturing process be risk-free and in line with the principles of waste minimization, reusability, and recycling. Pollution, emission of harmful gases, and contamination of soil and water resources can be reduced by adhering to environmental safety rules for the process.

The safety of pharmaceutical items relies on their thorough labelling and packaging. Manufacturing date, expiration date, dosage recommendations, potential adverse effects, components, chemical composition, warnings and precautions connected to use, etc. must all be included on product labels. The labels on medications assist doctors and patients prescribe and take the medication correctly, reducing the possibility of adverse effects and increasing patient safety (Wolfram et al., 2015).

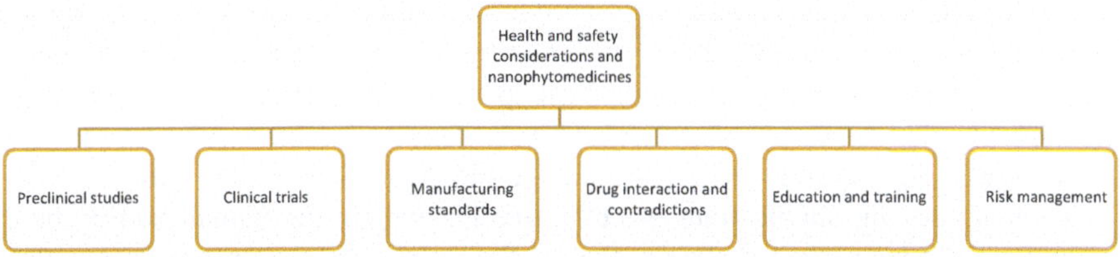

Fig. 16 Health and safety consideration of the use of nanothrepeutic techniques in herbal medicine

After being ingested, the nanophytomedicines could react with other molecules of medication. Pre-clinical and clinical trial dosages need to be carefully monitored and prescribed. Doctors should be familiar with their patients" medical histories so they can give appropriate dosages.

Proper training and education on the ingredients and usage of the nanophytomedicine pharmaceuticals is necessary for the healthcare professionals engaged in the treatment of patients, prescribing the medicines, and pharmacokinetic monitoring of the drug routes. Safer drug administration is promoted by training courses that cover topics including the medication"s make-up, how it works, where it goes in the body, what it does, what adverse effects it has, how it works in the body, and pharmacokinetics.

To ensure the safety of drug use, it is essential that the pharmaceutical industry and drug approval authority receive timely reports of the activity and side effects of the drugs in question. Risks associated with nanophytomedicines can be identified and mitigated with the use of publications like these and pharmacokinetic studies of the pharmaceuticals in various human bodies (Hua et al., 2018).

10 Quality Control for the Use of Nanophytomedicines

For nanophytomedicines to be used safely and effectively in the healthcare system, quality control measures must be put in place. The right measures must be adopted to guarantee the manufacturing of effective nanophytomedicines. Figure 17 depicts different quality control strategies for the use of nanophytomedicines.

Quality nanophytomedicine cannot be made without implementing good manufacturing practises. The quality of nanophytomedicines can be improved by the growth of high-grade herbal plants and the synthesis of effective nanomaterials. The use of cutting-edge technologies will guarantee the reliable production of high-quality pharmaceuticals at industrial scale.

The manufacturing of nanophytomedicines necessitates high-quality inputs. Good technology can be used to enhance the medicinal properties of herbal plants. Bio-based products are both efficient and environmentally friendly precursors to high-quality nanomaterials. Authenticity, purity, efficacy, and the absence of contaminants are only few of the aspects of raw material quality that can be checked (Hesari et al., 2021).

Before pharmaceuticals may be released to the public, they must undergo rigorous qualitative and quantitative analytical testing. The precise composition of pharmaceutical items can be evaluated with the help of cutting-edge analytical techniques as Superficial Fluid Chromatography, Fourier Transform Infrared Spectroscopy, Near Infrared Spectroscopy, Gas Chromatography Mass Spectrometry, etc.

To guarantee the medications are effective in vivo (in the human body), their potential must be evaluated. Depending on how the disease is progressing, it is important to conduct new clinical studies and improve the quality of the existing

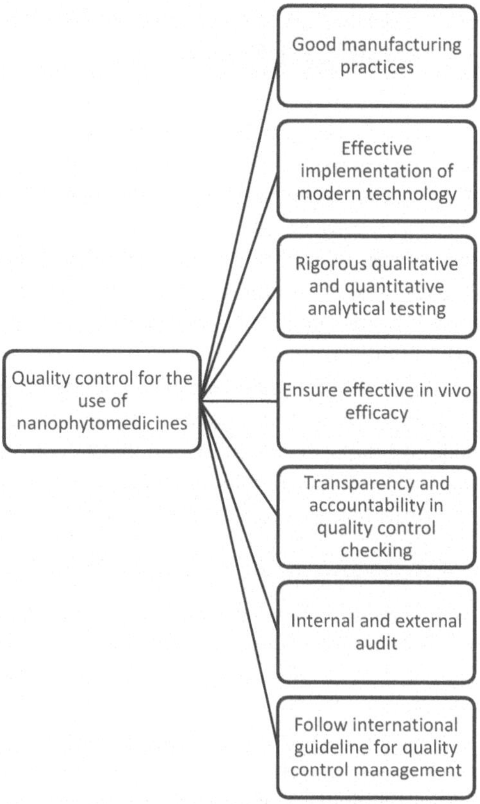

Fig. 17 Quality control for the use of nanophytomedicines

ones at frequent intervals. Newer technologies should be used to improve the drug"s therapeutic efficacy.

To guarantee responsibility, traceability, and openness in quality control, pharmaceutical manufacturers must keep detailed records of their quality management procedures. These can be things like batch records, specifications, quality control reports, and test techniques. Keeping detailed records of your actions helps you stay in line with the law.

Industrial production of nanophytomedicines requires both an internal and external audit. The specialised team is responsible for keeping a careful eye on the manufacturing process, quality of raw materials, quality of products, pre-clinical and clinical trial processes, and other components from manufacturing through distribution of products in the market. Audits can be conducted by internal quality assurance teams, government agencies, or impartial third parties (Tiwari et al., 2023).

It is possible that government regulatory authorities and industry standards will serve as guides for quality control management of nanophytomedicine production and delivery to the market. In order to provide patients with medicines that are both safe and effective, the pharmaceutical industry invests much in building quality control systems, undertaking extensive testing, and adopting continuous improvement processes (Paliwal & Paliwal, 2020).

11 Challenges of the Use of Nanophytomedicines in Current Healthcare System

Different challenges also needs to be faced for the implementation of nanophytomedicines in health care system as shown in Fig. 18. A key barrier to the therapeutic effectiveness of herbal plants is the poor bioavailability of their bioactive components. The primary function of nanomaterials is to facilitate the delivery of drugs into human cells. Many nanomaterials, however, have been questioned for their efficacy. The most efficient nanocarrier must be produced using cutting-edge techniques. Furthermore, certain nanomaterials are poisonous and may have unintended physiological effects.

The development of resistance to anticancer drugs is a major problem in modern medicine. There is a chance that resistance mechanisms will develop to nano-enabled herbal medications, reducing their effectiveness over time. Improving the long-term efficacy of nanophytomedicines requires a deeper understanding of resistance mechanisms and the development of measures to overcome or prevent resistance.

Herbal nanoparticles used to deliver medications to the damaged area of the body may cause side effects in healthy cells. When these chemically synthesised nanomaterials interact with human cells, they produce harmful effects on a large scale. To guarantee drug administration safety, it is crucial to prioritise the development of environmentally friendly methods for synthesising nanomaterials from biological components and agricultural wastes.

The commercialization of nanophytomedicine pharmaceuticals requires the protection of intellectual property rights. The success and monopoly of these drugs, however, might be threatened by factors like patent expiration, patent infringement,

Fig. 18 Challenges of the use of nanophytomedicines in current healthcare system

and the introduction of generic alternatives. The pharmaceutical industry must find a way to overcome these obstacles while maintaining a focus on both innovation and cost-effective distribution.

The nanophytomedicines"regulatory barriers are particularly convoluted. Developing countries lack the necessary resources to conduct high-quality preclinical and clinical research. Important obstacles to the efficient implementation of medications include a lack of resources for routine reporting on the clinical effects of drugs and for monitoring the pharmacokinetics of drug molecules.

Manufacturing, developing, and synthesising therapeutic compounds can be expensive, which is a problem for economies still getting their feet under them. Nanomaterial production and herbal plant culture are two distinct investment opportunities. Government and corporate actors working together must take this into account and provide adequate funding. It is difficult to find a middle ground between the high costs of drug research and development, manufacturing, and retail prices and the needs of patients and healthcare systems. Maintaining profitability and sustainability while expanding access to these medicines is a continuing challenge (Hua et al., 2018).

12 Future Prospective

Nanotechnology is an effective tool for improving the therapeutic efficiency of herbal drugs. Today's world is moving towards herbal drugs due to their efficiency and fewer side effects. However, most of the herbal extracts are not biocompatible and are unable to cross the blood-brain barrier. The high absorption characteristics and improved cellular uptake capacity increase the bioavailability of the herbal bioactive components. Figure 19 explains different strategies that may be adopted in future for the effective implementation of nanotherapeutic strategies in the field of herbal medicines. Nanoparticles can also be used as biomarkers for detecting the therapeutic action of drug molecules. Therefore, it will also help in the improvement of herbal treatment strategies. The application of smart nanopills will help in the pharmacokinetic study of herbal drugs, which will enable the improvement of herbal treatment strategies. The application of nanobiosensors facilitates the sensor-based delivery of drug molecules, which ensures the effective delivery of drugs in the target area. Nanocarriers also facilitate the effective delivery of plant-based chemotherapeutic agents, which mitigates the risk of side effects in chemotherapeutic treatments.

Nanomaterials facilitate the effective delivery of the genes to the affected site, which increases the capacity of gene therapy. T he application of nanofertilizers helps in the effective production of bioactive secondary metabolites in plant species. The nanotherapeutic approaches improve the quality of neurological and psychological disorders by facilitating the delivery of drug molecules to the affected neuron by crossing the blood-cerebrospinal fluid barrier effectively. The newly developed nanomaterials can also increase the immunological characteristics of the

Fig. 19 Future prospective of nanotherapeutic strategy of herbal medicines

body. The therapeutic characteristics of the phytochemicals can be improved by integrating nanomaterials with them.

Research needs to be conducted regarding the application of organometallic and organomineralic nanomaterials, i.e., bhasma, as an effective tool of nanotherapeutic herbal treatment strategy due to its ready availability and cost effectiveness in preparation.

The agricultural wastes, vegetable wastes, fruit wastes, and other plant-based wastes can be used as an effective component of herbal medicines by the application of nanotherapeutic strategies as they also contain a good amount of bioactive components.

The application of nanotechnology increases the possibility of the use of herbal extracts in the pharmaceutical industry. The use of nanomaterial-based plant extracts increases the therapeutic efficiency of existing pharmaceutical products.

The nanomaterials can be synthesised in an eco-friendly way from herbal extracts, which can be used effectively to improve the therapeutic efficiency of any pharmaceutical product.

The nanomaterials can be used for wound healing treatment and the treatment of bone transplant cases due to their effective self-repairing characteristics. Therefore, these self-healing nanosystems can be used along with calendula, pot marigold, and other plant extracts for the treatment of wound healing. The smart bandages prepared by nanotextile products can be used along with plant extracts for the speedy recovery of wounds and major fractures in bones.

13 Conclusion

The importance of herbal medicines is increasing day by day due to the rapidly changing nature of diseases. Nanotechnology becomes an important therapeutic strategy to improve the efficiency of herbal medicines as nanocarriers, nanobiosensors, and nanobiomarkers. It also increases the scope of research in the field of pharmacokinetics due to the application of smart pills. Nano-based neurophytomedicines enable the effective treatment of serious neurological diseases by effectively crossing the blood-cerebrospinal fluid barrier. The therapeutic efficiency of existing pharmaceutical products can also be improved by the application of biologically synthesised nanomaterials. Now, it is high time to return to our ancient therapeutic techniques based on the medicinal characteristics of herbal plants, as they have the capacity to increase the immune system and reduce the possibility of side effects. The naturally occurring bioactive compounds exhibit more therapeutic effects than the chemically synthesised ones. Further research is required to expand the application of nanotherapeutic strategies in herbal medicines.

References

Anjum, S., Ishaque, S., Fatima, H., Farooq, W., Hano, C., Bilal, H. A., & Anjum, I. (2021). Emerging applications of nanotechnology in healthcare systems: Grand challenges and perspectives. *MDPI, Pharmaceuticals, 14*, 707. https://doi.org/10.3390/ph14080707

Ansari, S. H., Islam, F., & Sameem, M. (2012). Influence of nanotechnology on herbal drugs: A review. *Journal of Advanced Pharmaceutical Technology & Research, 3*(3), 142–146. https://doi.org/10.4103/2231-4040.101006

Azharuddin, M., Zhu, G. H., Sengupta, A., Hinkula, J., Slater, N. K. H., & Patra, H. K. (2022). Nano toolbox in immune modulation and nanovaccines. *Cellpress, Trends in Biotechnology, 40*(10), 1195–1212. https://doi.org/10.1016/j.tibtech.2022.03.011

Beslay, M., Srour, B., Mejean, C., Alles, B., Fiolet, T., Debras, C., Chazelas, E., Deschasaux, M., Wendeu-Foyet, M. G., Hercberg, S., Galan, P., Monterio, C. A., Deschamps, V., Andrade, G. C., Kesse-Guyot, E., Julia, C., & Touvier, M. (2020). Ultra-processed food intake in association with BMI change and risk of overweight and obesity: A prospective analysis of the French NutriNet-Santé cohort. *PLoS Medicine, 17*(8), e1003256. https://doi.org/10.1371/journal.pmed.1003256

Bonifaclo, B. V., da Silva, P. B., dos Santos Ramos, M. A., Negri, K. M. S., Bauab, T. M., & Chorilli, M. (2014). Nanotechnology-based drug delivery systems and herbal medicines: A review. *International Journal of Nanomedicine, 9*, 1–15. https://doi.org/10.2147/IJN.S52634

Bukhari, S. N. A. (2022). Emerging Nanotherapeutic approaches to overcome drug resistance in cancers with update on clinical trials. *MDPI, Pharmaceutics, 14*(4). https://doi.org/10.3390/pharmaceutics14040866

Chenab, K. M., Eivazzadeh-Keihan, R., Maleki, A., Pashazadeh-Panahi, P., Hamblin, M. R., & Mokhtarzadeh, A. (2019). Biomedical applications of nanoflares: Targeted intracellular fluorescence probes. *Nanomedicine: Nanotechnology, Biology and Medicine, 17*, 342–358. https://doi.org/10.1016/j.nano.2019.02.006

Crichton, M. L., & Kendall, M. A. F. (2017). Nanopatches for vaccine delivery. *Micro- and Nanotechnology in Vaccine Development*, 343–356. https://doi.org/10.1016/B978-0-323-39981-4.00018-X

Demeke, H., Hasen, G., Sosengo, T., Siraj, J., Tatiparthi, R., & Suleman, S. (2022). Evaluation of policy governing herbal medicines regulation and its implementation in Ethiopia. *Dovepress, Journal of Multidisciplinary Healthcare, 15*, 1383–1394. https://doi.org/10.2147/JMDH.S366166

García-Quintero, A., & Palencia, M. (2021). A critical analysis of environmental sustainability metrics applied to green synthesis of nanomaterials and the assessment of environmental risks associated with the nanotechnology. *Science of the Total Environment, 793*, 148524. https://doi.org/10.1016/j.scitotenv.2021.148524

Gunasekaran, T., Haile, T., Nigusse, T., & Dhanaraju, M. D. (2014). Nanotechnology: An effective tool for enhancing bioavailability and bioactivity of phytomedicine. *Asian Pacific Journal of Tropical Medicine, 4*, S1–S7. https://doi.org/10.12980/APJTB.4.2014C980

Gupta, D., Boora, A., Thakur, A., & Gupta, T. K. (2023). Green and sustainable synthesis of nanomaterials: Recent advancements and limitations. *Environmental Research, 231*(3), 116316. https://doi.org/10.1016/j.envres.2023.116316

Hamdan, S., Pastar, I., Drakulich, S., Dikici, E., Tomic-Canic, M., Deo, S., & Daunert, S. (2017). Nanotechnology-driven therapeutic interventions in wound healing: Potential uses and applications. *ACS Central Science, 3*(3), 163–175. https://doi.org/10.1021/acscentsci.6b00371

Harilal, S., Jose, J., Parambi, D. G. T., Kumar, R., Mathew, G. E., Md Uddin, S., Kim, H., & Mathew, B. (2019). Advancements in nanotherapeutics for Alzheimer's disease: current perspectives. *Journal of Pharmacy and Pharmacology, 71*(9) Wiley. https://doi.org/10.1111/jphp.13132

Herranz, F., Almarza, E., Rodriguez, I., Sallinas, B., Rosell, Y., Desco, M., Bulte, J. W., & Ruiz-Cabello, J. (2011). The application of nanoparticles in gene therapy and magnetic resonance imaging. *Wiley Analytical Science, Microscopy Research & Technique, 74*(7), 577–591. https://doi.org/10.1002/jemt.20992

Hesari, M., Mohammadi, P., Khademi, F., Shackebaei, D., Momtaz, S., Moasefi, N., Farzaei, M. H., & Abdollahi, M. (2021). Current advances in the use of nanophytomedicine therapies for human cardiovascular diseases. *Dovepress, International Journal of Nanomedicine, 16*, 3293–3315. https://doi.org/10.2147/IJN.S295508

Hikal, W., Bratovcic, A., Baeshen, R., & Tkachenko, K. (2021). Nanobiotechnology for the detection and control of waterborne parasites. *Scientific Research Publishing, Open Journal of Ecology, 11*, 203–223.

Hua, S., de Matos, M. B. C., Metselaar, J. M., & Storm, G. (2018). Current trends and challenges in the clinical translation of nanoparticulate nanomedicines: Pathways for translational development and commercialization. *Frontiers in Pharmacology, 9*. https://doi.org/10.3389/fphar.2018.00790

Hunt, N. J., McCourt, P. A. G., Kuncic, Z., Le Couteur, D. G., & Cogger, V. C. (2022). Opportunities and challenges for Nanotherapeutics for the aging population. *Frontiers Nanotechnology, Section Biomedical Nanotechnology, 4*. https://doi.org/10.3389/fnano.2022.832524

Idrees, M. (2015). Role of nanotechnology in medical sciences: A review. *International Journal of Innovative Drug Discovery, 5*(1), 14–24.

India Brand Equity Foundation; December 17, 2020; *High demand for medicinal plant in India*; https://www.ibef.org/blogs/high-demand-for-medicinal-plants-in-india

Jain, K. K. (2009). Role of nanobiotechnology in the development of personalized medicine. *Future Medicine, Nanomedicine, 4*(3), 249–252. https://doi.org/10.2217/nnm.09.12

Jayaraj, R. L., Chandramohan, V., & Namasivayam, E. (2013). Nanomedicines for Parkinson disease: Current status and future perspective. *International Journal of Pharma and Bio Sciences, 4*(1).

Jiang, L., Chen, Y., Wang, X., Guo, W., Bi, Y., Zhang, C., Wang, J., & Li, M. (2022). New insights explain that organic agriculture as sustainable agriculture enhances the sustainable development of medicinal plants. *Frontiers in Plant Science, 13*. https://doi.org/10.3389/fpls.2022.959810

Kala, C. P., Dhyani, P. P., & Sajwan, B. S. (2006). Developing the medicinal plants sector in northern India: Challenges and opportunities. *BMC, Journal of Ethnobiology and Ethnomedicine, 2*, 32. https://doi.org/10.1186/1746-4269-2-32

Kenchegowda, M., Rahamathulla, M., Hani, U., Begum, M. Y., Guruswamy, S., Osmani, R. A. M., Gowrav, M. P., Alshehri, S., Ghoneim, M. M., Alshlowi, A., & Gowda, D. V. (2022). Smart nanocarriers as an emerging platform for cancer therapy: A review. *MDPI, Molecules, 27*(1), 146. https://doi.org/10.3390/molecules27010146

Kim, B., Park, J. E., Im, E., Cho, Y., Lee, J., Lee, H. J., Sim, D. Y., Park, W. Y., Shim, B. S., & Kim, S. H. (2021). Recent advances in nanotechnology with nano-phytochemicals: Molecular mechanisms and clinical implications in cancer progression. *MDPI, International Journal of Molecular Sciences, 22*(7), 3571. https://doi.org/10.3390/ijms22073571

Marcelino, S., Hamdane, S., Gaspar, P. D., & Paco, A. (2023). Sustainable agricultural practices for the production of medicinal and aromatic plants: Evidence and recommendations. *MDPI, Sustainability, 15*(19), 14095. https://doi.org/10.3390/su151914095

Moradi, S. Z., Momtaz, S., Bayrami, Z., Farzaei, M. H., & Abdollahi, M. (2020). Nanoformulations of herbal extracts in treatment of neurodegenerative disorders. *Frontiers in Bioengineering and Biotechnology, Section Nanobiotechnology, 8*. https://doi.org/10.3389/fbioe.2020.00238

Netscribes; October 27, 2020; *Four major ways nanotechnology is changing the future of healthcare*; https://www.netscribes.com/nanotechnology-in-healthcare/

Ngowi, E. E., Wang, Y. Z., Qian, L., Helmy, Y. A. S. H., Anyomi, B., Li, T., Zheng, M., Jiang, E., Duan, S., Wei, J., Wu, D., & Ji, X. (2021). The application of nanotechnology for the diagnosis and treatment of brain diseases and disorders. *Frontiers in Bioengineering and Biotechnology; Section, NanoBiotechnology, 9*. https://doi.org/10.3389/fbioe.2021.629832

Noore, S., Rastogi, N. K., O'Donnell, C., & Tiwari, B. (2021). Novel bioactive extraction and Nano-encapsulation. *MDPI, Encyclopedia, 1*(3), 632–664. https://doi.org/10.3390/encyclopedia1030052

Nouri, Z., Hajialyani, M., Izadi, Z., Bahramsoltani, R., Farzaei, M. H., & Abdollahi, M. (2020). Nanophytomedicines for the prevention of metabolic syndrome: A pharmacological and biopharmaceutical review. *Frontiers in Bioengineering and Biotechnology, 8*, 425. https://doi.org/10.3389/fbioe.2020.00425

Paliwal, R., & Paliwal, S. (2020). Quality control, scale-up, and regulatory aspects of herbal nanomedicine. *Advances in Nanochemoprevention*, 83–95. https://doi.org/10.1007/978-981-15-9692-6_7

Patra, J. K., Das, G., Fraceto, L. F., Campos, E. V. R., Rodriguez-Torres, M. P., Acosta-Torres, L. S., Diaz-Torres, L. A., Grillo, R., Swamy, M. K., Sharma, S., Habtemariam, S., & Shin, H. S. (2018). Nano based drug delivery systems: Recent developments and future prospects. *Springer, Journal of Nanobiotechnology, 16*, 71.

Press Information Bureau; Ministry of Chemicals and Fertilizers, Government of India; September 26, 2023; Dr Mansukh Mandaviya launches National Policy on Research and Development and Innovation in Pharma-MedTech Sector in India and Scheme for promotion of Research and Innovation in Pharma MedTech Sector (PRIP); https://pib.gov.in/PressReleasePage.aspx?PRID=1960812

Rajan, R., Robin, D. T., & Vandanarani, M. (2017). Biomedical waste management in Ayurveda hospitals ecurrentpractices & future prospectives. *Journal of Ayurveda and Integrative Medicine, 10*(3), 214–221. https://doi.org/10.1016/j.jaim.2017.07.011

Research and Markets; May 11, 2022; Nanotechnology in Medical Applications 2022: Analyses of Global Market Trends with Data from 2019, 2020, 2021 and Projection of CAGR through 2026; https://www.globenewswire.com/en/news-release/2022/05/11/2440527/28124/en/Nanotechnology-in-Medical-Applications-2022-Analyses-of-Global-Market-Trends-with-Data-from-2019-2020-2021-and-Projection-of-CAGR-through-2026.html

Ristow, M., & Schmeisser, K. (2014). Mitohormesis: Promoting health and lifespan by increased levels of reactive oxygen species (ROS). *Sage Journals, Dose Response, 12*(2), 12. https://doi.org/10.2203/dose-response.13-035.Ristow

Salinaro, A. T., Pennisi, M., Paola, R. D., Scuto, M., Crupi, R., Cambria, M. T., Ontario, M. L., Tomasello, M., Uva, M., Maiolino, L., Calabrese, E. J., Cuzzocrea, S., & Calabrese, V. (2018). Neuroinflammation and neurohormesis in the pathogenesis of Alzheimer's disease and Alzheimer-linked pathologies: Modulation by nutritional mushrooms. *BMC, Immunity & Ageing, 15*, 8. https://doi.org/10.1186/s12979-017-0108-1

Sarangi, M. K., & Padhi, S. (2018). Novel herbal drug delivery system: An overview. *Archives of Medicine and Health Sciences, 6*(1), 171–179. https://doi.org/10.4103/amhs.amhs_88_17

Singh, R., Goel, S., Bourgaede, P., Aleya, L., & Tewari, D. (2021). Ayurveda Rasayana as antivirals and immunomodulators: Potential applications in COVID-19. *Springerlink, Environmental Science and Pollution Research, 28*(40), 55925–55951. https://doi.org/10.1007/s11356-021-16280-5

Sutton, S. (2015 December 17). Smart Sensor Capsule; *The Medicine Maker*; https://themedicinemaker.com/manufacture/smart-sensor-capsule

The Business Global Company. (2023). Nanomedicine Global Market Report 2023; https://www.thebusinessresearchcompany.com/report/nanomedicine-global-market-report

The Economic Times. (2023 August 9). India exported Ayush, herbal products worth $1,240 mn in last two years: Govt to Rajya Sabha; https://economictimes.indiatimes.com/small-biz/trade/exports/insights/india-exported-ayush-herbal-products-worth-1240-6-million-dollars-in-last-two-years-govt-to-rajya-sabha/articleshow/102560037.cms

Tiwary, A., Joshi, M., Kenwat, R., Paliwal, S. R., Sulakhiya, K., & Paliwal, R. (2023). *Nanophytomedicine: Nanotechnology for herbal product development and value addition* (pp. 197–212). Elsevier, Phytopharmaceuticals and Herbal Drugs Prospects and Safety Issues in the Delivery of Natural Products. https://doi.org/10.1016/B978-0-323-99125-4.00004-4

Valikarimwala, M., Gadale, S., & Waghmode, S. (2021). Applications of nanotechnology in Ayurveda: A review. *Research & Reviews in Biotechnology & Biosciences, 8*(1), 163–169. https://doi.org/10.5281/zenodo.5118448

Wolfram, J., Zhu, M., Yang, Y., Shen, J., Gentile, E., Paolino, D., Fresta, M., Nie, G., Chen, C., Shen, H., Ferrari, M., & Zhao, Y. (2015). Safety of nanoparticles in medicine. *Bentham Science, Current Drug Targets, 16*(14), 1671–1681. https://doi.org/10.2174/1389450115666140804124808

Yadav, V. K., & Chaudhary, N. (2019). Fly ash as a source of natural nanomaterials. *Natural Nanostructured Materials*, 26–34.

Yang, G., Chen, S., & Zhang, J. (2019). Bioinspired and biomimetic nanotherapies for the treatment of infectious diseases. *Frontier Pharmacology, Section Translational Pharmacology, 10*. https://doi.org/10.3389/fphar.2019.00751

Yun, Y., Eteshola, E., Bhattacharya, A., Dong, Z., Shim, J., Conforti, L., Kim, D., Schulz, M. J., & Ahn, C. H. (2009). Tiny medicine: Nanomaterial-based biosensors. *Sensors, 9*, 9275.

Biosynthesis of Iron Oxide Nanoparticles (IONPs): Toxicity Evaluation and Applications for Magnetic Resonance Imaging and Magnetic Hyperthermia

Mohamed Abdelmonem, Emmellie Laura Albert, Nur Khalida Rahayu Zainon, Emma Ziezie Tarmizie, Samer H. Zyoud, and Che Azurahanim Che Abdullah

1 Introduction

The green biosynthesis of iron oxide nanoparticles (IONPs) and their potential biomedical applications are discussed in this chapter. Plant precursors are used to develop, manufacture, and improve green biosynthesized nanoparticles. The biosynthesized method for producing IONPs is less expensive and less harmful to the environment, and it expands their biological applications. This chapter describes biosynthesized iron nanoparticles that have been used or are being studied for biomedical applications. The course will cover IONP biosynthesis, synthesis, fabrication, and biomedical applications such as magnetic resonance imaging (MRI) contrast agents and magnetic hyperthermia. Because of their exceptional physicochemical properties, biocompatibility, nanoscale surface alteration, simplicity of synthesis, and other advances, green biosynthesized IONPs are appealing for biomedical applications. Green biosynthesis IONPs have been studied as MRI contrast agents and magnetic hyperthermia agents by our group and others. Subsections of the chapter cover biosynthesis, characterization, toxicity evaluation, MRI contrast agent applications, and magnetic hyperthermia. Future prospects and conclusions are included near the end of the chapter.

M. Abdelmonem · N. K. R. Zainon
Department of Physics, Faculty of Science, Universiti Putra Malaysia, Serdang, Selangor, Malaysia

E. L. Albert
Nanomaterial Synthesis and Characterization Laboratory, Institute of Nanoscience and Nanotechnology, Universiti Putra Malaysia, Serdang, Selangor, Malaysia

E. Z. Tarmizie
Centre of Foundation Studies for Agricultural Science, Universiti Putra Malaysia, Serdang, Selangor, Malaysia

S. A. Bansal et al. (eds.), *Emerging Applications of Novel Nanoparticles*, Lecture Notes in Nanoscale Science and Technology 37, https://doi.org/10.1007/978-3-031-57843-4_9

2 Synthesis of Iron Oxide Nanoparticles

Nanotechnology is the fastest-growing scientific trend and can fill the gaps in current technology. Over the past two decades, the latter science has made significant strides, particularly in the field of nanomedicine. Nanomedicine focuses on improving the biological and physicochemical properties of nanoscale structures in order to target difficult-to-reach sites with enhanced solubility, greater bioavailability, and minimal side effects (Ray et al., 2014). These particles are known as nanoparticles (NPs). Typical NP diameters range between 1 and 100 nm (Darezereshki et al., 2010). Due to their ability to be synthesised in a variety of shapes, sizes, and lattices, NPs can be used in numerous applications, such as electronic devices, biosensors, textile fabrication, and drug delivery, etc. (LewisOscar et al., 2015).

Metal oxide nanoparticles are one of the most important types of nanoparticles, with applications in numerous fields, including environmental sciences, electrochemistry, and magnetism biology (Saratale et al., 2018). IONPs are notable and unique metal oxides due to their biocompatibility and strong magnetic properties, which aid in the treatment and diagnosis of a wide range of diseases (Vasantharaj et al., 2019).

Iron oxide is an abundantly occurring compound in nature. In addition, it can have a variety of crystal structures and magnetic properties. The three most common forms of these naturally occurring or synthetic mineral structures are hematite, magnetite, and maghemite. The interstitial positions of iron cations are octahedral or tetrahedral. These three oxides have a crystal structure characterised by closely packed planes of oxygen anions (Campos et al., 2015). Depending on the desired nano size and shape, nanoparticles can be synthesised utilising either a top-down or bottom-up approach. Both approaches result in the production of nanomaterials in the desired nanoscale range, but their underlying principles are distinct. Using milling, sputtering, grinding, and photolithography, the top-down method converts bulk materials into nanomaterials. Mechanical processes are used to transform bulk materials into nanomaterials under the top-down methodology (Meyers et al., 2006).

S. H. Zyoud
Department of Mathematics and Science, Ajman University, Ajman, UAE

Center of Medical and Bio-Allied Health Sciences Research, Ajman University, Ajman, UAE

C. A. C. Abdullah (✉)
Department of Physics, Faculty of Science, Universiti Putra Malaysia, Serdang, Selangor, Malaysia

Nanomaterial Synthesis and Characterization Laboratory, Institute of Nanoscience and Nanotechnology, Universiti Putra Malaysia, Serdang, Selangor, Malaysia

UPM-MAKNA Cancer Research Laboratory, Institute of Bioscience, Universiti Putra Malaysia, Serdang, Selangor, Malaysia

Centre for Diagnostic Nuclear Imaging, Universiti Putra Malaysia, Serdang, Selangor, Malaysia
e-mail: azurahanim@upm.edu.my

Top-down approach is renowned for its practicability, which results in the formation of a large mass of NPs; however, it can damage the surface of the produced NPs, and in most cases, the NPs themselves can be damaged (Thakkar et al., 2010). Due to the fact that the physicochemical optical properties of nanoparticles depend primarily on their surface architecture, the top-down approach is limited in certain applications. The bottom-up method, on the other hand, produces NPs by assembling atoms into molecules and molecules into clusters until the formation of NPs of various sizes (Decarolis et al., 2020).

Bottom-up approach techniques include molecular self-assembly, electrochemical nanostructure precipitation, laser pyrolysis, sol-gel method, and biosynthesis. These techniques have rendered the bottom-up method controllable, thereby enabling the fabrication of nanostructures with a variety of applications (Dhand et al., 2015). As shown in Fig. 1, the researchers have categorised the synthesis methods into three broad categories: physical, chemical, and environmentally friendly green synthesis.

Various techniques, including ball milling, sputtering, deposition, and plasma-based techniques, are used in the physical synthesis of nanoparticles. Unfortunately,

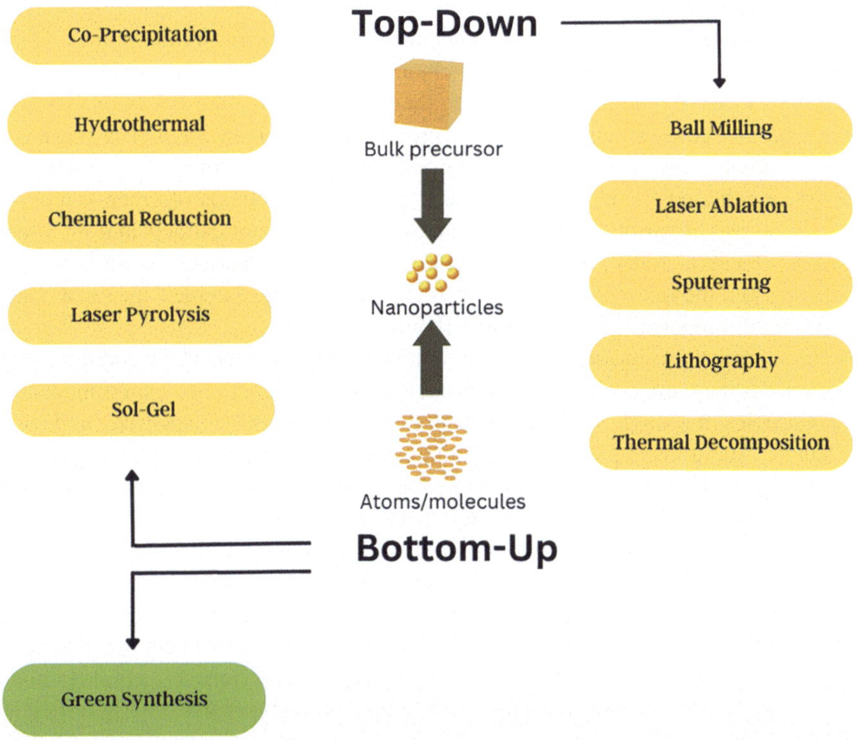

Fig. 1 A diagram summarizing the synthesis approaches used in the synthesis of nanoparticles namely top-down and bottom-up

it is impossible to ignore the disadvantages of the latter techniques. One of them is that the rate of synthesis for the majority of these physical processes is considered to be very slow. For example, ball milling yields only 50% nanomaterials (Yadav et al., 2012). In addition, some techniques, such as the sputtering technique, can produce nanomaterials with a wide particle size distribution, and only 6–8% of the formed NPs are less than 100 nm in size. In addition, laser ablation and plasma-based techniques necessitate a substantial amount of energy. Large size distribution, extensive energy consumption, slow production time, and the formation of by-products make physical methods costly to use and inapplicable to a variety of applications (Seetharaman et al., 2018).

In addition to physical method, numerous chemical strategies have been employed in the synthesis of nanoparticles. These techniques include pyrolysis, sol-gel, the hydrothermal method, chemical reduction, microemulsion, and polyol synthesis, among others (Darroudi et al., 2013). These chemical techniques have several disadvantages. These include the use of a toxic substance, the production of environmentally hazardous by-products, and the contamination of precursor compounds during the synthesis process. To create nanoparticles while preserving their functionality and optical, electrical, and magnetic properties, researchers are currently attempting to develop novel, green, nontoxic, and environmentally friendly strategies (Zhang et al., 2019).

The formulation of nanoparticles using eco-friendly synthesis techniques offers numerous advantages over hazardous chemical methods. Positive aspects of the green synthesis of nanoparticles include their safety; during the synthesis process, no hazardous components are used or produced. The green method is the most straightforward, cost-effective, and environmentally friendly alternative to the previously mentioned chemical and physical techniques. Methods of green synthesis permit the use of various biological sources, including plant extracts and microorganisms (bacteria, algae, fungi, yeast, etc.). Figure 2 summarises the biosynthesis pathways and the role of biological substrate, whereas Fig. 3 depicts the most commonly used biological substrate in IONP biosynthesis.

Among all these biological sources, plant extracts are the most common and efficient, as they are capable of both reducing and stabilising the desired magnetic nanoparticles in a single step. In addition, they are harmless, inexpensive, and easy to handle. The utilised plant extract contains both stabilising and reducing agents. Figure 4 provides a summary of green synthesis method focusing on the procedure of the synthesis of IONPs using different parts of the plant.

According to the scientific literature, the utilization of toxic agents in both physical and chemical procedures has the potential to generate nanoparticles that may have adverse effects on human health. As a result, it became necessary to develop a technique for synthesizing nanoparticles that is not only economical but also eco-friendly. To biologically synthesise nanoparticles, "green synthesis" employs non-toxic, non-hazardous reagents. These procedures are inexpensive, energy-efficient and eco-friendly.

The formulation of nanoparticles through environmentally friendly synthesis techniques has numerous advantages over hazardous chemical methods. Positive

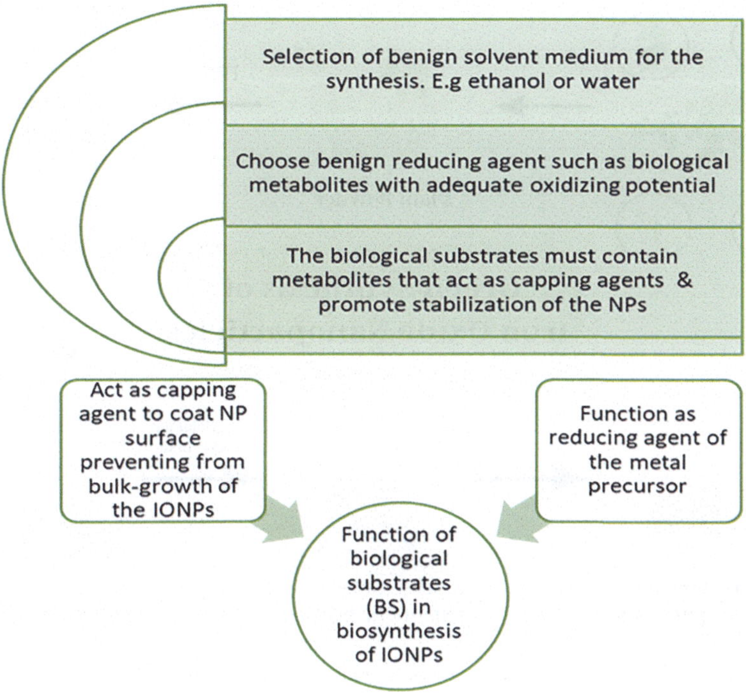

Fig. 2 Biosynthesis pathways and the role of biological substrate IONPs using different plant parts

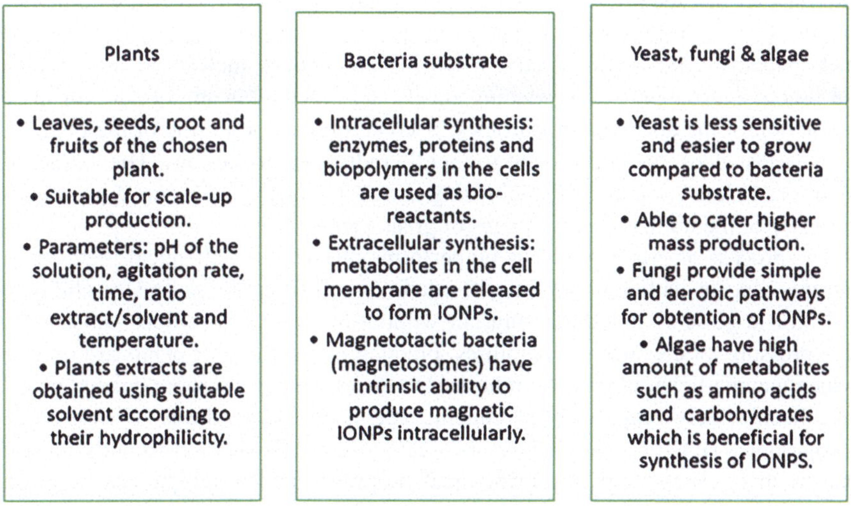

Fig. 3 Different types of biological substrates used in biosynthesis of IONPs

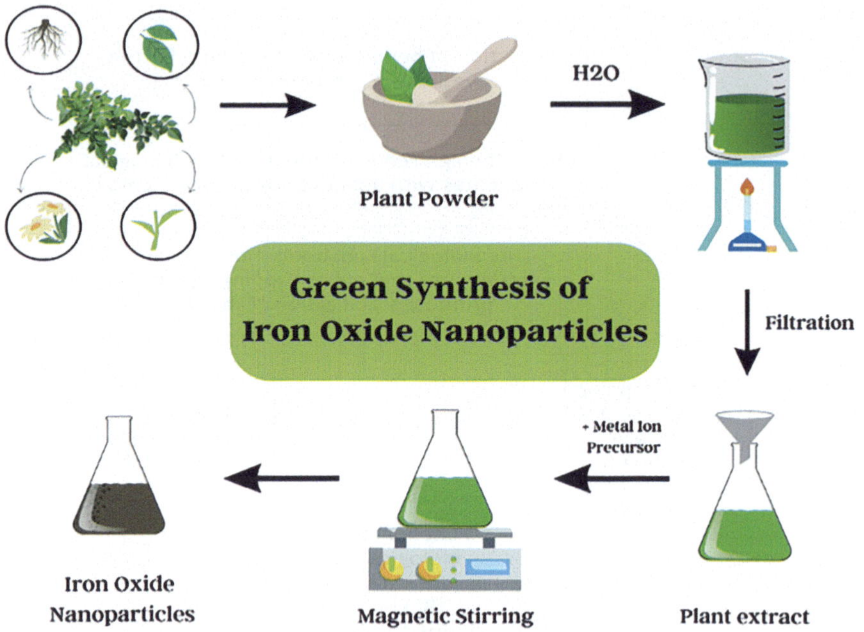

Fig. 4 An Illustration to demonstrate the procedure of the synthesis of IONPs using different plant parts

aspects of the green synthesis of nanoparticles include their safety; no hazardous materials are utilised or generated during the synthesis process. There are numerous examples of environmentally friendly nanoparticle synthesis techniques. The first method involves the use of an active biological component, such as enzymes, which can function as capping and reducing agents. This method is appropriate for mass production of the desired nanoparticles. Second, utilising a portion of the plant or, in some instances, the entire plant to rapidly reduce the metal ions. The extract of the utilised plant contains both stabilising and reducing agents. Figure 5 presents a summary of the numerous advantages of green synthesis.

The process of green synthesis outperforms physical and chemical methods in multiple aspects, including rapidity, affordability, eco-friendliness, and the ability to scale for large-scale manufacturing, all without the use of hazardous substances, high pressure, or elevated temperatures. Frequently, microbes or plants are used to reduce nanoparticles in green synthesis. In comparison to microorganisms, plants have been widely employed as reducing agents in the synthesis of iron oxide nanoparticles (IONPs) due to their abundant, renewable, and eco-friendly nature. Due to their exceptional physiochemical properties and developments in green chemistry and plant-based research, plants are able to synthesise metal nanostructures from every part. From plant components such as root, stem, seed, leaves, peel, fruit, callus, pod, bark, petals, gum, and rhizome, the green synthesis produced IONPs of various shapes and sizes.

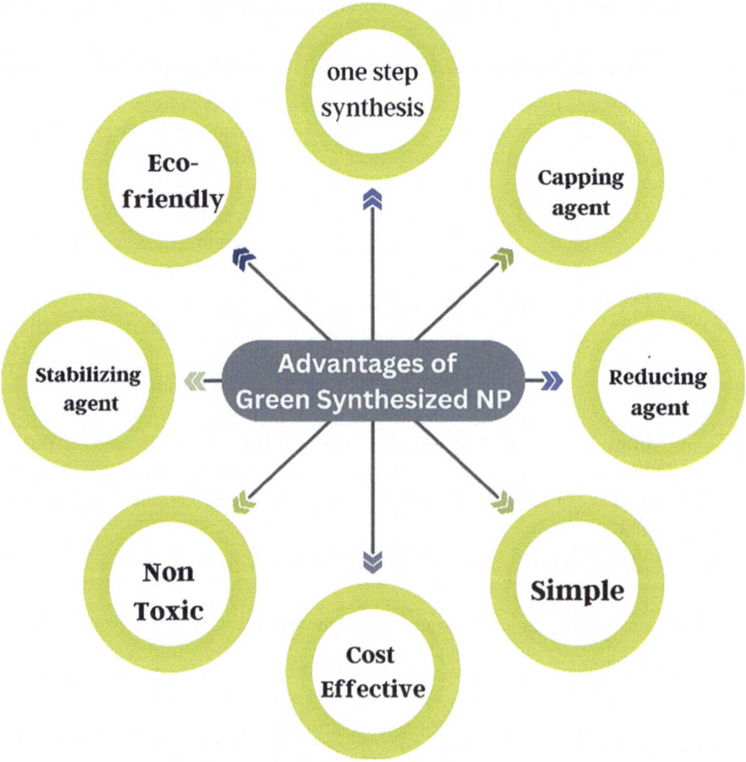

Fig. 5 Positive aspects of adopting green synthesis approach

Phytochemicals found in plants, such as alkaloids, ketones, terpenoids, polysaccharides, phenols, vitamins, amines, amides, aromatic dicarboxylic acids, enzymes, and tannins, are not always essential for plant growth, but they can serve as reducing agents to convert metals from bulk to ionic forms. In green synthesis, phytochemicals and secondary metabolites typically serve as capping and reducing agent stabilisers. Table 1 demonstrates that numerous plants have successfully produced IONPs.

3 Characterization of Biosynthesis Iron Oxide Nanoparticles

Characterization of biosynthesized IONPs is comprised of two types: physical and chemical (see Table 2). To aid in formulation development, scanning electron microscope (SEM) and transmission electron microscope (TEM) are used to examine morphology and particle size, dynamic light scattering (DLS) is used to determine particle size in solution, and zeta potential (ZP) is used for IONPs interaction measurements. The Brunauer–Emmett–Teller (BET) isotherm is typically used to

Table 1 Summary of IONPs synthesized from different parts of plant

Plant	Plants parts	Type of IONPs	Shape	Size (nm)	Applications	References
Vanilla planifolia Cinnamonum verum	Leaves	Fe_3O_4-γFe_2O_3	Core -shell	10–14	Hyperthermia treatment in BV2 microglial cells	Ramirez-Nuñez et al. (2018)
Pimenta dioica	Leaves	Fe_3O_4	Spherical	15	Deep tissue imaging, hyperthermia	Kharey et al. (2019)
Gardenia resinifera	Leaves	α-Fe_2O_3	Spherical	5	Magnetic hyperthermia	Karade et al. (2019)
Punica granatum	Fruit Peel	Fe_3O_4	Spherical	14.38	Magnetic hyperthermia, MRI	Yusefi et al. (2021a)
Garcinia mangostana	Fruit Peel	Fe_3O_4	Spherical	13.42	Magnetic hyperthermia	Yusefi et al. (2021b)
Artemisia absinthium L	Leaf and stem	Fe_3O_4	Spherical	Less than 10	Hyperthermia	Moacă et al. (2022)

Table 2 The Physical and Chemical Characterisation of IONPs

Physical characterisation	Chemical characterisation
Scanning electron microscopy (SEM)	Fourier-transform infrared spectroscopy (FTIR)
Dynamic Light Scattering (DLS)	Raman spectroscopy (RS)
Zeta potential	Nuclear Magnetic Resonance (NMR)
Brunauer–Emmett–Teller (BET) isotherm	Liquid Chromatography-Mass Spectrometry (LC-MS)
Transmission Electron Microscopy (TEM)	Inductively Coupled Plasma Mass Spectroscopy (ICPMS)
X-Ray Diffraction (XRD)	Time-of-Flight Secondary Ion Mass Spectrometry (ToF-SIMs)
UV-Visible (UV-Vis)	X-Sray photoelectron spectroscopy (XPS)

measure surface area and porosity. Powder X-ray diffraction (XRD) will be used to confirm the crystalline nature of the samples.

FTIR, Raman, and NMR are used to examine the chemical information of biosynthesized IONPs. ICPMS is utilised for the analysis of inorganic elements, while LCMS is utilised for the analysis of organic compounds. ToF-SIMs and XPS were used for surface chemistry and analysis.

Raman, AFM, UV-Vis, TEM, XRD, FTIR, XPS, SEM, EDS, and DLS are the most commonly used characterization techniques, as shown in Fig. 6. A variety of methods are available to analyze the characteristics of IONPs, which facilitate the determination of parameters such as particle size, shape, crystallinity, and surface properties. The results of various characterizations of biosynthesis IONPs are displayed in Table 3.

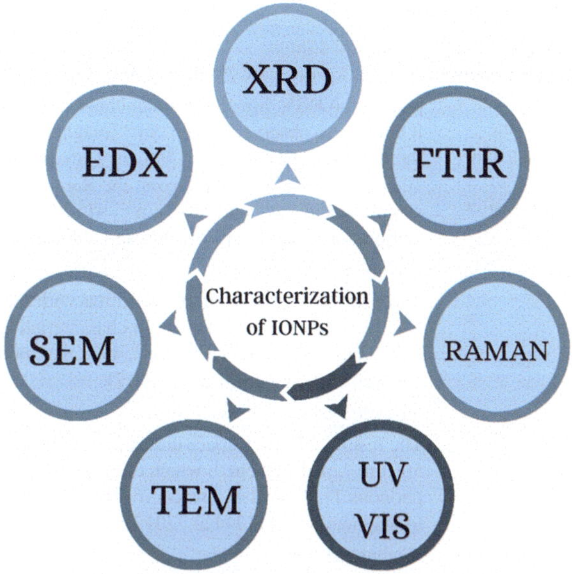

Fig. 6 Characterization methods usually employed used to validate the biosynthesis IONPS

Table 3 Summary and highlights of biosynthesized IONPs characterization

X-Ray Diffraction Analysis (XRD) -to assess the nanoparticles' crystalline structure and average size	*Pimenta dioica*
	At 2θ = 30.41°, 35.75°, 43.55°, 57.3°, and 62°, biosynthesis IONPs exhibited a distinct diffraction peak. The peaks corresponded to the (220), (311), (400), (511), and (440) crystal planes, respectively. It is discovered that the resulting diffraction peaks match the standard for Fe_3O_4 literature data (JCPDS card no. 00-003-0045) (Kharey et al., 2019).
	Gardenia resinifera
	The X-ray spectrum of α-Fe_2O_3 exhibited peaks at 2θ = 33.1°, 35.53°, and 62.53°, which were found to correspond to the lattice planes of (104), (110), and (214), respectively (Karade et al., 2019).
	Garcinia mangostana
	The synthesized IONPs displayed diffraction peaks at 2θ = 30.44°, 35.82°, 43.47°, 53.92°, 57.45°, 63.01°, and 74.69°. Based on the standard diffraction data for magnetite (JCPDS file no. 00-003-0863), these peaks were found to correspond to the crystal planes (220), (311), (400), (422), (511), (440), and (533), respectively, indicating the presence of the pure Fe_3O_4 crystal phase in the magnetite (Yusefi et al., 2021b).
	Punica granatum
	The presence of intense peaks at 2θ = 18.38°, 30.18°, 35.55°, 43.25°, 53.60°, 57.11°, 62.76°, 71.30°, and 74.50° indicated the existence of crystal planes (111), (220), (311), (400), (422), (511), (440), (620), and (533), respectively, confirming the successful synthesis of IONPs and verifying the formation of Fe_3O_4 based on the standard magnetite diffraction patterns (JCPDS file no.: 00-003-0863) (Yusefi et al., 2021a).

(continued)

Table 3 (continued)

Raman Spectroscocpy-used to determine nanoparticle elemental composition	*Ceratonia siliqua*
	Raman spectroscopy can be employed to identify the type of oxide present in prepared IONPs. Specific Raman bands corresponding to the formation of IONPs were observed at 206 cm^{-1} (magnetite), 953 cm^{-1} (akageneite), 594 cm^{-1} (hematite), 1312 cm^{-1} and 1568 cm^{-1} (maghemite), 1113 cm^{-1} (goethite), and 1029 cm^{-1} (hematite). Moreover, the Raman spectrum exhibited additional signals at 124 cm^{-1} and 1688 cm^{-1}, which were attributed to phenolic compounds, such as epicatechin and chlorogenic acid (Demirezen et al., 2019).
	Psoralea corylifolia
	The resulted bands found to be similar with the characteristic modes of α-Fe_2O_3 nanoparticles. The main characteristics of the wavenumbers are found at 222 cm^{-1}, 288 cm^{-1}, 399 cm^{-1}, 492 cm^{-1}, 605 cm^{-1} and a broad peak at 1300 cm^{-1} (Nagajyothi et al., 2017).
	Sageretia thea
	The resultant Raman spectrum exhibited three distinct shifts at 388 cm^{-1}, 500 cm^{-1}, and 772 cm^{-1}, which corresponded to the three specific Raman active phonons $T_{2g,}$ E_g and A_{1g}, respectively. These three Raman frequencies are unique to maghemite (Khalil et al., 2017).
Scanning and Transmission Electron Microscopy (SEM and TEM) and Atomic Force Microscopy (AFM)	*Pimenta dioica*
	The morphology of the biosynthesized IONPs was confirmed by means of field emission scanning electron microscopy (FESEM) analysis. The results demonstrated that the nanoparticles had a spherical shape and a narrow size distribution, with an average size of 15 nm (Kharey et al., 2019).
	Gardenia resinifera
	High-resolution transmission electron microscopy (HR-TEM) was employed to evaluate the size and morphology of biogenic IONPs. The findings revealed that the IONPs had a spherical shape, and the size distribution was normal, ranging from 3 to 8 nm (Kharey et al., 2019).
	Garcinia mangostana
	TEM was utilised to observe the shape, size, and dispersion of IONPs. The results indicated that the nanoparticles were spherical with a high degree of dispersion and minimal agglomeration. The size distribution ranged from 6 to 20 nm, with an average of 14 nm (Yusefi et al., 2021b).
	Punica granatum
	TEM analysis demonstrated that the biosynthesized IONPs were spherical with some minor agglomeration, likely due to Van Der Waals forces and hydroxyl group interactions. The average particle size of the IONPs synthesized using Punica granatum extract was 14.38 nm, and the size distribution ranged from 6 to 24 nm (Yusefi et al., 2021a).
	Artemisia absinthium L
	The SEM results indicated the formation of phase uniform IONPs. It was discovered that nanoparticles have spherical shape with particle sizes ranging from 1 to 10 nm and a slight agglomeration. Agglomeration is caused by the large ratio of surface area to volume and the interactions between molecules adsorbed to the surface of nanoparticles (Moacă et al., 2022).

(continued)

Table 3 (continued)

UV-vis spectroscopy	*Pimenta dioica*
	The sample exhibited its peak absorbance at 435 nm, accompanied by a significant absorption band at 1065 nm. This latter feature is ascribed to an inter-valence charge transfer mechanism, involving the electron transition between Fe^{+2} and Fe^{+3} ions within the octahedral sites of the inverse spinel structure of magnetite (Kharey et al., 2019).
	Avicennia marina
	At room temperature, an analysis of the UV-Vis spectrum of FeO-NPs was conducted. The presence of an absorption peak between 298 and 301 nm may indicate the formation of biogenic IONPs (Karpagavinayagam & Vedhi, 2019).
Fourier transform infrared spectroscopy (FTIR)	*Pimenta dioica*
	Fourier-transform infrared spectroscopy (FTIR) was utilized to verify functional groups on biosynthesized IONPs. The FTIR spectrum showed a notable band at 3325 cm^{-1}, indicative of aliphatic or phenolic hydroxyl groups. Additionally, a 2840 cm^{-1} peak signified -CH2 group C-H stretching, and a 1625 cm^{-1} band pointed to C=C alkene bond stretching in eugenol. Methoxy groups on eugenol's benzene ring, specific to Pimenta dioica, were confirmed by a 1307 cm^{-1} peak. The IONPs' spectrum featured distinct peaks at 3420, 2851, 1629, 1382, 868, 634, and 574 cm cm^{-1}. The 868 cm cm^{-1} peak indicated a benzene ring C–H bending mode, while peaks between 400 and 650 cm^{-1} were consistent with Fe_3O_4 (Kharey et al., 2019).
	Gardenia resinifera
	The FTIR analysis revealed an Fe-O bond vibration in IONPs at 622 cm cm^{-1}. A broad band around 3300–3400 cm^{-1}, attributed to phenolic hydroxyl groups in polyphenols, was observed in both IONPs and plant extracts. Additional bands at 1550–1600 cm^{-1}, 1400 cm^{-1}, and 1026.60 cm^{-1} indicated the presence of phenolic acids, terpenoids, and aliphatic amines, corresponding to C=O stretching, C-H bending, and C-N stretching vibrations, respectively (Karade et al., 2019).
	Garcinia mangostana
	In the FTIR spectra of IONPs synthesized using Mangostana extract, a prominent band ranging from 1621 to 1649 cm^{-1} suggests C=O stretching vibrations. Peaks between 3442 and 3596 cm^{-1}, attributed to O-H stretching, were more intense in IONPs with higher extract ratios, indicating the presence of hydroxyl groups. These observations, along with the detection of O-H, C=O, and C-O functional groups, are likely due to the presence of gartanin, a xanthone from G. Mangostana fruit peel. Additionally, bands between 636 and 721 cm^{-1} in all synthesized IONPs signified the Fe-O bond, suggesting the extract's efficacy as both a stabilizing and capping agent in the green synthesis of IONPs (Yusefi et al., 2021b).
	Psoralea corylifolia
	Using FTIR spectroscopy, the chemical constituents and functional groups of Psoralea corylifolia were analyzed. The presence of alcohols and phenols was inferred from the O-H stretching band at 3400 cm^{-1} . Peaks at 1736 and 1364 cm^{-1} indicated C=O stretching in carboxylic acids. Moreover, bands at 1212 and 1040 cm^{-1} suggested C-N stretching in aliphatic amines, while peaks at 526 and 450 cm^{-1} in Fe_2O_3 confirmed the Fe-O bond vibrations (Nagajyothi et al., 2017).
	Sageretia thea
	Prior studies have identified that the Fe-O bond exhibits a distinctive vibration mode, evident in the FTIR absorption peak at 500 cm $^{-1}$. The broad OH group stretching observed around 3400 cm $^{-1}$ typically signifies the presence of phenolic compounds in the analyzed sample. Furthermore, IR absorption peaks at 1100–1200, 1600, and 2200 cm $^{-1}$ correspond to -C-O, -C=O, and -CN functional groups, respectively. These absorptions are associated with phytochemicals that contribute to the reduction, stabilization, and capping processes in the synthesis of IONPs (Khalil et al., 2017).

4 Applications of Iron Oxide Nanoparticles

Magnetic iron oxide nanoparticles are used in versatile applications and have been verified to be useful as drug delivery carriers, contrast agents, and in magnetic hyperthermia, as their unique magnetic as well as electrical properties make these nanoparticles ultimately efficient in photothermal therapy and a powerful diagnostic and imaging tool. Magnetism is typically associated with the ability to respond to an external magnetic field (Montiel Schneider et al., 2022).

4.1 Iron Oxide Nanoparticles as Contrast Agent for Magnetic Resonance Imaging

Positron emission tomography, computed tomography, magnetic resonance imaging, and ultrasound imaging are examples of in vivo imaging techniques. Among these, MRI is considered the most versatile. The latter technique can provide functional and anatomical information in addition to a high-quality image. In addition, the use of non-ionizing radiation in MRI is the primary advantage of its versatility, as it has allowed longitudinal studies to be conducted without the risk of adverse effects (Avasthi et al., 2020). Contrast agents (CAs) and Gadolinium (Gd)-chelates, which are commonly used in clinical practise, can be used in an MRI to further enhance the contrast of the images. However, these CAs lack selectivity, and it was recently discovered that they are linked to some severe toxicities caused by the release of free Gd during the process (Spencer et al., 1997). The characteristics of an ideal positive and negative contrast agent are summarised in Fig. 7.

Fortunately, a superior alternative to these CAs, IONPs with vastly improved selectivity and biocompatibility, has been identified. During the past two decades, numerous studies have been conducted to develop new IONPs that can improve the diagnosis of various diseases and, as a result, lead to the selection of an appropriate treatment. IONPs were the most researched magnetic nanoparticles for use as CAs in MRI. It is distinguished by its superparamagnetism, which results in high relaxivity and compatibility, and its surfaces are easily functionalizable with highly targeted molecules for molecular imaging (Xie et al., 2018). Due to their biocompatibility, ability to decompose easily in vivo, and low toxicity in comparison to other nanoparticles, magnetite and maghemite are the two iron oxide types that are most commonly used. IONP labels are currently approved for use in clinical settings (Yeh et al., 1993).

4.2 Iron Oxide Nanoparticles for Magnetic Hyperthermia

Cancer is regarded as one of the most significant global health concerns. Consequently, over the past ten years, new measures have been developed to diagnose, monitor, and eradicate cancer. The diagnosis of cancer is regarded as

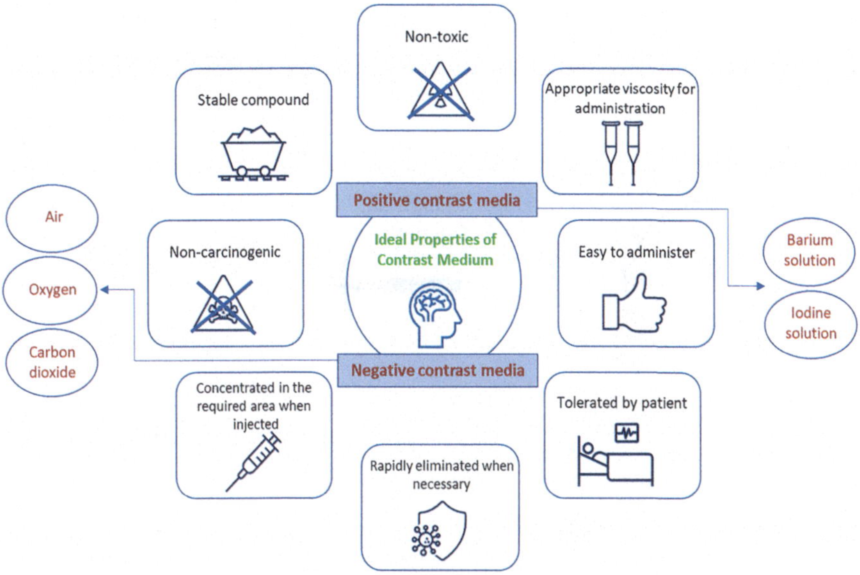

Fig. 7 Characteristics of and ideal contrast agent for MRI

fundamental, as the early detection of tumours influences the selection of the most effective cancer treatment (van der Meel et al., 2019). Standard cancer treatments include radiation therapy, chemotherapy, and surgery. However, conventional interventions have numerous disadvantages and side effects (de Oliveira et al., 2021). Researchers have recently implemented hyperthermia as a means to combat tumour growth.

Hyperthermia is an antitumor therapy with potential benefits for the treatment of a variety of cancer types. It can also enhance the efficacy of other treatments, such as chemotherapy (Yang et al., 2020), radiotherapy (Van Vulpen et al., 2004), or both (Anscher et al., 2000). Hyperthermia is characterised by a rise in body temperature from 37 °C to 45 °C–46 °C (Roti Roti, 2008). As a result, cancer cells will be eradicated via apoptosis, causing the tumour to shrink. The primary disadvantage of hyperthermia is that it does not target the tumour mass selectively, meaning that normal cells may also be affected (Kobayashi, 2011). However, this disadvantage can be overcome by employing heat sources designed for hyperthermia. This will allow specific body zones to be heated and the surrounding cells in the infected area to be killed without harming normal cells, thereby reducing side effects and achieving selectivity (Kok et al., 2020).

IONPs are the most extensively researched and reliable heat sources, this process is referred to as magnetic hyperthermia, which involves the conversion of electromagnetic energy into heat by the nanoparticles through exposure to an external alternating magnetic field (Peeken et al., 2017). Figure 8 depicts the processes involved in magnetic hyperthermia. In addition, magnetic hyperthermia is considered a noninvasive cancer treatment method. In fact, it is a viable alternative for

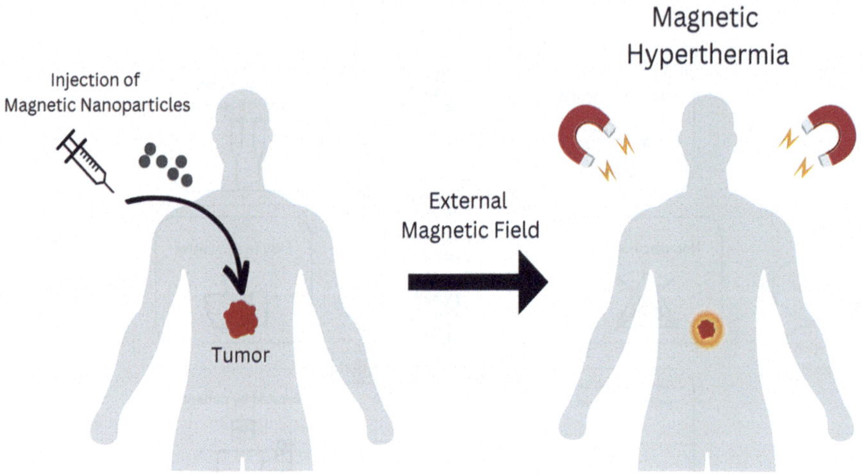

Fig. 8 Magnetic hyperthermia processes

certain types of cancer, particularly solid tumours, because the external magnetic field can penetrate the depth of the targeted area (Perigo et al., 2015). In addition, this method permits the introduction of a high concentration of IONPs that can selectively reside at the tumour site or within the tumour mass for multiple treatment sessions (Moacă et al., 2021).

Maghemite, magnetite, and hematite are regarded as the most frequently utilised IONPs. Due to their distinct physicochemical properties and extraordinary biological characteristics, these three forms are utilised often. The particle size of superparamagnetic materials is typically less than 100 nm, resulting in a high ratio of surface area to volume. In addition, IONPs are biocompatible, easily controlled, have a high heating efficiency, and can be functionalized by modifying their surface. Maghemite and magnetite are the best candidates for magnetic hyperthermia as a cancer treatment.

5 Toxicity Evaluation of Biosynthesis Iron Oxide Nanoparticles

Currently, in vitro studies are regarded as a crucial factor when introducing IONPs as a diagnosis or treatment for specific diseases in the human body. Initial in vitro research on the use of IONPs revealed alterations in cell membrane integrity, metabolic activity, and genetic material in response to IONPs. There are numerous aspects of in vitro nanotoxicity testing, such as providing reproducible and reliable results without the use of animals, being economical, quick, and straightforward (Patil et al., 2015). MTT (3-[4,5-dimethylthiazol-2-yl]-2,5 diphenyl tetrazolium bromide) assay and lactate dehydrogenase (LDH) assay are the most common cell-based assays utilised to demonstrate IONPs' toxicity. MTT assay is used to examine mitochondrial function, whereas LDH assay is primarily used to evaluate the integrity of the cell membrane. IONPs also referred to as magnetic nanoparticles, are globally recognized for their superparamagnetic properties. IONPs composed of ferro- or ferromagnetic materials can exhibit superparamagnetic behaviour if the particle size is less than 20 nm (Marghussian, 2015). Numerous applications can benefit from their superior specific surface area and electrical and thermal conductivity (Ganapathe et al., 2020).

IONPs have multiple potential applications such as targeted drug delivery, magnetic separation, biological labelling, MRI contrast agents, and hyperthermia therapy (Pankhurst et al., 2003). Recently, the demand for IONPs generated via the biogenic pathway has grown. Biogenic synthesis involves the utilisation of microorganisms or plant extract as the reducing agent (Khandel et al., 2018). Green tea extract was used to produce spherical IONPs with an average size of less than 15 nm in 2010. A human epidermal keratinocyte (HaCat) cell line was utilised to assess the toxicity of IONPs. After 48 hours of incubation with biogenic IONPs, cells are still viable (Nadagouda et al., 2010). In a study published in 2019, *Centella asiatica* (CA) was utilised to produce biogenic IONPs with spherical shapes and sizes smaller than 40 nm. As prepared IONPs were evaluated for their in vivo toxicity by exposing 2000 mg/kg IONPs on Swiss albino mice. Two weeks post exposure, it was determined that mice were not vulnerable and prone to IONPs (Pravallika et al., 2019).

The grey mangrove *Avicennia marina* was utilised for biogenic synthesis of IONPs. The synthesised IONPs have a particle size between 10 and 25 nm and are spherical. Furthermore, IONPs were less toxic to human breast cancer cells (HBL100). The results revealed that increasing the IONP concentration to 200 g/ml had little effect on the health of the cells. Therefore, IONPs are either non-toxic or minimally cytotoxic (Pravallika et al., 2019). Aloe Vera leaf and flaxseed aqueous extracts have been synthesised, according to research conducted in 2020. For both plants, the IONP particle size is less than 50 nm. The in vitro cytotoxicity of a substance on MCF-7 breast cancer cell lines was measured using the MTT assay. Low cytotoxicity was observed at concentrations below 4.7 ug/mL (Rahmani et al., 2020). IONPs with a size in the range of 10–100 nm are non-toxic at low concentrations, according to earlier research; however, the toxicity depends on the type of cells present (Lu et al., 2021). Moreover, the toxicity of IONPs is proportional to the amount administered to the specimen or cells. Few studies have been conducted to investigate the in vitro and in vivo cytotoxicity of biogenic IONPs. For practical applications, it is crucial to analyse and explain the mechanisms underlying the toxicological evaluation of biogenic IONPs materials. The results of the cytotoxicity of several IONPs synthesised using the biogenic method are presented in Table 4.

6 Challenges and Future Prospect

The study of metallic nanoparticles, particularly IONPs for medical and environmental applications, is thriving. IONPs cannot currently be mass-produced. To synthesise large amounts of IONPs for commercial applications, it is necessary to

Table 4 Summary related to the biogenic IONPs toxicity assessment and their significant outcome

Green source of reducing/capping agent	Type of toxicity test	Cell lines	IC_{50} /Significant Outcome	References
Tea	MTS assay LDH	HaCaT cells	At 24 and 48 hours, the MTS assay revealed no reduction in cell proliferation in any of the samples tested. There was an increase in LDH leakage when particle concentrations exceeded 50 mg/mL.	Nadagouda et al. (2010)
Centella Asiatica	In vivo test using mice	NA	At a dose of 2000 mg/kg/day orally, no clinical signs of toxicity were observed in the test animals, nor was there any mortality up to 14 days. No notable physiological variations were observed between the control group and the group treated with IONPs.	Pravallika et al. (2019)
Avicennia Marina	MTT assay	HBL100 cells	Human HBL100 cells were less susceptible to IONP toxicity. The results indicated that increasing the concentration of IONPs to 200 g/ml had little effect on the viability of the cells.	Ramalingam et al. (2019)
Flaxseed	MTT assay	MCF-7 cells	The cell treated with the green IONPs depicted low cell viability.	Rahmani et al. (2020)
Cocos nucifera	MTT assay	Human liver carcinoma cells (HepG2 cells)	The activity against HepG2 cell lines showed a dose-dependent pattern, as the viability of cancer cells was considerably reduced at higher concentrations of IONPs (400 and 500 g/mL), with values of 71.48% and 68.87%, respectively.	Rajendran et al. (2021)

(continued)

Table 4 (continued)

Green source of reducing/capping agent	Type of toxicity test	Cell lines	IC_{50} /Significant Outcome	References
Aloe Vera	MTT assay	MCF-7 cells	At a concentration of less than 4.7ug/mL, the results indicated low cytotoxicity that was dose dependent.	Rahmani et al. (2020)
Plumeria obtusa	MTT assay	AU565 and HeLa cell	According to the results, the IC_{50} for AU565 is (47.51 ± 0.08 ug/mL) and for HeLa cell lines it is (55.36 ± 0.07 ug/mL).	Perveen et al. (2022)
Enterococcus faecalis	MTT assay	A549 cells	The data showed cytotoxicity at a minimal dosage of IC_{50} = 50 μg/L	Samuel et al. (2021)
Spinacia Oleracea	In vivo test using Wistar albino mouse	NA	Multiple tests on organs and the rat body showed signs of toxicity at both 10 mg/kg and 100 mg/kg, suggesting that these doses are hazardous.	Kulkarni et al. (2020)
Couroupita guianensis	MTT	HepG2 cell lines	IONP produce significantly increased cytotoxicity at a minimal concentration of IC_{50} = 44.5 ug/mL.	Sathishkumar et al. (2018)
Gardenia resinifera	NA	Human mesenchymal cells	After incubation of 1 mg/mL NPs with human mesenchymal cells for 24 and 48 hours, the cell viability is 82% and 81%, respectively.	Karade et al. (2019)

NA not available

investigate plant biomaterials and reaction parameters in depth. IONPs have only been synthesised and tested in the lab for research purposes to date. These biosynthesis IONPs must be commercialised immediately for biomedical applications, including cancer treatment and MRI contrast agent, because conventional chemotherapeutic methods have numerous side effects. Using a magnetic field, the IONPs-coated drug could be targeted to the site of the tumour without harming normal healthy cells, allowing for the site-specific destruction of cancer cells. Due to their biocompatibility and magnetic properties, IONPs may replace conventional MRI contrast agent. The simplicity of functionalizing IONP surfaces with various ligands will facilitate the development of new molecular MRI techniques. In addition to cancer, IONPs can detect brain inflammation and thrombosis. Increasing the use of IONPs for MRI applications requires synthesis of the inorganic core and proper functionalization to make them biocompatible, targeting of specific tissues or cells, and in vivo tumour models.

7 Conclusions

This section explores the biomedical applications of IONPs derived from green biosynthesis. In recent years, a green approach to metallic nanoparticles has garnered a great deal of interest, particularly due to their vast array of applications. Various nanoparticle sizes, structures, dispersions, and surface modifications have been manufactured utilising eco-friendly synthesis techniques. Since the parameters of biosynthesis vary widely, it is impossible to compare the results directly. Prior to in vivo testing, it is essential to refine and standardise synthesis protocols and parameters in order to produce the most effective nanoparticles. In spite of growing scientific data on theranostic IONPs, biosynthesis IONPs that can deliver multiple anticancer drugs with controlled release and specific tissue targeting may be needed in a variety of clinical settings. Disease prevention, treatment, and diagnosis are included. MRI has been the primary focus of IONPs; as a result, clinical evaluations of other applications are imminent. To apply biosynthesized IONPs on a large scale, it is necessary to enhance the limitations of biogenic synthesis and gather more data on their biological properties. Improving reaction parameters and developing strategies for bulk production is crucial. The potential of biogenic IONPs from renewable resources for biomedical applications requires further investigation through in vitro and in vivo experiments. Future research will focus on advancing and optimizing biogenic synthesis for large-scale production and expanding the knowledge base of their biological properties.

References

Anscher, M. S., Lee, C., Hurwitz, H., Tyler, D., Prosnitz, L. R., Jowell, P., Rosner, G., Samulski, T., & Dewhirst, M. W. (2000). A pilot study of preoperative continuous infusion 5-fluorouracil, external microwave hyperthermia, and external beam radiotherapy for treatment of locally advanced, unresectable, or recurrent rectal cancer. *International Journal of Radiation Oncology Biology Physics, 47*(3), 719–724.

Avasthi, A., Caro, C., Pozo-Torres, E., Leal, M. P., & García-Martín, M. L. (2020). Magnetic nanoparticles as MRI contrast agents. *Surface-modified Nanobiomaterials for Electrochemical and Biomedicine Applications*, 49–91.

Campos, E. A., Pinto, D. V. B. S., Oliveira, J. I. S. D., Mattos, E. D. C., & Dutra, R. D. C. L. (2015). Synthesis, characterization and applications of iron oxide nanoparticles—A short review. *Journal of Aerospace Technology and Management, 7*, 267–276.

Darezereshki, E., Ranjbar, M., & Bakhtiari, F. (2010). One-step synthesis of maghemite (γ-Fe2O3) nano-particles by wet chemical method. *Journal of Alloys and Compounds, 502*(1), 257–260.

Darroudi, M., Sabouri, Z., Oskuee, R. K., Zak, A. K., Kargar, H., & Abd Hamid, M. H. N. (2013). Sol–gel synthesis, characterization, and neurotoxicity effect of zinc oxide nanoparticles using gum tragacanth. *Ceramics International, 39*(8), 9195–9199.

de Oliveira, S. A., Borges, R., dos Santos Rosa, D., de Souza, A. C. S., Seabra, A. B., Baino, F., & Marchi, J. (2021). Strategies for cancer treatment based on photonic nanomedicine. *Materials, 14*(6), 1435.

Decarolis, D., Odarchenko, Y., Herbert, J. J., Qiu, C., Longo, A., & Beale, A. M. (2020). Identification of the key steps in the self-assembly of homogeneous gold metal nanoparticles produced using inverse micelles. *Physical Chemistry Chemical Physics, 22*(34), 18824–18834.

Demirezen, D. A., Yıldız, Y. Ş., & Yılmaz, D. D. (2019). Amoxicillin degradation using green synthesized iron oxide nanoparticles: Kinetics and mechanism analysis. *Environmental Nanotechnology, Monitoring & Management, 11*, 100219.

Dhand, C., Dwivedi, N., Loh, X. J., Ying, A. N. J., Verma, N. K., Beuerman, R. W., Lakshminarayanan, R., & Ramakrishna, S. (2015). Methods and strategies for the synthesis of diverse nanoparticles and their applications: A comprehensive overview. *RSC Advances, 5*(127), 105003–105037.

Ganapathe, L. S., Mohamed, M. A., Mohamad Yunus, R., & Berhanuddin, D. D. (2020). Magnetite (Fe3O4) nanoparticles in biomedical application: From synthesis to surface functionalisation. *Magnetochemistry, 6*(4), 68.

Karade, V., Parit, S., Dawkar, V., Devan, R., Choudhary, R., Kedge, V., Pawar, N., Kim, J., & Chougale, A. (2019). A green approach for the synthesis of α-Fe2O3 nanoparticles from Gardenia resinifera plant and it's in vitro hyperthermia application. *Heliyon, 5*(7), e02044.

Karpagavinayagam, P., & Vedhi, C. (2019). Green synthesis of iron oxide nanoparticles using Avicennia marina flower extract. *Vacuum, 160*, 286–292.

Khalil, A. T., Ovais, M., Ullah, I., Ali, M., Shinwari, Z. K., & Maaza, M. (2017). Biosynthesis of iron oxide (Fe2O3) nanoparticles via aqueous extracts of Sageretia thea (Osbeck.) and their pharmacognostic properties. *Green Chemistry Letters and Reviews, 10*(4), 186–201.

Khandel, P., Yadaw, R. K., Soni, D. K., Kanwar, L., & Shahi, S. K. (2018). Biogenesis of metal nanoparticles and their pharmacological applications: Present status and application prospects. *Journal of Nanostructure in Chemistry, 8*, 217–254.

Kharey, P., Dutta, S. B., Manikandan, M., Palani, I., Majumder, S., & Gupta, S. (2019). Green synthesis of near-infrared absorbing eugenate capped iron oxide nanoparticles for photothermal application. *Nanotechnology, 31*(9), 095705.

Kobayashi, T. (2011). Cancer hyperthermia using magnetic nanoparticles. *Biotechnology Journal, 6*(11), 1342–1347.

Kok, H. P., Cressman, E. N., Ceelen, W., Brace, C. L., Ivkov, R., Grüll, H., Ter Haar, G., Wust, P., & Crezee, J. (2020). Heating technology for malignant tumors: A review. *International Journal of Hyperthermia, 37*(1), 711–741.

Kulkarni, S., Mohanty, N., Kadam, N. N., Swain, N., & Thakur, M. (2020). Green synthesis to develop iron-nano formulations and its toxicity assays. *Journal of Pharmacopuncture, 23*(3), 165–172.

LewisOscar, F., MubarakAli, D., Nithya, C., Priyanka, R., Gopinath, V., Alharbi, N. S., & Thajuddin, N. (2015). One pot synthesis and anti-biofilm potential of copper nanoparticles (CuNPs) against clinical strains of Pseudomonas aeruginosa. *Biofouling, 31*(4), 379–391.

Lu, D., Wu, X., Wang, W., Ma, C., Pei, B., & Wu, S. (2021). Synthesis and application of iron oxide nanoparticles in bone tissue repair. *Journal of Nanomaterials, 2021*, 1–14.

Marghussian, V. (2015). Magnetic properties of nano-glass ceramics. *Nano-Glass Ceramics*, 181–223.

Meyers, M. A., Mishra, A., & Benson, D. J. (2006). Mechanical properties of nanocrystalline materials. *Progress in Materials Science, 51*(4), 427–556.

Moacă, E.-A., Watz, C.-G., Socoliuc, V., Racoviceanu, R., Păcurariu, C., Ianoş, R., Cîntă-Pînzaru, S., Tudoran, L. B., Nekvapil, F., & Iurciuc, S. (2021). Biocompatible magnetic colloidal suspension used as a tool for localized hyperthermia in human breast adenocarcinoma cells: Physicochemical analysis and complex in vitro biological profile. *Nanomaterials, 11*(5), 1189.

Moacă, E.-A., Socoliuc, V., Stoian, D., Watz, C., Flondor, D., Păcurariu, C., Ianoş, R., Rus, C. I., Barbu-Tudoran, L., & Semenescu, A. (2022). Synthesis and characterization of bioactive magnetic nanoparticles from the perspective of hyperthermia applications. *Magnetochemistry, 8*(11), 145.

Montiel Schneider, M. G., Martín, M. J., Otarola, J., Vakarelska, E., Simeonov, V., Lassalle, V., & Nedyalkova, M. (2022). Biomedical applications of iron oxide nanoparticles: Current insights progress and perspectives. *Pharmaceutics, 14*(1), 204.

Nadagouda, M. N., Castle, A. B., Murdock, R. C., Hussain, S. M., & Varma, R. S. (2010). In vitro biocompatibility of nanoscale zerovalent iron particles (NZVI) synthesized using tea polyphenols. *Green Chemistry, 12*(1), 114–122.

Nagajyothi, P., Pandurangan, M., Kim, D. H., Sreekanth, T., & Shim, J. (2017). Green synthesis of iron oxide nanoparticles and their catalytic and in vitro anticancer activities. *Journal of Cluster Science, 28*, 245–257.

Pankhurst, Q. A., Connolly, J., Jones, S. K., & Dobson, J. (2003). Applications of magnetic nanoparticles in biomedicine. *Journal of Physics D: Applied Physics, 36*(13), R167–R181.

Patil, U. S., Adireddy, S., Jaiswal, A., Mandava, S., Lee, B. R., & Chrisey, D. B. (2015). In vitro/in vivo toxicity evaluation and quantification of iron oxide nanoparticles. *International Journal of Molecular Sciences, 16*(10), 24417–24450.

Peeken, J. C., Vaupel, P., & Combs, S. E. (2017). Integrating hyperthermia into modern radiation oncology: What evidence is necessary? *Frontiers in Oncology, 7*, 132.

Perigo, E. A., Hemery, G., Sandre, O., Ortega, D., Garaio, E., Plazaola, F., & Teran, F. J. (2015). Fundamentals and advances in magnetic hyperthermia. *Applied Physics Reviews, 2*(4).

Perveen, S., Nadeem, R., Ur Rehman, S., Afzal, N., Anjum, S., Noreen, S., Saeed, R., Amami, M., Al-Mijalli, S. H., & Iqbal, M. (2022). Green synthesis of iron (Fe) nanoparticles using Plumeria obtusa extract as a reducing and stabilizing agent: Antimicrobial, antioxidant and biocompatibility studies. *Arabian Journal of Chemistry, 15*(5), 103764.

Pravallika, P. L., Mohan, G. K., Rao, K. V., & Shanker, K. (2019). Biosynthesis, characterization and acute oral toxicity studies of synthesized iron oxide nanoparticles using ethanolic extract of Centella asiatica plant. *Materials Letters, 236*, 256–259.

Rahmani, R., Gharanfoli, M., Gholamin, M., Darroudi, M., Chamani, J., Sadri, K., & Hashemzadeh, A. (2020). Plant-mediated synthesis of superparamagnetic iron oxide nanoparticles (SPIONs) using aloe vera and flaxseed extracts and evaluation of their cellular toxicities. *Ceramics International, 46*(3), 3051–3058.

Rajendran, A., Alsawalha, M., & Alomayri, T. (2021). Biogenic synthesis of husked rice-shaped iron oxide nanoparticles using coconut pulp (Cocos nucifera L.) extract for photocatalytic degradation of Rhodamine B dye and their in vitro antibacterial and anticancer activity. *Journal of Saudi Chemical Society, 25*(9), 101307.

Ramalingam, V., Dhinesh, P., Sundaramahalingam, S., & Rajaram, R. (2019). Green fabrication of iron oxide nanoparticles using grey mangrove Avicennia marina for antibiofilm activity and in vitro toxicity. *Surfaces and Interfaces, 15*, 70–77.

Ramirez-Nuñez, A., Jimenez-Garcia, L., Goya, G., Sanz, B., & Santoyo-Salazar, J. (2018). In vitro magnetic hyperthermia using polyphenol-coated Fe3O4@ γFe2O3 nanoparticles from Cinnamomun verum and Vanilla planifolia: The concert of green synthesis and therapeutic possibilities. *Nanotechnology, 29*(7), 074001.

Ray, G., Shetti, D., Yadav, A., & Kundu, G. (2014). Nanomedicine: Therapeutic applications and limitations. In S. Soni, A. Salhotra, & M. Suar (Eds.), *Handbook of research on diverse applications of nanotechnology in biomedicine, chemistry, and engineering* (pp. 64–89). IGI Global.

Roti Roti, J. L. (2008). Cellular responses to hyperthermia (40–46 C): Cell killing and molecular events. *International Journal of Hyperthermia, 24*(1), 3–15.

Samuel, M. S., Datta, S., Chandrasekar, N., Balaji, R., Selvarajan, E., & Vuppala, S. (2021). Biogenic synthesis of iron oxide nanoparticles using enterococcus faecalis: Adsorption of hexavalent chromium from aqueous solution and in vitro cytotoxicity analysis. *Nanomaterials, 11*(12), 3290.

Saratale, R. G., Karuppusamy, I., Saratale, G. D., Pugazhendhi, A., Kumar, G., Park, Y., Ghodake, G. S., Bharagava, R. N., Banu, J. R., & Shin, H. S. (2018). A comprehensive review on green nanomaterials using biological systems: Recent perception and their future applications. *Colloids and Surfaces B: Biointerfaces, 170*, 20–35.

Sathishkumar, G., Logeshwaran, V., Sarathbabu, S., Jha, P. K., Jeyaraj, M., Rajkuberan, C., Senthilkumar, N., & Sivaramakrishnan, S. (2018). Green synthesis of magnetic Fe3O4 nanoparticles using Couroupita guianensis Aubl. Fruit extract for their antibacterial and cytotoxicity activities. *Artificial Cells, Nanomedicine, and Biotechnology, 46*(3), 589–598.

Seetharaman, P. K., Chandrasekaran, R., Gnanasekar, S., Chandrakasan, G., Gupta, M., Manikandan, D. B., & Sivaperumal, S. (2018). Antimicrobial and larvicidal activity of eco-friendly silver nanoparticles synthesized from endophytic fungi Phomopsis liquidambaris. *Biocatalysis and Agricultural Biotechnology, 16*, 22–30.

Spencer, A. J., Wilson, S. A., Batchelor, J., Reid, A., Pees, J., & Harpur, E. (1997). Gadolinium chloride toxicity in the rat. *Toxicologic Pathology, 25*(3), 245–255.

Thakkar, K. N., Mhatre, S. S., & Parikh, R. Y. (2010). Biological synthesis of metallic nanoparticles. *Nanomedicine: Nanotechnology, Biology and Medicine, 6*(2), 257–262.

van der Meel, R., Sulheim, E., Shi, Y., Kiessling, F., Mulder, W. J., & Lammers, T. (2019). Smart cancer nanomedicine. *Nature Nanotechnology, 14*(11), 1007–1017.

Van Vulpen, M., De Leeuw, A., Raaymakers, B., Van Moorselaar, R., Hofman, P., Lagendijk, J., & Battermann, J. (2004). Radiotherapy and hyperthermia in the treatment of patients with locally advanced prostate cancer: Preliminary results. *BJU International, 93*(1), 36–41.

Vasantharaj, S., Sathiyavimal, S., Senthilkumar, P., LewisOscar, F., & Pugazhendhi, A. (2019). Biosynthesis of iron oxide nanoparticles using leaf extract of Ruellia tuberosa: Antimicrobial properties and their applications in photocatalytic degradation. *Journal of Photochemistry and Photobiology B: Biology, 192*, 74–82.

Xie, W., Guo, Z., Gao, F., Gao, Q., Wang, D., Liaw, B.-S., Cai, Q., Sun, X., Wang, X., & Zhao, L. (2018). Shape-, size-and structure-controlled synthesis and biocompatibility of iron oxide nanoparticles for magnetic theranostics. *Theranostics, 8*(12), 3284–3307.

Yadav, T. P., Yadav, R. M., & Singh, D. P. (2012). Mechanical milling: A top down approach for the synthesis of nanomaterials and nanocomposites. *Nanoscience and Nanotechnology, 2*(3), 22–48.

Yang, S.-J., Huang, C.-H., Wang, C.-H., Shieh, M.-J., & Chen, K.-C. (2020). The synergistic effect of hyperthermia and chemotherapy in magnetite nanomedicine-based lung cancer treatment. *International Journal of Nanomedicine, Volume 15*, 10331–10347.

Yeh, T. C., Zhang, W., Ildstad, S. T., & Ho, C. (1993). Intracellular labeling of T-cells with superparamagnetic contrast agents. *Magnetic Resonance in Medicine, 30*(5), 617–625.

Yusefi, M., Shameli, K., Hedayatnasab, Z., Teow, S.-Y., Ismail, U. N., Azlan, C. A., & Rasit Ali, R. (2021a). Green synthesis of Fe3O4 nanoparticles for hyperthermia, magnetic resonance imaging and 5-fluorouracil carrier in potential colorectal cancer treatment. *Research on Chemical Intermediates, 47*, 1789–1808.

Yusefi, M., Shameli, K., Su Yee, O., Teow, S.-Y., Hedayatnasab, Z., Jahangirian, H., Webster, T. J., & Kuča, K. (2021b). Green synthesis of Fe_3O_4 nanoparticles stabilized by a Garcinia mangostana fruit peel extract for hyperthermia and anticancer activities. *International Journal of Nanomedicine, 16*, 2515–2532.

Zhang, M., Yang, J., Cai, Z., Feng, Y., Wang, Y., Zhang, D., & Pan, X. (2019). Detection of engineered nanoparticles in aquatic environments: Current status and challenges in enrichment, separation, and analysis. *Environmental Science: Nano, 6*(3), 709–735.

Effect of Annealing Temperature on Structural, Morphological and Optical Properties of CdZnTe Thin Films

Sudeshna Surabhi, Kumar Anurag, and S. R. Kumar

1 Introduction

The discrete energy levels of quantum dots, a type of low-dimensional semiconductor, have drawn attention because of their potential uses in the fabrication of optoelectronic devices, quantum information processing, and terahertz devices (Chakraborty, 1999). II-VI semiconductors, out of all the materials, have drawn interest lately due to their higher quantum efficiency and exciton binding energies, which can be used for optoelectronic devices in the middle of the spectrum. Because of the lattice mismatch between its counter constituents, CdTe/ZnTe material is regarded as a significant semiconductor. With the most recent developments in growing technology, CdZnTe quantum dots of superior quality are produced. These days, CdZnTe-based yellow and green laser diodes have been developed (Kryshtab et al., 2008). Bagaev et al. (2000) examined the temperature dependence of integral photoluminescence (PL) in a variety of CdZnTe/ZnTe quantum dots generated by molecular beam epitaxy. They also examined the impact of exciton level energy depth on these PL spectra. The optical characteristics and the mechanism of activation energy augmentation in the CdTe/ZnTe nanostructure have been studied using temperature-dependent photoluminescence measurement (Lee et al., 2006; Walter

S. Surabhi (✉)
Brindavan College of Engineering, Bengaluru, Karnataka, India

K. Anurag
School of Energy Materials, Mahatma Gandhi University, Kottayam, Kerala, India

S. R. Kumar
Thin Film Laboratory, National Institute of Advanced Manufacturing Technology (Formerly NIFFT), Ranchi, India

S. A. Bansal et al. (eds.), *Emerging Applications of Novel Nanoparticles*, Lecture Notes in Nanoscale Science and Technology 37, https://doi.org/10.1007/978-3-031-57843-4_10

et al., 2001). Temperature change has also been explored in relation to the optical characteristics of CdZnTe. By using electrical current deep level transient spectroscopy and low temperature cathodoluminescence, the band offsets in strained ZnTe/CdZnTe square quantum well structures were found (Kozlovsky et al., 2000a).

Using pseudo potential formalism, a thorough analysis of energy gaps, dielectric constants, and the composition ranging from 0 to 1 has just been completed (Bouarissa, 2007). It is believed that the influence of strain distribution in CdTe/ZnTe material is a fascinating topic for achieving unusual optical and electrical features. In strained CdZnTe/ZnTe heterostructures, the impact of strain on the electronic energy band structure has been discussed (Hong, 2010). Any low dimensional semiconductor system"s linear and non-linear optical properties are influenced by the quantum dot"s composition (Zn in CdTe quantum dot). Based on superlattice theory, the electrical structure and optical characteristics of a few III-V and II-VI semiconductor superlattices were examined using the super lattice representation formalism (Johnson et al., 1990). Either the zinc-blend phase or the wurtzite phase can crystallize in CdZnTe. On the other hand, systems based on zinc blends are more advantageous than wurtzite systems (Novikov et al., 2008). Any semiconductor heterostructure"s potential barrier height is dictated by the band lineups and offsets at the interface. The CdZn/ZnTe heterostructure exhibits a wide range in the band offset ratio as a result of mismatched crystal properties, significant intrinsic elastic strain, and electrostatic fields. Deep level transient spectroscopy using electrical current and low temperature cathodoluminescence have been used to study the band offsets of some II-VI materials (Kozlovsky et al., 2000b).

In our work, we employed a non-aqueous chemical bath deposition (CBD) method to synthesize CdZTe thin films and investigated their structural properties under varying annealing temperatures. CdZTe can be deposited using both physical and chemical techniques, each with its own set of advantages and limitations. While physical techniques offer high-quality films, they come at a significant cost and involve complex vacuum chamber operations. On the other hand, chemical techniques such as CVD offers high quality film such as adhesion and resistance to defects, however, CVD also has some disadvantages, such as strict temperature control, hazardous precursors etc. Therefore, establishing a balance between efficiency and cost is crucial for optimizing the deposition process. Figure 1 illustrates the broad categorization of deposition techniques. Advantages of chemical bath deposition technique is illustrated in Fig. 2.

2 Experimental Procedure

The study examines CdZnTe films that were produced on nickel substrate using a chemical bath process in a non-aqueous solution and have varying annealing temperatures (Table 1).

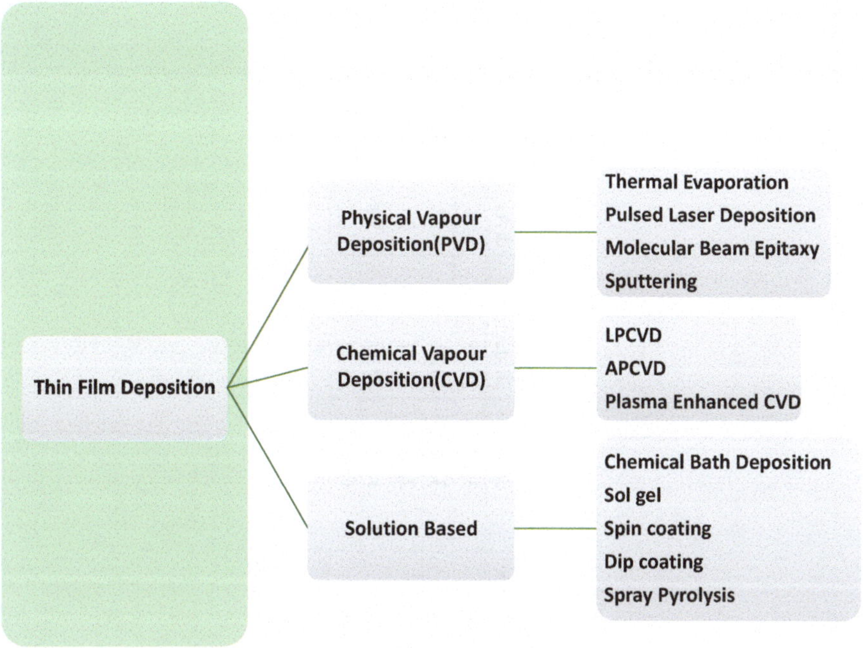

Fig. 1 Various deposition techniques

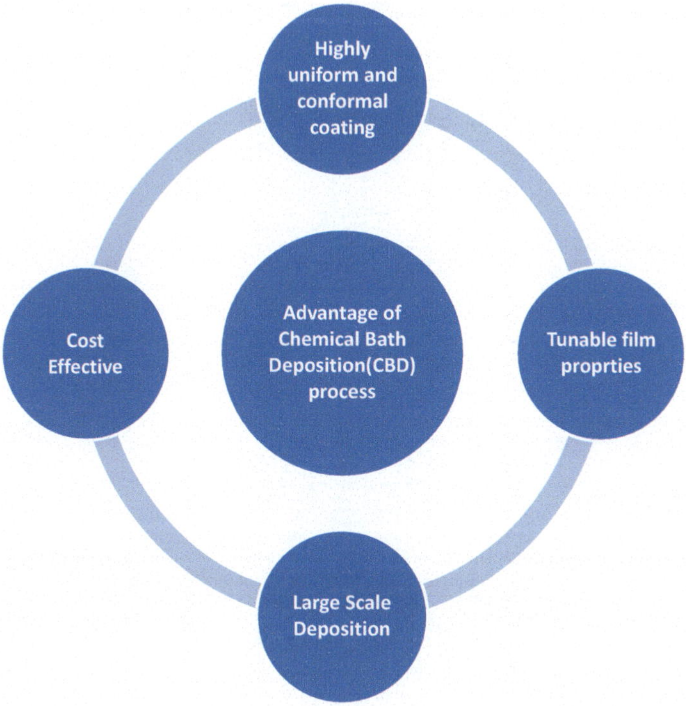

Fig. 2 Advantages of CBD technique

Table 1 2θ, ‘d’ and miller-indices of CdZnTe thin films with varying annealing temperature

Varying annealing temperature (300 °C, 350 °C, 400 °C and 450 °C)								Standard values		
300 °C		350 °C		400 °C		450 °C				
2θ(degree)	d(Å)	2θ(degree)	d(Å)	2θ(degree)	d(Å)	2θ(degree)	d(Å)	2θ(degree)	d(Å)	Miller indices
23.02	3.86	24.27	3.64	24.30	3.66	25.67	3.47	23.75	3.74	(111)
29.32	3.04	28.65	3.12	29.38	3.04	29.35	3.04	28.51	3.13	(200)
40.46	2.23	39.76	2.27	39.80	2.26	41.88	2.16	41.01	2.20	(220)
47.01	1.93	46.91	1.94	46.96	1.93	46.60	1.95	46.87	1.94	(311)
56.90	1.62	57.32	1.61	57.27	1.61	56.99	1.61	56.83	1.62	(400)

2.1 Sample Preparation and Experimental Setup

It is made up of an electrolyte containing Borosil glass reaction cell. To agitate the electrolyte, a magnetic stirrer/hot plate was employed. Nickel plate and glass were the substrates employed in this project. A stand was used to secure the substrate vertically. The substrates were effectively cleansed in distilled water, hot soap solution, and pure water before the deposition. The substrate was allowed to dry at ambient temperature (Fig. 3).

2.2 Preparation of Bath

A magnetic stirrer was used to help swirl 40 milliliters of ethylene glycol in a beaker until the temperature reached 160 °C. A 0.05 M TeO2 AR grade was added to the solvent. The dimensions of the nickel substrate were 2 cm × 1 cm × 0.1 cm. It was polished with 600 grit carborandum paper and then thoroughly cleaned with boiling distilled water, hot soap solution, and distilled water again. With the assistance of firm support, the nickel substrate was dipped within the electrolyte. Following sample loading, the solution was agitated for 10 min with the addition of 0.05 M tellurium dioxide. Next, for two hours, 0.05 M cadmium acetate is added and mixed. Finally, 0.5 M zinc chloride was added to the mixture, and it was agitated for the following two hours. To get rid of counter ions and organic contaminants, the film

Fig. 3 Experimental setup

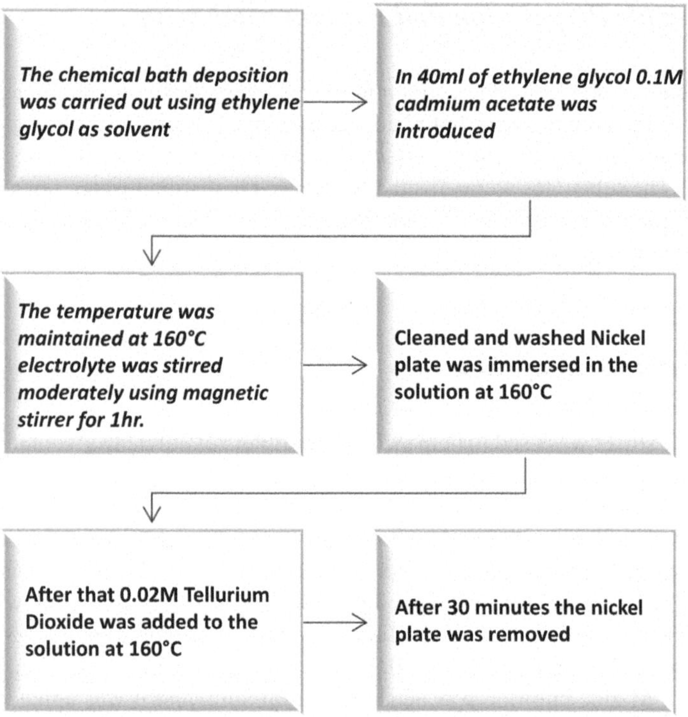

Fig. 4 Synthesis of CdZnTe through CBD process

was rinsed with pure water for ten seconds following its thirty-minute deposition. The CdZnTe nanocrystalline films that were deposited on nickel plates showed good adhesion and physical stability. The CdZnTe film that had been as deposited was annealed in air at 300°, 350°, and 400 °C. In order to observe the solid-state characteristics of the deposited films, their optical properties and structural form and composition were studied. The electrolyte preparation steps is shown in Fig. 4.

3 Results and Discussions

3.1 Structural Analysis

Figure 5 displays the XRD spectra of annealed CdZnTe films that were deposited over nickel substrate at different temperatures (300 °C, 350 °C, 400 °C, and 450 °C). At a scan rate of 3°/min, the films are scanned from 20° to 80° of 2θ. Every peak in the spectrum has a label and an identification. The annealed CdZnTe films" phase, crystallite size, miller indices, and structure are examined. Following sample annealing at 300°, 350°, 400°, and 450 °C, all film properties remain the same. The

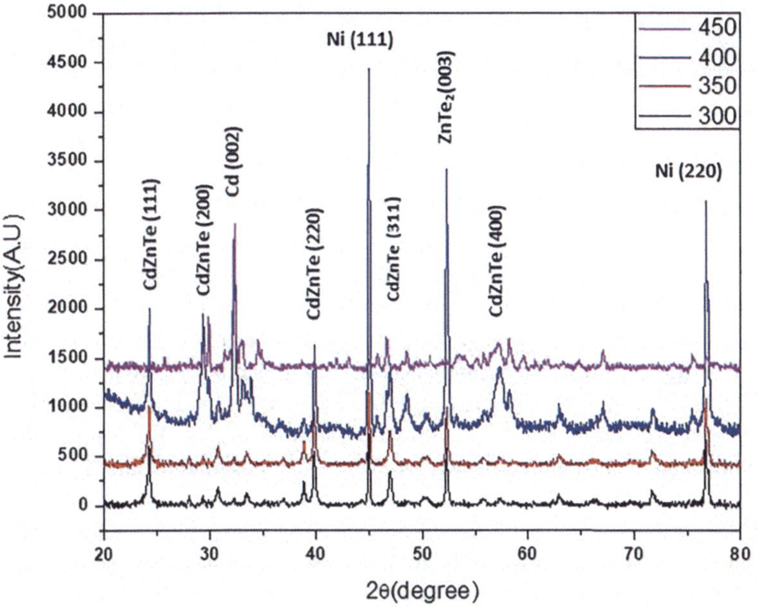

Fig. 5 XRD spectra of CdZnTe film with varying annealing temperature

(111), (200), (220), (311), and (400) planes are responsible for the CdZnTe reflection peaks, which are detected at 24.27°, 29.38°, 39.80°, 46.59°, and 57.27°, respectively. The (111) diffractive peak of the cubic zinc blende annealed CdZnTe films is responsible for the prominent peak at 24.27° (Chander & Dhaka, 2017). The intensity of preferred diffraction peak (111) is observed to increase with annealing temperature which may be attributed to the annealing process that provides the freedom movement of cadmium, zinc and tellurium atoms to form crystalline structure and also revealed to the improvement in crystallinity. The angular position of preferred diffraction peak (111) is found to shift slightly upper side due to decrease in corresponding lattice constant (Chander & Dhaka, 2017). Additionally, reflection from the nickel substrate is noted at 44.96° (111) and 76.77° (220). Because annealing increases the adherence between the deposits and substrate, the resulting annealed films have good stability and adherence. All of these are shown in Table 2, including the lattice constant (a) of the cubic phase and the interplanar spacing (d), which were determined using the relevant relation and the Scherrer formula (Chander & Dhaka, 2016a) for the average crystallite size (D). The effect of stress, vacancies, and defects on the surface of deposited thin films may be the reason for the slightly higher lattice constant of pristine CdZnTe thin films, which is determined to be 6.7 Å, compared to the usual JCPDS results of 6.4 Å. Due to a tiny increase in the corresponding angular position, it is seen to diminish with thermal annealing (Chander & Dhaka, 2017). Due to a reduction in the corresponding lattice constant, the inter-planner spacing is observed to decrease from 3.860 Å to 3.467 Å

Table 2 The crystallographic parameters of annealed CdZnTe thin films

CdZnTe film	Lattice constant(a) (Å)	Crystallite size D (nm)	Micro-strain (ε X 10^{-3})	Dislocation density ($\delta \times 10^{10}$ cm^{-2})
300 °C	6.68	30.76	64.60	105.60
350 °C	6.34	54.15	36.60	3.41
400 °C	6.33	102.9	29.30	0.94
450 °C	6.00	153.4	28.00	0.42

increasing post-annealing temperature. Table 2 presents a summary of the findings. The full width at half maximum (FWHM, β) of the diffraction peak (111) was utilized to compute the average crystallite size (D), dislocation density (δ), and lattice strain (ε) using typical mathematical relations (Chander & Dhaka, 2016b). The CdZnTe films that were annealed at 300°, 350°, 400°, and 450° degrees Celsius have been observed to have an average crystallite size of 30.76 nm, 54.15 nm, 102.9 nm, and 153.4 nm, respectively. Following thermal annealing, it is noticed that the average crystallite size increases. This could be because the associated FWHM decreases from 0.264 to 0.115, indicating a stronger contact between the substrate and atoms as a result of heat treatment, as evidenced by increased crystallinity. Larger grains are produced when the smaller grains on the substrate"s surface begin to coalesce with one another after absorbing thermal energy (Chander & Dhaka, 2016c). The lattice mismatch and decrease in strain within the deposits might potentially be the cause. While the dislocation density is defined as lines per unit volume, it is well known that lattice strain can emerge in the deposited thin films due to variation in atom displacement with respect to their reference lattice positions. Due to an increase in grain size brought about by thermal annealing, the dislocation density and lattice strain are reduced. This phenomenon may be explained by the strong interaction between the vapor atom and substrate (Zeng et al., 2007). Cd and ZnTe2 reflections are also noted at 2θ = 32.36^ (002) and 52.15^ (003), where phase inclusions cause ZnTe2 reflections. Table 1 displays the preferred planes and measured "d" values of annealed CdZnTe films at different temperatures.

3.2 *Surface Morphology Studies of CdZnTe Thin Films with Varying Annealing Temperature*

SEM images of CdZnTe films annealed at various temperatures (300 °C, 350 °C, 400 °C, and 450 °C) are presented in Fig. 6a–d. Using a 25KX magnification, the SEM study is conducted. Comparing the grains to the CdZnTe thin film that was deposited, they are larger, more homogeneous, spherical, tightly packed, and free of cracks. Grains are closely packed in the film that was annealed at 300 °C. The film exhibits some voids but the grains are evenly dispersed throughout the surface with no charging impact. But the 350 °C-annealed film is smooth, with uniformly sized

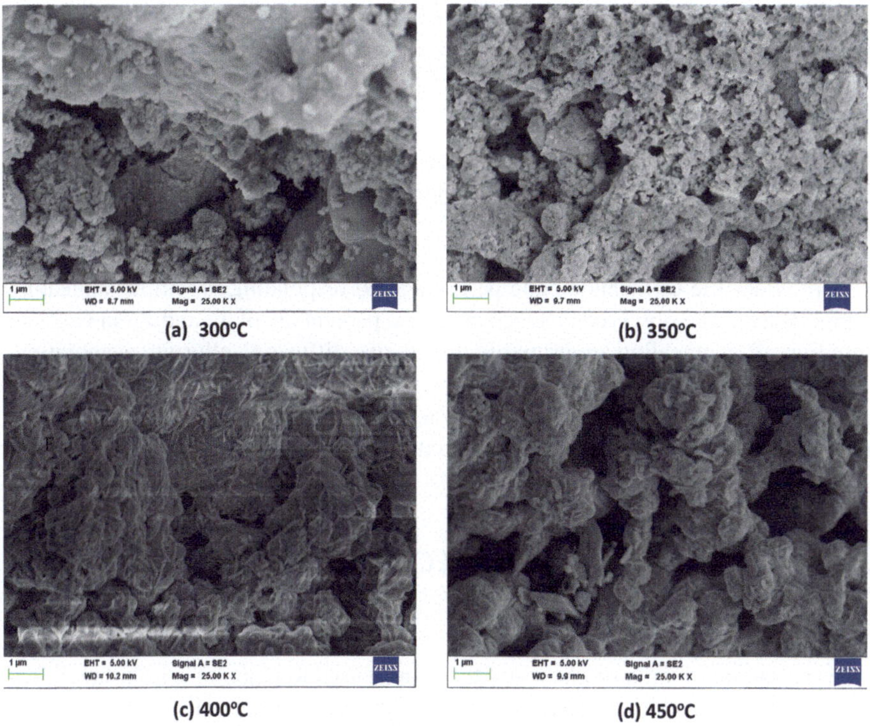

Fig. 6 (**a–d**) SEM image of annealed CdZnTe thin films annealed with varying temperature

spherical grains arranged randomly, and shows no signs of a hazy surface. It also shows some sort of voids and pits. On further increasing the annealing temperature to 400 °C, it is observed that the film surface is homogeneous, grains are distinct in single as well as cluster form throughout the analyzed area. Additionally, it depicts pits and voids of some sort. As the annealing temperature is raised to 400 °C, it is noted that the film surface is uniform and that the grains are unique both in single and cluster forms across the area under study. The figure also shows that randomly distributed nanorods were present over the surface of CdZnTe films after they were annealed at 400 °C (Ali et al., 2020). When the annealing temperature is raised to 450 °C, it is noticed that micron-sized grains take the form of circles. Ultimately, we can state that upon annealing, the surface topology demonstrates the individuality of the grains and the creation of crystalline particle clusters. The crystallization process accelerated as a result of annealing, forming the cluster to reduce the fault. Along with the XRD data, the post-heat treatment results show an increase in grain size (Fig. 6c, d). This improvement in surface attributes may be attributed to an increase in crystallinity and packing density.

3.3 Compositional Studies of CdZnTe Thin Films with Varying Annealing Temperature

Energy dispersive spectroscopy (EDS) is used to perform the elemental analysis of CdZnTe thin films that have been annealed at various temperatures, as seen in Fig. 7a–d. EDS patterns in the binding energy range of 0–20 KeV were noted. The peaks in the spectrum shown in Fig. 7a–d indicated that the deposited CdZnTe thin films contained elements of tellurium (Te), zinc (Zn), and cadmium (Cd). Table 3 also provides the elements atomic percentage corresponding to various annealing temperatures. The observed changes in atomic percentage of Te and Zn in CdZnTe thin film after annealing are primarily due to zinc diffusion, tellurium evaporation, ZnTe formation, and CdTe decomposition. These processes can significantly alter the composition and properties of the CdZnTe film, making it important to carefully control the annealing temperature and duration.

3.4 FTIR Analysis of Zn Doped CdTe Films with Varying Temperature

One crucial tool in organic chemistry that can be used to pinpoint certain functional groups in a molecule is infrared spectroscopy. Furthermore, the distinct set of absorption bands can be used to quickly identify particular contaminants as well as determine the properties of pure chemicals. The idea behind infrared spectroscopy is that molecules vibrate at particular frequencies (Meier, 2005). These frequencies (between about 4000 and 200 cm^{-1}) are in the infrared region of the electromagnetic spectrum. When infrared radiation strikes a sample, it broadcasts other frequencies while absorbing radiation at frequencies comparable to its molecular vibration frequencies. An infrared spectrometer is used to find the frequencies of absorbed radiation, and the resulting plot of absorbed energy against frequency is known as an infrared spectrum. Different materials have different vibrations and produce different infrared spectra, which makes it possible to identify a particular compound. A nonlinear molecule with N atoms can be described as having 3 N–6 fundamental vibrations, also known as normal modes, for each of its atoms. The infrared spectrum only displays molecular vibrations that are infrared active, meaning they absorb infrared light. A change in the vibrating molecule"s dipole moment is necessary for a mode to be IR active. Thus, symmetric vibrations are absent from the infrared spectrum. The vibrations in a molecule that has a center of symmetry are symmetric with regard to the center and IR passive. All of these asymmetric molecule vibrations, however, are IR active and manifest themselves in the spectrum. As a result, every chemical group included in the sample is detected simultaneously. It"s interesting to note that this method easily finds water molecules and ammonia acids, both of which are challenging to find using other spectroscopies. In the infrared spectrum, chemical groups with persistent dipoles such as polar bonds show

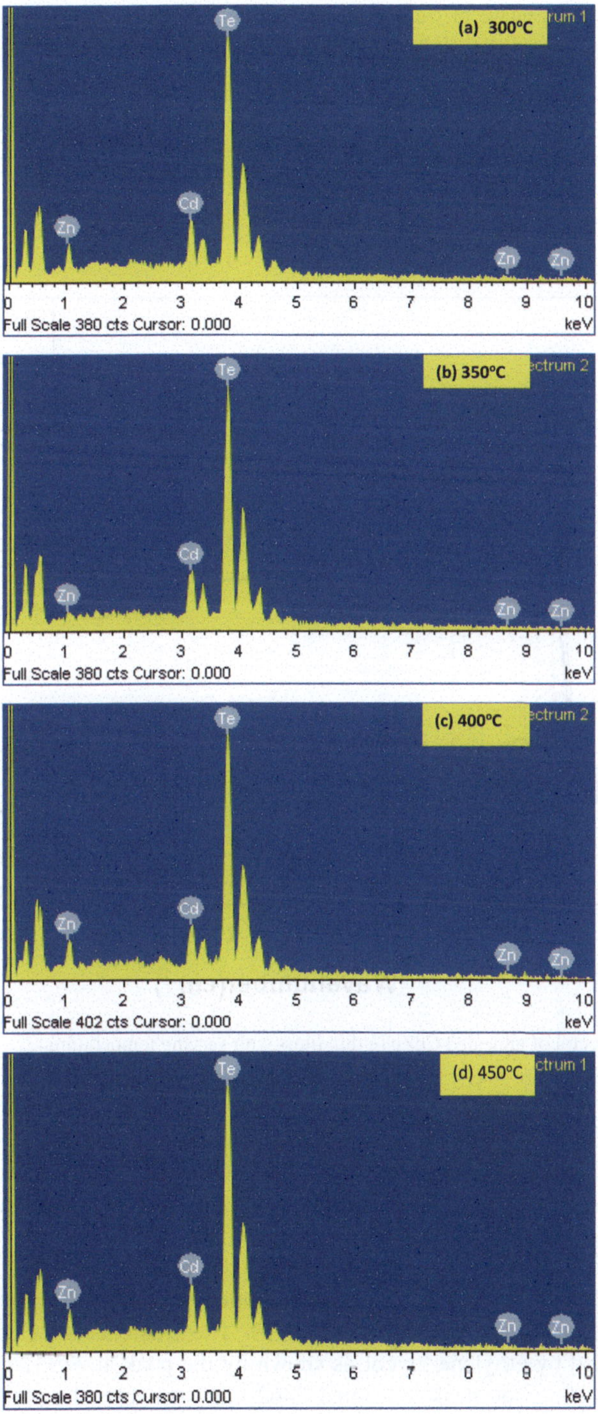

Fig. 7 (**a–d**) EDS spectra of CdZnTe thin films annealed at different temperature

Table 3 Atomic percentage of Cd, Zn and Te in CdZnTe thin films with varying annealing temperatures

Temperature	Atomic% Cd	Atomic %Zn	Atomic % Te
300 °C	16.64	6.61	19.26
350 °C	17.28	07.87	18.91
400 °C	19.18	08.19	17.98
450 °C	19.82	08.51	17.05

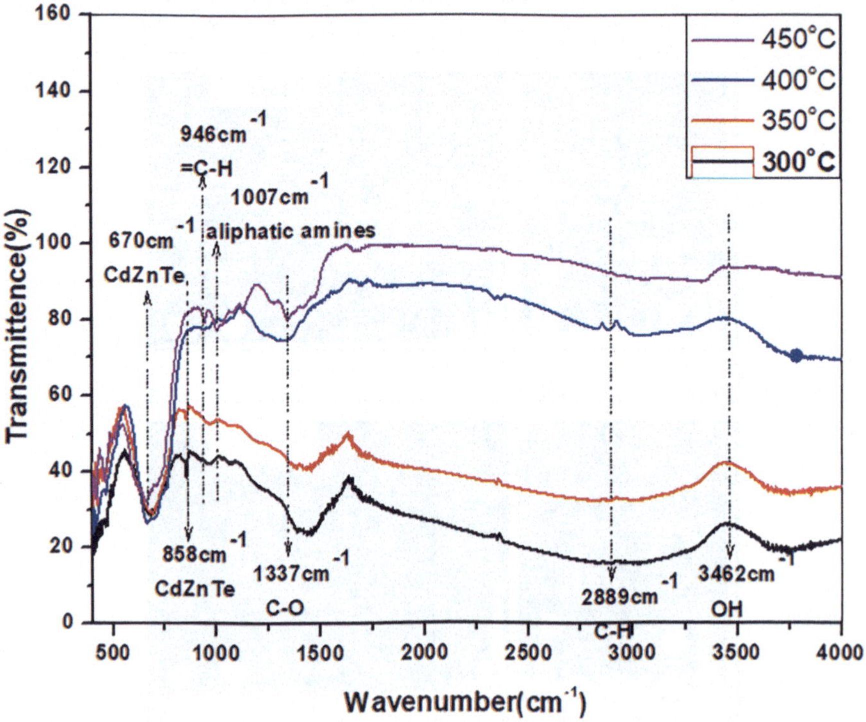

Fig. 8 FTIR spectra of annealed CdZnTe thin films with varying temperature

high absorption. The FTIR transmission spectra of annealed CdZnTe thin films at different temperatures (300°, 350°, 400°, and 450 °C) within a wavenumber range of 400–4000 cm^{-1} is displayed in Fig. 8. The CdZnTe IR spectra in the sample are situated at around 670 cm^{-1} and 858 cm^{-1} (Chander & Dhaka, 2017). The presence of alkene bending vibrations is indicated by the peak at ~946 cm^{-1} (Bakri & Jayamani, n.d.). The stretching mode of aliphatic amines is shown by the peak at ~1007 cm^{-1} (Janakiraman & Johnson, 2015). The existence of C-O stretching, which is caused by ethylene glycol, is shown by the peak at ~1337 cm^{-1} (Kumar et al., 2017). The wide peak at around 3462 cm^{-1} suggests the presence of OH extension, which is the cause of the atmosphere (Kumar et al., 2015). The presence

Table 4 Different functional groups appeared in CdZnTe thin films with varying annealing temperatures

S.No.	Functional groups	Wavenumber(cm^{-1})
1.	CdZnTe	~670
2.	CdZnTe	~858
3.	=C-H	~946
4.	Aliphatic amines	~1007
5.	C-O	~1337
6.	C-H	~2889
7.	O-H	~3462

of C-H stretching vibration is shown by the peak that appears at ~2889 cm^{-1} (Tang et al., 2005). The picture also shows that the IR transmittance increases with post-annealing temperature, which may be related to the annealing process"s reduction of crystal defects and the improvement in crystalline behavior that the XRD results indicate. Table 4 shows different functional groups appeared in annealed CdZnTe thin films with varying temperature.

3.5 Optical Analysis of CdZnTe Thin Films with Varying Annealing Temperatures

3.5.1 Photoluminescence Analysis of CdZnTe Thin Films with Varying Annealing Temperature

Using a spectrophotometer with a wavelength range of 200–800 nm, the optical behavior of annealed CdZnTe thin films is investigated. The photoluminescence spectra of CdZnTe films at different annealing temperatures (300 °C, 350 °C, 400 °C, and 450 °C) are displayed in Fig. 9. In this study, excitation is provided via a 250 nm wavelength Xenon lamp source. All of the CdZnTe crystals" free carriers are excited by the photon absorption. The strongly luminescent excited CdZnTe crystal"s band gap edge is shown by the excited PL spectra. The emission peaks are measured at wavelengths of 372, 448, and 471 nm, in that order. The spectra provide evidence that the visible emission peak, which corresponds to the crystalline thin layer of CdZnTe, is detected at 470 nm. The emission intensity increases sharply as the annealing temperature varies. When compared to the emission peak at ambient temperature, the emission peaks exhibit a significant red shift at higher temperatures, specifically from 300° to 450 °C. This results from the combined effects of the band gap"s temperature-induced red shift and exciton-induced blue shift. The narrower and more symmetric band edge PL peak that appears at a different wavelength suggests that annealing can remove stress and flaws from within the crystal lattice. Defect elimination improves the annealed films" crystallinity and reduces deep trap emission (Jin et al., 2009).

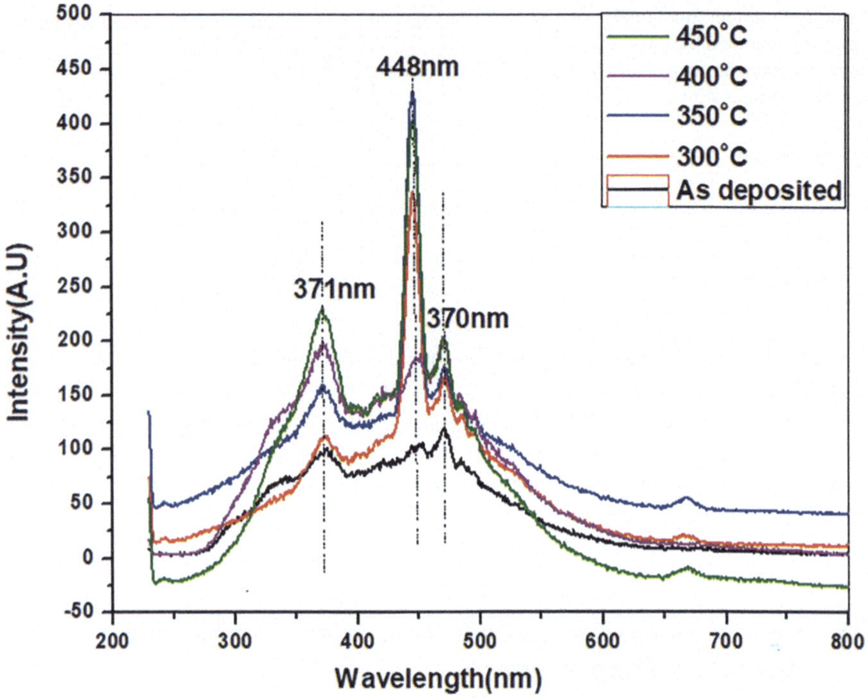

Fig. 9 PL spectra of annealed CdZnTe thin films with varying temperature

4 Conclusion

Successful preparation of the CdZnTe thin films on nickel substrate was achieved using the chemical bath deposition approach in a non-aqueous solution. Using various characterization approaches, the solid state and optical properties of annealed films at different temperatures (350 °C, 400 °C, 450 °C) were effectively examined. Every film that has been deposited at a different temperature shows a peak that matches the (111) plane in the cubic zinc blende structure. The intensity of the diffraction peak associated with the (111) plane rises as the annealing temperature rises. After annealing, the average crystallite size and crystallinity of the deposited CdZnTe films improve. After annealing, the SEM images display varied film orientation and homogenous, compact, void-free surfaces. Te, Zn, and Cd are confirmed to be present by EDS spectra. These thin films" photoluminescence characteristics might reveal important details about the surface states that could account for the surface fluorescence"s size dependency. Following annealing, a narrower, more symmetric band edge PL peak appears at various wavelengths. Higher temperatures cause a significant red shift in the emission peak. The annealed CdZnTe film that is being described has good characteristics for usage in sensor applications and optoelectronics for solar cells.

Acknowledgements The authors express their gratitude to the Central Instrumentation Facility (CIF) at BIT-Mesra and the Centre for Research Facilities (CRF) at the Indian Institute of Science (IISc) Bengaluru for providing valuable characterization support.

References

Ali, M., Ahmed, S., Younus, F., & Ali, Z. (2020). Electrical, charge transients and photo response study of as-deposited and phosphorus implanted Cd1-xZnxTe devices for PV applications. *Radiation Physics and Chemistry, 166*, 108498.

Bagaev, V. S., Zaitsev, V. V., Onishchenko, F. E., & Sadofyev, Y. G. (2000). Photoluminescence of various $Zn_{1-x}Cd_xTe$/ZnTe heterostructures grown by MBE on GaAs(0 0 1) and ZnTe(0 0 1) substrates: temperature dependences. *Journal of Crystal Growth, 214*, 250–254.

Bakri, M. K. B., & Jayamani, E. (n.d.). Comparitive Study of Functional Groups In Natural Fibers: Fourier Transform Infrared Analysis (FTIR). In *In Book: Futuristic trends in engineering, science, humanities, and technology* (pp. 167–174). Edition: 1, Chapter: 30, Publisher: TRO India.

Bouarissa, N. (2007). Pseudopotential calculations of $Cd_{1-x}Zn_xTe$: Energy gaps and dielectric constants. *Physica B, 399*, 126–131.

Brun, M., Haunt, S., Woehl, J. C., Motte, J. F., Marshal, L., & Mariette, H. (2002). Excitons and multi-excitons in single CdTe quantum dots probed by near-field spectroscopy. *Solid State Communications, 121*, 407–410.

Chakraborty, T. (1999). *Quantum dots-a survey of the properties of artificial atoms*. Elsevier.

Chander, S., & Dhaka, M. S. (2016a). Effect of thickness on physical properties of electron beam vacuum evaporated CdZnTe thin films for tandem solar cells. *Physica E, 84*, 112–117.

Chander, S., & Dhaka, M. S. (2016b). Influence of thickness on physical properties of vacuum evaporated polycrystalline CdTe thin films for solar cell applications. *Physica E, 76*, 52–59.

Chander, S., & Dhaka, M. S. (2016c). Thermal evolution of physical properties of vacuum evaporated polycrystalline CdTe thin films for solar cells. *Journal of Materials Science: Materials in Electronics, 27*, 11961–11973.

Chander, S., & Dhaka, M. S. (2017). Thermal annealing induced physical properties of electron beam vacuum evaporated CdZnTe thin films. *Thin Solid Films, 625*, 131–137.

Hong, W. P. (2010). Optical properties of strained CdTe/ZnTe pyramidal quantum dots. *Japanese Journal of Applied Physics, 49*, 012801.

Janakiraman, N., & Johnson, M. (2015). Functional Group of Tree Ferns (Cyathea) using FTIR: Chemotaxonomic implications. *Romanian Journal of Biophysics., 25.*

Jin, C. Q., Zhong, W., Zhang, X., Deng, Y., Au, C., & Du, Y. (2009). Crystal growth & design, *9*(11), 4602–4606.

Johnson, N. F., Ehrenreich, H., Hui, P. M., & Young, P. M. (1990). Electronic and optical properties of III-V and II-VI semiconductor superlattices. *Physical Review B41*, 3655.

Kozlovsky, V. L., Sadofyev, Y. G., & Litvinov, V. G. (2000a). Band alignment in ZnCdTe/ZnTe and ZnCdSe/ZnSe SQW structures grown on GaAs(100) by MBE. *Nanotechnology, 11*, 241.

Kozlovsky, V. I., Sadofev, Y. G., & Litvinov, V. G. (2000b). 8th International Symposium Nanostructures. *Physics and Technology, 185.*

Kryshtab, T., Andrace, J. A., Borkorska, L. V., Korsunska, N. O., Ye, F., & Venger, Y. G. (2008). Sadofyev, effect of CdTe monolayer insertion on CdZnTe/ZnTe quantum well characteristics. *Microelectronics Journal, 39*, 418–422.

Kumar, S. R., Kumar, S., Sharma, S. K., & Roy, D. (2015). Structure, composition and optical properties of non-aqueous deposited ZnCdS nanocrystalline film. *Materials Today: Proceedings, 2*, 4563–4568.

Kumar, S., Rajpal, S., Sharma, S. K., Roy, D., & Kumar, S. R. (2017). Effect of Zn concentration on the structural, morphological and optical properties of ternary ZnCdS Nanocrystalline thin films. *Digest Journal of Nanomaterials and Biostructures, 12*, 339–347.

Lee, D., Johnson, A. M., Zucher, J. E., Feldman, R. D., & Austin, R. F. (1991). Room temperature excitonic absorption in CdZnTe/ZnTe quantum wells: Contributions to exciton linewidth. *Journal of Applied Physics, 69*, 6722–6724.

Lee, H. S., Park, H. L., & Kim, T. W. (2006). Optical properties of CdTe/ZnTe sandwiched between two quantum wells with ZnTe separation barriers. *Applied Physics Letters, 89*, 181929.

Meier, R. J. (2005). Vibrational spectroscopy: A 'vanishing' discipline? *Chemical Society Reviews, 34*, 743–752.

Novikov, S. V., Stanton, N. M., Campion, R. P., Morris, R. D., Geen, H. L., Foxon, C. T., & Kent, A. J. (2008). Molecular beam epitaxy as a method for the growth of freestanding zinc-blende (cubic) GaN layers and substrates. *Semiconductor Science and Technology, 23*, 015018.

Pan, E. (2002). Elastic and piezoelectric fields around a quantum dot: Fully coupled or semi-coupled model? *Journal of Applied Physics, 91*, 3785–3796.

Stebe, B., Essaoudi, I., Alinanc, A., & Saber, M. (2001). *Phys. Rev* (p. 085304). B64.

Tang, H., Yan, M., Zhang, H., Xia, M., & Yang, D. (2005). Preparation and characterization of water soluble CdS nanocrystals by surface modification of ethylene diamine. *Materials Letters, 59*, 1024–1027.

Walter, G., Holonyak, N., Jr., Ryou, J. H., & Dupis, R. D. (2001). Room-temperature continuous photopumped laser operation of coupled InP quantum dot and InGaP quantum well InP–InGaP–In(AlGa)P–InAlP heterostructures. *Applied Physics Letters, 79*, 1956–1958.

Yang, J.-H., Chen, S., Wan-Jian Yin, X. G., Walsh, G. A., & Wei, S. H. (2009). *Electronic structure and phase stability of MgTe, ZnTe, CdTe, and their alloys in the B3, B4, and B8 structures* (p. 245202). Physical Review B79.

Zeng, D., Jie, W., Zha, G., Wang, T., & Yang, G. (2007). Effect of annealing on the residual stress and strain distribution in CdZnTe wafers. *Journal of Crystal Growth, 305*, 50–54.

Two-Dimensional Molybdenum Disulfide Nanosheets Based Optoelectronic Devices

Arun Singh Patel, Praveen Mishra, Anirban Chakraborti, and Prianka Sharma

1 Introduction

In 2004, a revolutionary advancement was made in the two-dimensional (2D) field of materials by exfoliating single layered Graphene from bulk graphite (Novoselov et al., 2004). Two-dimensionality of this wonder element attributes to its exotic and exciting properties like higher electrical conductivity of ~10^4 Ω^{-1} cm^{-1} due to zero band gap, specific surface area high upto 2630 m^2/g, higher transparency etc. Monolayer of graphene exhibits Young"s modulus as high as 1.1 TPa also (Allen et al., 2010; Soldano et al., 2010). sp_2 -hybridized orbitals in graphene leads to unique properties like anomalous quantum Hall effect, excellent thermal conductivity, high mobility of charge carriers, chirality etc. (Meyer et al., 2006; Kumar et al., 2015; Li et al., 2011). Semi-metallic nature of graphene is due to the presence of $\pi - \pi^*$ bonds. Graphene's electronic property arises due to symmetry of valence and conduction band near the Dirac point and explained by Dirac equation instead of Schrodinger equation. There Fermi surface arises at the valence band and the conduction band intersection of this zero band gap material. Thus, practical applications of graphene in solar cells, devices for energy storage, transparent electrodes, catalysis etc. arises due to these remarkably excellent properties (Zang et al., 2014; Bae et al., 2010; Li et al., 2014a, b). But the uniqueness of graphene i.e. it's zero

A. S. Patel
Department of Physics, Dyal Singh College, University of Delhi, New Delhi, India

P. Mishra · A. Chakraborti
School of Computational & Integrative Studies, Jawaharlal Nehru University, New Delhi, India

P. Sharma (✉)
Department of Physics, School of Basic & Applied Sciences, Maharaja Agrasen University, Solan, HP, India

S. A. Bansal et al. (eds.), *Emerging Applications of Novel Nanoparticles*, Lecture Notes in Nanoscale Science and Technology 37, https://doi.org/10.1007/978-3-031-57843-4_11

band gap poses limitation to its many applications in low power logical circuits of electronic switching. Another 2D material that exhibit similar properties like graphene but is an insulator is hexagonal Boron nitride (hBN). It finds wide application as electrical insulator in gate dielectrics of capacitors (Hui et al., 2016).

Hence, to overcome the limitations of layered 2D graphene, lot of investigations have been focussed on the search of other layered 2D materials, so that the range of applications of these materials can be widely extended to next generation semiconductor devices. In this search, a new class of 2D layered materials, Transition Metal Dichalcogenides (TMDCs), an isostructural analogous of graphene have redeveloped interest in diverse applications in various fields (Tang & Zhou, 2013; Butler et al., 2013; Lv et al., 2015). These materials are chemically stable and inert at room temperature and possess high melting point about 1000 °C. These newly discovered materials include molybdenum disulfide (MoS_2), molybdenum diselenide ($MoSe_2$), tungsten disulphide (WS_2), tungsten diselenide (WSe_2) etc. Since molybdenite is abundantly available in nature, MoS_2 is the most deeply investigated layered transition metal dichalcogenides (TMDCs). 1.8 eV direct band gap of monolayer of MoS_2 and its flexible tunability from direct to indirect band gap serve as the most important property that can overcome the limitation of zero band gap of graphene. This makes MoS_2 an efficient 2D material for applicability in next level optoelectronic and switching devices. Hence, MoS_2 can be efficiently utilized in hydrogen evolution reaction (HER), energy conversion, lithium ion batteries etc. Additionally, odd numbered layers of MoS_2 show alternating current outputs and piezoelectric voltage that leads to its prospective applications in the field of energy storage nanodevices and stretchable electronics (Kumar et al., 2015; Li et al., 2011; Wi et al., 2014; Ding et al., 2012; Stephenson et al., 2014; Wu et al., 2014; Jariwala et al., 2014).

In present time, optoelectronic devices have become an integral part of electronic era of data-driven technologies due to the requirement of high-speed optical communication. Though graphene has been extensively utilized in many electronic devices, however its semi-metallic behaviour and zero band gap puts limit to its possible applications in semiconducting optical devices. In this context, recently MoS_2 has proved to be a promising semiconducting material in optoelectronic devices. MoS_2 demonstrates good light-matter interaction characteristics that can be efficiently tuned with external physical factors like magnetic and electric field, strain, pressure etc. Improved thermal stability, good mechanical stability, flexible and sizeable bandgap of MoS_2 in the visible and infrared region gives MoS_2 favourable edge for efficient optoelectronic photodetectors and light-emitting devices in optical communication of short range over traditional silicon based technology.

In this chapter, we have attempted to present a broad view of 2D TMDCs large family, highlight composition of MoS_2 and its physical properties. We shall be discussing few opto-electronic devices like photodetectors, photovoltaics, and solar cells which are based on 2D MoS_2 nanosheets, along with the device structure and architecture and their working mechanisms.

2 Properties of TMDCs

2.1 Composition and Phase Change

TMDCs are made up of group 4 ~ 10 transition metals combined with chalcogens as shown in Fig. 1. General formula MX_2 represents TMDCs where M represents the transition metals (Ti, Zr, Hf, V, Nb, Ta, Mo, W, Tc, Re, Co, Rh, Ir, Ni, Pd, Pt) while X position is occupied by atoms of chalcogen like Sulphur (S), Selenium (Se), or Tellurium (Te). TMDCs exhibit strong anisotropy in chemical, physical, electrical, thermal and mechanical properties and they are easily exfoliated into layers like graphite. Layered structure is dominant in Group 4–7 TMDCs, however, TMDCs of group 8–10 show non-layered structure. MoS_2 is a group 6 TMDC with layered structure, having 6 ~ 7 Å thickness of each layer typically. Layers are hexagonally packed with transition metal (TM) atoms of molybdenum (Mo) sandwiched between chalcogen atoms of sulphur (S) in each layer as S-Mo-S (Chhowalla et al., 2014; Li & Zhu, 2015; Ganatra & Zhang, 2014). Strong intra-layer covalent bond exists between TM (Mo) and chalcogen (S) atoms while weak Van-der Waals forces couple the layers that form a sandwich. This weaker Van-der Waals coupling provides the advantage of easy cleavage of the surface layers of MoS_2 (Bertolazzi et al., 2011). +4 oxidation state is attained by TM atoms since the bond states are filled by four electrons supplied by TM atoms. On the other hand −2 oxidation state is attained by the chalcogen atoms which supply the lone-pair electrons to the surface layers that show stability towards environmental reactions due to the absence of dangling bonds. The length of the M-X atoms bond influences the size of the chalcogen and metal atoms that vary between 3.15 Å and 4.03 Å. However, the bond length of M-X atoms is 15–25% greater than that of elemental transition metals that poses limitations over the TMDCs d orbitals spatial overlapping.

Two types of metallic co-ordinations are exhibited by layered TMDCs i.e. either octahedral or trigonal prismatic depending upon thermodynamic stability and combination of M –X elements (Fig. 2). Unlike graphite, stacked polytypes and various polymorphs are exhibited by bulk TMDCs. Two phases of three individual X-M-X atom layers form monolayer MX_2. Commonly existing polymorphs are represented

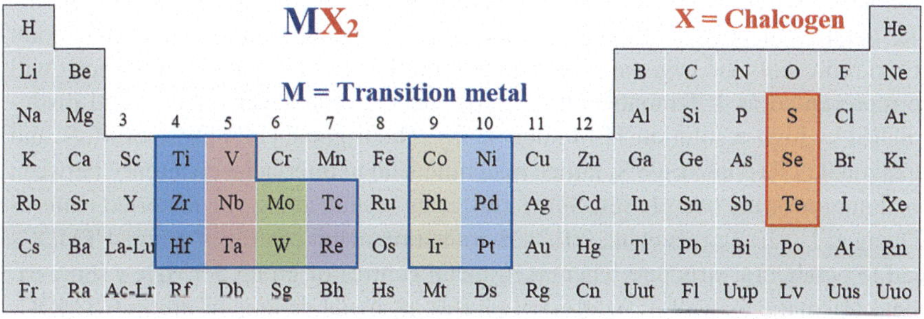

Fig. 1 Position of transition metals and chalcogen atoms commonly utilized forTMDCs

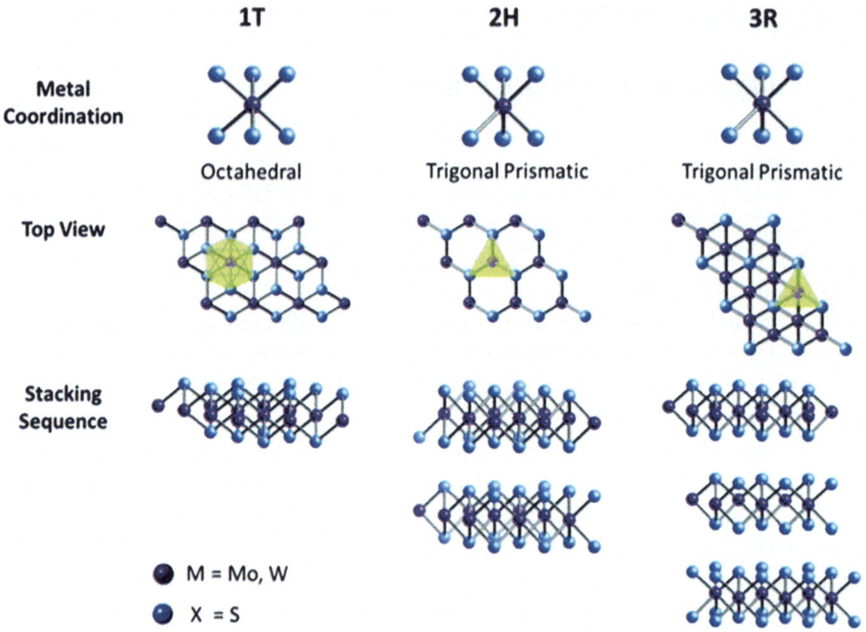

Fig. 2 Representations of 1 T octahedral coordination, 2H or 3R triangular prismatic coordination of MoS_2 (Toh et al., 2017). (Reprinted Figure under @CCC (Royal Society of Chemistry))

by 1 T(trigonal), 2H(hexagonal) and 3R(rhombohedral) forms with stacking sequence of sandwiched layers of X-M-X structures. Existence of multiple polymorphs and polytypes depend upon the formation of TMDC. For example, '2H phase' exists for naturally occurring MoS_2 shown by AbABaB structure. Here the upper case in the stacking sequence represents the chalcogen atoms and the metallic atoms are shown by the lower case. However, synthetically fabricated MoS_2 is represented by '3R phase' showing stacking sequence AbACaCBcB (Ganal et al., 1993). Trigonal prismatic phase is exhibited by the metal atom in both the cases. Group 4 TMDC like TiS_2 shows AbCAbC stacking sequence with 1 T phase and octahedral metal coordination. In simpler case focus is on monolayer TMDC. For monolayer TMDCs two polymorphs phases exist: octahedral and trigonal prismatic. The octahedral polymorph comes under the D_{3d} point group whereas D_{3h} point group represents the trigonal prismatic polymorph. Further, these polymorphs will be represented as 1 T monolayer (or D_{3d}) and 1H monolayer (or D_{3h}) and distinguished by high-resolution TEM microscopy. The trigonal (1 T) phase of MoS_2 and hexagonal (1H) phase of WS_2 lattice matching lead to interfacial coherence between the domains of the two phases. Some clustering results in zig-zag distorted patterns (Zhang & Li, 2020; Eda et al., 2012). d_0 transition metal centres of group 4 TMDCs lead to octahedral structure whereas d_1 metal centres of group 5 TMDCs show trigonal prismatic as well as octahedral phases. d_2 metal centres of group 6 TMDCs show trigonal prismatic geometry. Distorted octahedral structure is exhibited by d_3

metal centres of group 7 TMDCs. However, octahedral structure is represented by d_6 metal centres of group 10 TMDCs. It has been observed that TMDC intercalation with alkali metals can bring alteration in the TMDCs phase. For example, lithium intercalation of MoS_2 transforms 2H polymorph to 1 T (Py & Haering, 1983). Reversely, Li intercalation of TaS_2 leads to transition of 1 T polymorph to 2H. By transferring alkali metal s orbital valence electron to TM d orbital, there is destabilization in the original phase due to d orbital electrons change and free energy change of the two phases as well. Presently, 2H-1 T phase hybrid structure can be obtained with partial transformation. Seifert et al. developed an effective technique of controlling local phase transformation by doping with 1 T stabilized phase leading hybrid structure of metal–semiconductor (Lorenz et al., 2012). Metallic character of MoS_2 is shown by 1 T phase, thus electronic hetero-junction is shown by 1H-1 T interfaces across chemically homogeneous layer. This feature of MoS_2 can be effectively utilized as potential way of fabricating electronic devices at molecular level. However, instability in the electronic structure of MoS_2 give rise to certain periodic distortions in the crystal lattice besides phase transformations (Castro Neto, 2001). It has been observed that at comparatively lower temperature (<40 K for $NbSe_2$ and < 120 K for $TaSe_2$) there is stability in phases and with room temperature lattice distortions starts appearing for intercalated MoS_2. This lattice distortion in MoS_2 can be explained by splitting of partially occupied degenerate orbitals arising due to free energy reduction. This phenomenon is known as 'Jahn–Teller instability' (Heising & Kanatzidis, 1999). At room temperature, metal atoms form super lattice of Li intercalation in MoS_2 by formation of chain clusters. As a result, atoms are shifted from their equilibrium position and intercalant is removed to maintain the distorted phase to form the superlattice. However, properties of undistorted TMDCs are significantly different from that of distorted TMDCs (Castro Neto et al., 2009).

2.2 *Electronic Properties*

TMDCs are a large family exhibiting electronic nature from metal to semiconductor and superconducting properties as well (Lee et al., 2010). The electronic and semiconducting nature of TMDC arises from the d-electrons of TM and their coordination environment given in Table 1. The M–X bonds of TMDCs of 1 T and 1H phases exhibit band gap are of two types: bonding (σ) and anti-bonding (σ*) existing between non-bonding d bands. The D_{3d} TM centres lead to formation of degenerate orbitals of $d_{yz,xz,xy}$ (t_{2g}) and d_{z^2,x^2-y^2} (e_g) with octahedral coordination that can accommodate maximum 6 d electrons for group 10 TMDCs. On contrary, TM d orbital splits in case of D_{3h} trigonal prismatic coordination in three groups $d_{x^2-y^2,xy}$ (e), d_{z^2} (a1) and $d_{xz,yz}$ (e') having ~1 eV gap among orbitals of first two groups.

The electronic properties of group 4 to group 10 TMDCs strongly depend upon continuous filling of d bands. Metallic conductivity of $NbSe_2$-2H phase and ReS_2-1 T arises due to partially filled orbitals. The semiconducting behaviour is exhibited by

Table 1 Electronic characteristics of layered TMDCs

Group	M	X	Properties
4	Ti, Zr, Hf	S, Te, Se	Semiconducor with ($E_g = 0.2 \sim 2$ eV)
5	V, Ta, Nb	S, Te, Se	Metals with narrow energy band metals i.e. semi-metals
6	W, Mo	S, Te, Se	Selenides and sulfides are semiconductor with semi-metallic properties of telurides
7	Re, Tc	S, Te, Se	Narrow band gap semiconductors
10	Pt, Pd	S, Te, Se	Selenides and sulfides are semiconductor with Metallic telurides and superconducting $PdTe_2$

2H-MoS_2, 1 T-PtS_2 and 1 T-HfS_2 TMDCs due to completely filled orbitals. The chalcogen atoms electronic structure is affected little in comparison to TM atoms due to the orbital filling. As the chalcogen atoms atomic number increases, there is decrement in the band gap with broadening of d band. For example, as the chalcogen atoms atomic number increases (S < Se < Te) in 2H phase of MoS_2, $MoSe_2$ and $MoTe_2$, the band gap decreases smoothly from 1.3 to 1.0 eV (Lee et al., 2010; Wilson & Yoffe, 1969). Thus, it has been observed that number of *d*-electrons in TM determines the phase of TMDC. Table 1 presents summarization of the differently layered TMDCs electronic character.

2.3 Optical Properties of Layered MoS_2

Optical properties of MoS_2 plays significant role in many device applications (Frindt & Yoffe, 1963). Raman spectroscopy significantly helps us to understand the structural evolution of 2D layered TMDCs from 3D bulk material (Buscema et al., 2014). It helps to go deep into the layered TMDC and gives information about layer number and crystal quality (Eda et al., 2011). Lee and co-workers explained transition from few layered to single layer MoS_2 by Raman spectroscopy (Lee et al., 2010). In general, there are two typical peaks in Raman spectra for pristine MoS_2 depending upon the frequency difference and position of the peaks. These peaks are the E^1_{2g} and the A_{1g} where the former peak represents the opposite direction vibration of atoms of Mo and S giving rise to the in-planer vibration mode that arises due to built-in strain in 2D MoS_2. The other peak A_{1g} arises from interlayer van der Waals interaction that lead to S atoms vibrate out-of plane (Zhan et al., 2012). As the number of layer reduces from bulk to single layer, the A_{1g} shows a red shift while the E^1_{2g} peak exhibits a blue shift for MoS_2 monolayer. This trend of peaks shift in opposite direction arises from induced charge stacking possibility in intra-layer bond due to Columbic interaction partially. Also, the peak frequency decreases between A_{1g} and E^1_{2g} as one goes from many to few layered MoS_2. Another observation shows linear increase in peak intensity for both the modes while moving from bulk to few layers. But as the layer number reduces below four, the peak intensity also falls and becomes broader for MoS_2 monolayer (Zhan et al., 2012; Molina-Sánchez et al., 2015) (Fig. 3).

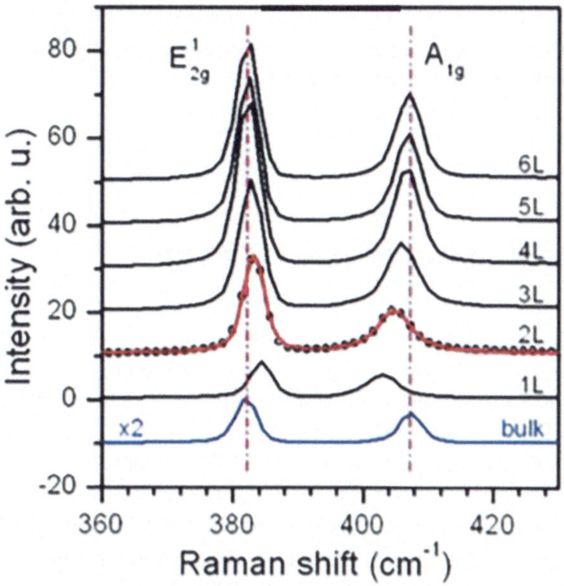

Fig. 3 Variation in peak intensity in Raman spectra of thin and bulk MoS_2 films (Lee et al., 2010). (Reprinted figure with permission Copyright (2010) by the American Physical Society)

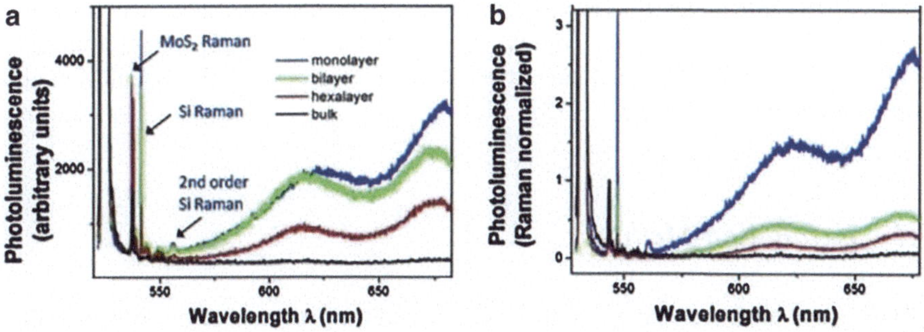

Fig. 4 (**a**) Pl spectra of the MoS_2 NSs, (**b**) Raman spectra of thin and bulk MoS_2 films (Splendiani et al., 2010). (Reprinted figure with permission Copyright (2010) by the American Physical Society)

Other optical properties depending on MoS_2 layers are the photoluminescence (PL) and band structure engineering. Splendiani et al. studied the PL response of 1 L (monolayer), 2 L (bilayer), 6 L (hexalayer) and bulk MoS_2 (Splendiani et al., 2010). It has been observed that negligible PL signals are shown by bulk MoS_2. However, significant PL signals are received as the number of layers decreases Fig. 4a, b. Strong PL signals are received from monolayer MoS_2 giving absorption peaks of A1 and B1 excitons at 670 nm and 627 nm (Mak et al., 2010). Spin-orbit coupling leads to splitting of valence band energy giving rise to excitonic peaks. PL resonances become significant due to direct transitions of the excitonic peaks at the Brillouin zone K point in MoS_2 monolayer. This explains well the indirect (1.2 eV) to direct (1.8 eV) transition of band gap while moving from bulk to single layer MoS_2 (Splendiani et al., 2010). But there disappearance of PL absorption peaks for bulk MoS_2. Thus, it can be well concluded that the qualitative band structure manipulation can be observed by altering the layered engineering from multilayer to monolayer MoS_2.

3 Synthesis Methods of TMDCs

In the recent years of evolution of TMDCs, specific morphological fabrication of MoS_2 has attracted interest of many research groups. Over these works, it has been understood that surface morphologies plays crucial in understanding the TMDCs role in various applications. Hence, it is necessary to develop high quality scalable methods for synthesis of 2D TMDCs that are cost effective too. Thus, different methods have been categorized into bottom-up and top-down synthesis approaches of MoS_2. Chemical vapour deposition (CVD) and sulphurization method and hydrothermal synthesis (Seravalli & Bosi, 2021; Wang et al., 2013; Hong et al., 2018; Guan et al., 2021; Hyun et al., 2018; Yue et al., 2019; Jin et al., 2014; Kim et al., 2023; Duraisamy et al., 2021; Li et al., 2016; Muralikrishna et al., 2015; Chaudhary et al., 2018) fall under bottom-up approaches, whereas, mechanical exfoliation method and liquid exfoliation methods come under top-down approaches (Ottaviano et al., 2017; Li et al., 2014c; Huang et al., 2020; Zhang et al., 2021; Nicolosi et al., 2013; Niu et al., 2016; Dines, 1975).

3.1 Exfoliation Method

In MX_2 exfoliated nanosheets, there is exposure of the prismatic edges as well as the basal planes. The terminating edges of M or X atoms are controlled by the growth environment chemical potential. By exfoliating the MX_2 adjacent layers, they are made free from s-p_z orbital interaction thus widening the bandgap. This reformation of orbital hybridization helps to transform indirect band gap TMDCs to direct band gap monolayered TMDCs. Hence, transformation of the band gap and its widening enhances the TMDCs photoluminescence property compared with bulk counterpart. Thus, coordination bonding at the edges of exfoliated TMDCs governs the surface chemistry of TMDCs. The surface chemistry is also governed by controlled shape and size of sheets. When X–M–X tri-layer sheets size laterally reduced, then low-coordination step-edges develop in the nanosheets. This also results in development of additional local chemical effects which arises due to corners and kinks at the edges. Once, density of edges in MX_2 nanosheets drop below certain critical threshold size, the nanosheets equilibrium shape is controlled by adjustment of the edge atoms which dominates the basal planes. This leads to open sites or missing edges in MX_2 nanosheets which give rise to metallic edge states preferably for catalytic applications.

3.2 Mechanical Exfoliation

Graphene successfully exfoliated from bulk graphite by simple "Scotch tape method" encourages fabrication of other 2D materials like (Yi & Shen, 2015). monolayered 2D MoS_2 was synthesized from their bulk conterparts (Ottaviano et al., 2017; Li et al., 2014c). Mechanical exfoliation is an efficient method for preparation of mono-layered TMDCs. This method furnishes high quality monolayers for production of intrinsic nanosheets for fundamental research. The basic properties of these layered materials are studied by scanning tunneling microscopy (STM), atomic force microscopy (AFM), transmission electron microscopy (TEM) and optical microscopy etc. However, the large scale industrial or device applications this method of exfoliation does not find much suitability since the number of layers and size of the sheet (tens of micron scale) is quite difficult to control in this method. Hence, low yield is an important disadvantage of the mechanical exfoliation. This poses many limitations to the applications in real devices.

3.3 Liquid Exfoliation by Ultrasonication

For exploiting the potential application of TMDCs, nanosheets are required in large quantities which can be obtained by adopting a high yielding top-down approach i.e. liquid exfoliation by ultrasonication in solvent or ion intercalation for preparation of MoS_2 nanosheets (Fig. 5). This is a simple chemical approach (Nicolosi et al., 2013; Niu et al., 2016; Gupta et al., 2015; Rani et al., 2020; Rani et al., 2021). This route initiates from bulk MoS_2 that exfoliates into flakes of layers of various shapes and sizes. This method of exfoliation gives high yield but quality is lower. In this ultrasonication, rapid rarefaction and compression are created by sound waves which forms cavities of vacuum between the medium that collapses due to high pressure jets which peels off the layer. Cavity formation is lower in organic liquids compared to water and varies with types of solvents. Solvents like xylene, styrene, toluene, etc. which have cyclic structure preferably form more cavities in compared to aliphatic amines and carboxylic acids (Korn et al., 2003; Zhang et al., 2015). Thus, sonication in presence of suitable aqueous solvent or solution exfoliates nanosheets and limits the re-aggregation of nanosheets that may happen due to electrostatic repulsion arising from adsorption of molecules from solution. Thus, selection of appropriate solvent helps effective exfoliation. Liquid exfoliation in solvent has three major roles: (1) firstly, the acoustic power of the ultrasound bath sonicator is effectively transferred to the solvent medium, (2) secondly, it helps in minimizing the enthalpy mixing of the solvent and the crystals of interfacial layers permits solvent intercalation and (3) lastly, it inhibits the re-aggregation for stabilization of exfoliated sheets by formation of a steric barrier. Specifically, nitrogen-functionalised liquids are found to be effective solvents for high performance exfoliation of nanosheets of TMDCs as they have more density than water. N-alkylpyrrolidones based solvents, like

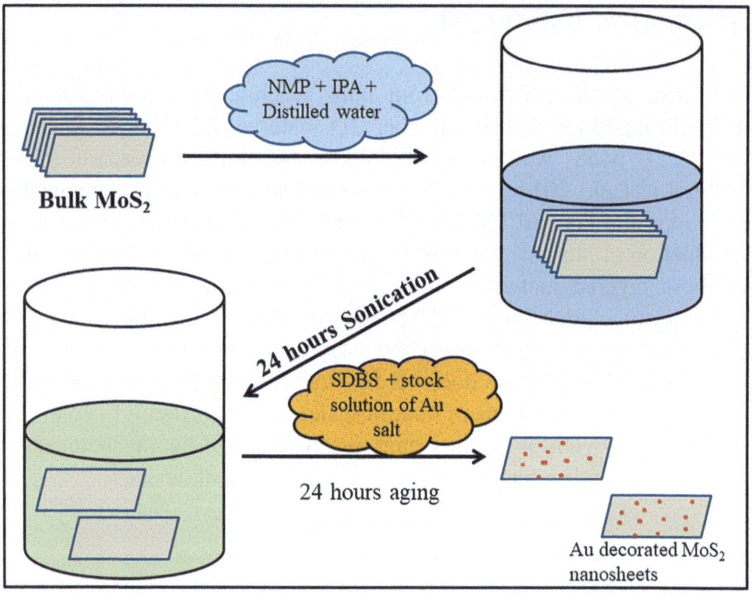

Fig. 5 Schematic representation of synthesis of MoS_2 sheets by liquid exfoliation method

N-cyclohexylpyrrolidone (CHP) and N-methylpyrrolidone (NMP) are quite useful, as they lead to dispersion of nanosheets for high concentration which effectively remain stable for quite longer period. However, these nitrogen based solvents might be hazardous as their boiling point is high, which limits their use in large scale practical applications. Zhang et al. used another alternative to these high boiling point solvents: ethanol and water for the formation of stable dispersions of MoS_2 (Zhang et al., 2015). Yao et al. performed MoS_2 nanosheets sonication for stable dispersion in ethanol/water solution on large scale for promoting practical applications (Yao et al., 2012). Similarly, Savjani et al. directly exfoliated Mo minerals in NMP that are unrefined but could be source of low-dimensional MoS_2 (Savjani et al., 2014). Nguyen et al. used liquid sonication in NVP solvent for explaining the size-dependent properties of two-dimensional MoS_2 and showed their application in various field (Nguyen et al., 2016). Mukherjee et al. synthesized MoS_2 nanocrystals of different sizes by controlled sonication with gradient centrifugation for investigation of optical emission characteristics alongwith tunable direct band gap (Mukherjee et al., 2015).

3.4 *Liquid Exfoliation by Lithium Intercalation*

Another way of inhibiting re-aggregation of nanosheets in liquid phase exfoliation is intercalation by lithium ions (Dines, 1975) (Fig. 6). Ramakrishna Matte et al. firstly used *n*-butyllithium in hexane for intercalation to be inserted into MoS_2 and

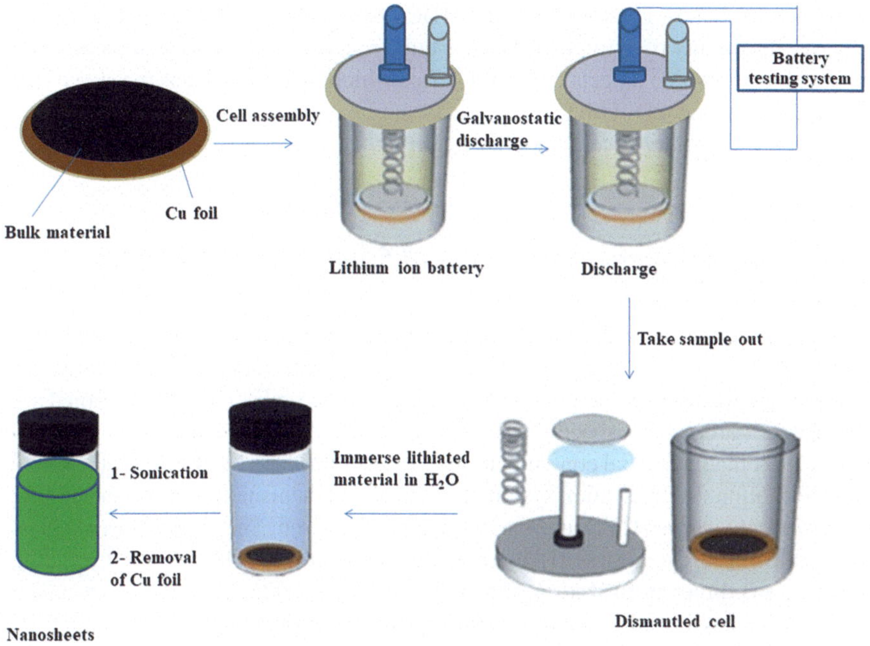

Fig. 6 Schematic representation of synthesis of MoS_2 sheets by liquid exfoliation by Lithium intercalation

WS_2 which can be subsequently exfoliated in water to get high yield single-layered materials (Ramakrishna Matte et al., 2010). However, feasibility of this interacalation method is limited as quantity of lithium insertion is not controlled easily and it is quite time consuming. Zeng et al. adopted another method—electrochemical lithiation to control the insertion of lithium ion content yielding high single layered TMDC nanosheets (Zeng et al., 2015). Layered bulk TMDCs like MoS_2, WS_2, ZrS_2 and TiS_2 were incorporated into electrochemical cell as cathode. In this process of lithium intercalation Li_xXS_2 compound is formed which effectively controls the yield of the exfoliated layers. 1 T phase of MX_2 is influenced by the degree of lithium intercalation. Intercalation of lithium can be maintained during the discharge process. The compounds which are intercalated are ultra-sonicated in ethanol or water to get enhanced quality of mono layered exfoliated TMDCs in large amounts (Garah et al., 2018). The key importance of this process of Li intercalation lies in getting 1 T metallic phase which can be induced by charge transfer from Li ion to TMDC. As a result, the atomic structure of TMDC gets locally rearranged from 2H to 1 T phase. With stability of the 1 T phase, the negative charge on the layered nanosheets becomes stable by residual Li ions or water (Feng et al., 2009; Azhagurajan et al., 2016). This method yields 100% of exfoliated TMDCs yet it has some limitations. Firstly, high temperature for long hours is required to carry out the experiment. Also, the intercalation process must carried out effectively so that formation of metal nanoparticles as well as Li_2S precipitates is avoided. Though, chemical exfoliation gives high yield TMDCs compared to mechanical exfoliation however the quality of exfoliated sheets is quite reduced. This is arises as sonication process induces some defects in the 2D lattice which thereby reduces the size of the sheet, limiting the 2D TMDC nanosheets applications in the area of electronic devices as well as large-scale integrated circuits.

3.5 Chemical Vapour Deposition of Single Layered TMDCs

Recently, preparation of 2D TMDCs by CVD has attracted wide attention. By this method wafer-scale 2D TMDCs can be synthesized with large-area uniformity for potential applications in integrated electronics at large scale. The layers formed are uniform in nature with specific thickness and hetrostructures layers are directly grown avoiding interfacial contamination that gets introduced at the time of layer by layer formation. In this section systematic growth mechanism of monolayer- or few-layer MoS_2 by various CVD approaches will be discussed. The approaches are classified into three categories: (i) thermal decomposition of TMDC on insulating substrate (ii) direct sulfurization of Mo or Mo based oxides and (iii) MoO_3 metal oxide is converted to MS_2 metal disulfide by sulfurization.

3.6 Synthesis of MoS_2 Film by Thermal Decomposition of $(NH_4)_2MoS_4$

Thermal decomposition of $(NH_4)_2MoS_4$ is an effective way of synthesizing MoS_2 wafer scale film by controlled formation of layer (Baby & Kumar, 2021; Wu et al., 2022) (Fig. 7). Liu et al. synthesized bi- or tri MoS_2 layered uniform films on insulating substrates by thermolysis approach (Liu et al., 2012). In this method, an insulating substrate of silica or sapphire was dipped in ammonium thiomolybdates [$(NH_4)_2MoS_4$] and the substrate was annealed at 500 °C and undergoes sulfurization with sulphur vapours at 1000 °C for formation of thin MoS_2 coating on the insulating substrate. Following chemical reaction leads to the formation of MoS_2 layers.

$$(NH_4)_2 MoS_4 + H_2 \rightarrow 2NH_3 + 2H_2S + MoS_2.$$

With secondary annealing at higher temperature (1000 °C) is done to improve the crystallinity of MoS_2 domains. Such samples grown under sulphur vapours exhibit high electrical performance and mobility of electron upto 6 $cm^2\ V^{-1}\ s^{-1}$ for n-type field effect transistors (FET) and ON/OFF current ratio ~ 105.

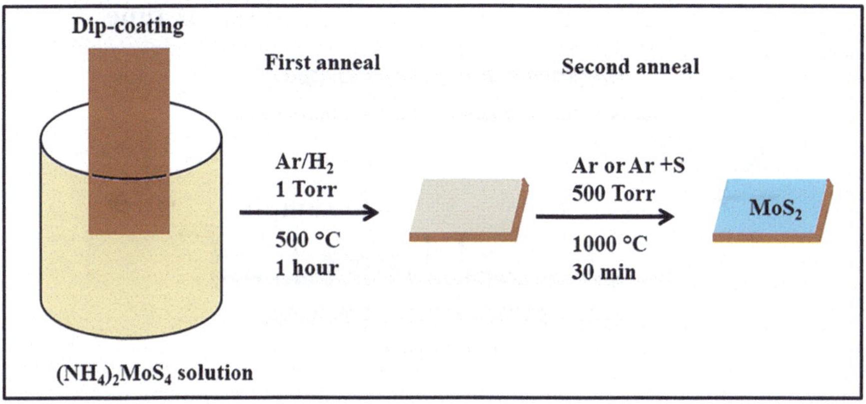

Fig. 7 Schematic diagram of synthesis of MoS_2 sheets by thermal decomposition method

3.7 *Sulfurization of Mo and Mo Based Oxides*

A different strategy was adopted by Zang et al. for deposition of single layered MoS_2. A thin Mo layer of ~1–5 nm is priory deposited by e-beam evaporation on SiO_2 substrate which was placed at 750 °C in tube furnace for reaction with sulphur vapours. This leads to deposition of ordered phase of sulphur atoms formed with Mo crystal (Zhan et al., 2012). Thus, MoS_2 nuclei in conjunction with sulphur atoms appear and under appropriate temperature and pressure thin film is formed by diffusion of sulphur. In this way, two or three layers of MoS_2 with spacing of ~6.6 ± 0.2 Å between layers were obtained. The electrical measurements confirm the resistor-like behavior the MoS_2 samples with sheet resistance of 1.46×10^4–2.84×10^4 Ω/m and typical mobility of 0.004–0.04 cm^2/(V s). Another way of thermal deposition was adopted by Lin et al. for formation of MoO_3 thin films on sapphire substrate. With successful sulfurization of MoO_3 for the formation of few layered MoS_2 by thermal decomposition process, the layer thickness of 2 nm could be obtained (Lin et al., 2012). Electrical performance of such MoS_2 thin films was studied for mobility of electron (~0.8 cm^2/V s) and n-type bottom-gate FET device with on/off current ratio ~ 10^5.

3.8 *Sulfurization of Mo Based Compounds*

For large area synthesis of mono layered MoS_2, Lee et al. adopted an alternative method was by using solid reactants: molybdenum trioxide (MoO_3) and sulphur (S) powders (Lee et al., 2012). In an ambient environment, thin film of MoS_2 was deposited on insulating SiO_2 substrate (Fig. 8). However, to enhance nucleation points for the growth of MoS_2 film, this substrate must be pre- treated with aromatic molecules of reduced graphene oxide etc. The reaction mechanism follows.

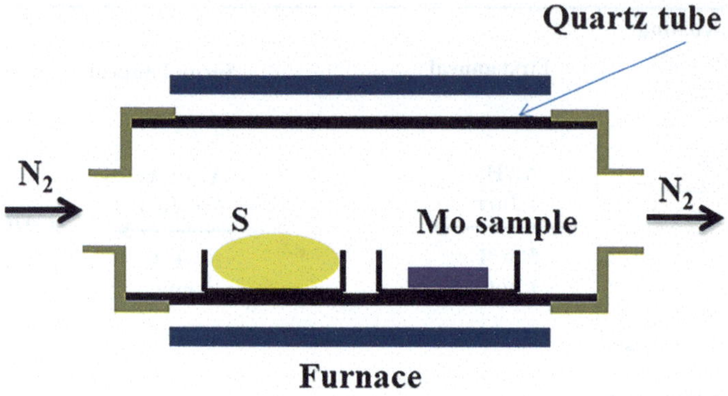

Fig. 8 Schematic diagram of synthesis of MoS_2 sheets by sulfurization of Mo based compounds

$$MoO_3 + x/2S \rightarrow MoO_{3-x} + x/2SO_2.$$

$$MoO_{3-x} + (7-x)/2S \rightarrow MoS_2 + (3 - x/2SO_2.$$

Here, an intermediary phase MoO_2 is formed at the time of reaction at $x = 1$. This gives MoS_2 layer of 0.72 nm thickness quite close to the layer obtained by mechanical exfoliation. The FET based on these MoS_2 films shows n-type behaviour with current on/off ratio of 10^4. Later, a CVD method was reported by Yu et al. under pressure of 2 Torr for the preparation of MoS_2 thin layers by using molybdenum chloride ($MoCl_5$) as precursor. Quality of the films improves and FET based on these films exhibit competing results with that of Lee et al.

3.9 *Hydrothermal Method*

Among various 2D TMDC nanostructures synthesizing methods, hydrothermal method is considered to be an efficient synthesis technique (Miao et al., 2016; Luo et al., 2019; Zhang et al., 2016, 2019). In hydrothermal method, ammonium molybdate tetrahydrate and thiourea are dissolved separately in distill water and undergoes sonication for 30 mins. The sonicated solution is then transferred into a 100 mL autoclave of teflon-stainless steel coating. The reaction between precursors is carried out at particular temperature in a muffle furnace for required hours. After cooling of the furnace naturally, the solution is filtered and precipitates of dark brown colour are obtained. Washing of these brown precipitates with deionized water for many times lead to removal of any residual traces. Finally, the MoS_2 nanostructures undergo drying for 7 hrs at 60 °C (Fig. 9). By this method many reactive edge states of MoS_2 appears that can be exposed to different environmental conditions for getting enhanced chemical activities. These reactive edge states of MoS_2 are essential for practical application like hydrodesulphurization, lithium ion battery, hydrogen

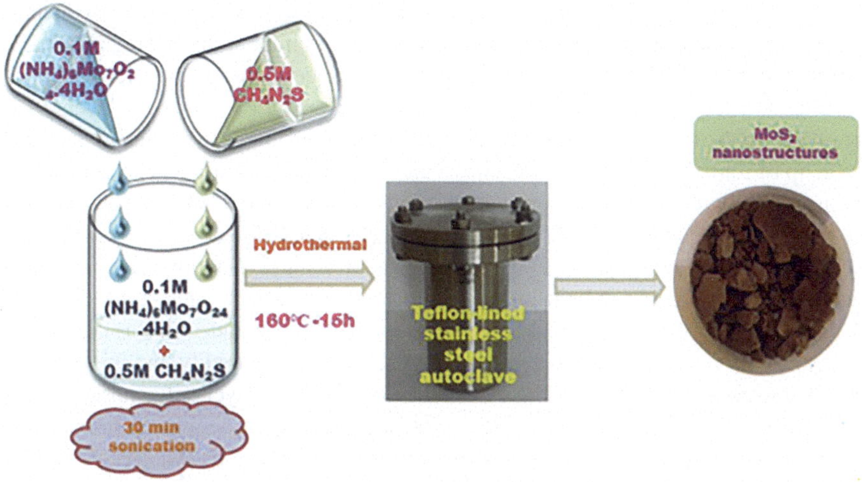

Fig. 9 Schematic diagram of synthesis of MoS_2 sheets by hydrothermal method

evolution reaction etc. (Rani et al., 2022, 2023). Hydrothermal method is preferred due to the advantage that by this method TMDC can form functional nanocomposites with other nanomaterials like graphene, semiconducting materials like TiO_2, carbon nanotubes (CNTs) etc. depending upon the need of application. Additionaly, hydrothermal method can be utilized to produce 2D polycrystalline TMDCs. Till date, different morphologies of 2H-MoS_2 nanocomposites have been reported such as nanorods, nanosheets, nanoflowers, nanoplates, nanowalls, and hollow nanoparticles. All these structures depend strongly on phase, size, crystallinity, interfaces and morphology of MoS_2. Hydrothermal method is quite flexible in controlling these properties of MoS_2 by tuning the various parameters of the process. Hence, to get MoS_2 nanostructure of desired morphology for particular application, it is essentially necessary to understand effects of different parameters like temperature, solvents, precursor, reaction time, additives and substrates for the growth mechanism of 2H- MoS_2. The most important parameter influencing the morphology and crystallinity of desired MoS_2 nanostructures in the hydrothermal synthesis process, is the temperature of reaction process.

The hydrothermal method is most favoured to get different morphologies of MoS_2 since by the effect of reaction temperature has been investigated by Tian et al. on morphologies of MoS_2 nanostructures. At hydrothermal temperature of 160 °C amorphous nanosheets were obtained which appeared as MoS_2 mussy fragments. However, with increase in temperature to 180 °C, three more morphologies of MoS_2 appeared. These are MoS_2 nanotubes of 200 nm diameter, MoS_2 nanorods obtained of diameter of 60–100 nm and lastly, 'rag-like' microstructure. To better understand the formation of these different MoS_2morphologies, we basically need to understand the formation of MoS_2 nanosheets by formed hydrothermal process. During this hydrothermal process, MoS_2 nanosheets are active but quite unstable due to the presence of unsaturated Mo or

S atoms dangling bonds. The atoms appear at the edge sites which then tend to interact in between the layer with each other for the purpose of elimination of dangling bonds which reduces to lowest energy level. Here, the nanosheets appear thin, crinkled, having irregular shape. The edges are not only highly active rather within the layers they try to roll up the layers of nanosheet and form nanotubes. Since, MoS_2 nanotubes are made up of folded and crinkled nanosheets they appear as caterpillar. When the layers are more tightly rolled up, then there is formation of nanorods. However, these nanorods are limpsy and pliable, as a result of which they entangle themselves with each other. At further higher temperature, 200 °C lesser nanorods and nanotubes are comparatively observed. Further, when the temperature reaches 220 °C, nanotube diameter increases upto ~500 nm with thicker wall. Thus, it can be concluded that as the hydrothermal reaction temperature increases to 200 °C and above rigidity and thickness of MoS_2 nanosheets increases which poses difficulty to roll the layers of nanosheets further into nanotubes and nanorods. However, increase in temperature improves the samples crystallinity. Crystallinity of MoS_2 samples was studied by Wang et al. studied within temperature 160–240 °C for 12 h. In this temperature range, nanoflower morphology of MoS_2 sample is exhibited with 100–500 nm diameter (Wang et al., 2014). Interlayer distance of MoS_2 nanosheets expands from 0.62 nm to 0.8 nm which indicates lattice expansion. As the surface energy is high and Van der waals interaction is attractive, MoS_2 layers tend to aggregate during the process of fabrication. Hence, many of the active sites present in MoS_2 are lost and this property of MoS_2 makes these nanostructures of 2D TMDC different from other materials. Hence, major attention is towards prevention of layer aggregation in 2D MoS_2 nanostructures thereby aiming at enhanced electrochemical and catalytic performances. This shortcoming of MoS_2 layers can be reduced by introducing or adding additives like acids, solvents and surfactants during hydrothermal process with the precursor. However, inspite of these advantages, 2D MoS_2 nanostructures prepared by hydrothermal method exhibit poor crystallinity and formation of non-uniform number of layers which renders low damage threshold to practical applications like fast optically modulated devices. Therefore, though controlled morphology nanocomposites of 2D MoS_2 can be prepared by hydrothermal process yet other alternative method is needed for improving the crystallinity of these nanosheets at lower reaction temperature. This will maintain control over the 2D materials layers and make it suitable for high performance applications.

3.10 Opto-Electronic Applications of MoS_2

MoS_2 nanosheets exhibit flexibility in bandgap depending upon the number of layers which can be effectively utilized for fabrication of varied optoelectronic devices. In case of MoS_2 monolayer, direct electronic transition between K" and K points of the Brillouin zone change the structure of the electronic bands. With change in layer number there is variation in thickness. The direct excitonic states K-point does not change but the Γ-point transitions shift while moving from indirect to direct band gap. Recent, experimental and theoretical investigations conclude that out-of-plane vibrational direction of MoS_2 single layer due to quantum confinement effect leads to decrease in thickness of MoS_2 layers, phonon dispersion relations, exciton binding energies and flexible band gap (Castellanos-Gomez et al., 2012). The d-orbitals in Mo atoms strengthen the spin orbit coupling in MoS_2. As a result of this coupling, the valence band degeneracy break and splitting of band is of the order of 160 meV. This creates two valence band maximas at the Brillouin zone K and K' points. Hence, degeneracy is broken that can be exploited at different levels for varied optoelectronic applications of MoS_2.

MoS_2 nanocrystals have been investigated for silicon based compatible CMOS technology (Xiong et al., 2018). This opens up a new horizon of MoS_2 NC's that can be explored deeply for future sensing, memory, energy conservation and optoelectronics devices. Mukherjee et al. investigated the next generation photonic devices by exploiting the tunable band gap of MoS_2 nanocrystals in their switching characteristics and optical responsivity (Mukherjee et al., 2015; Yang & Li, 2020). Higher on/off ratio, carrier mobility and photoresponsivity is exhibited by MoS_2 monolayer based phototransistors. These phototransistors based on MoS_2 monolayer exhibit improved green light detection characteristics due to 1.82 eV direct band gap for the monolayer. While red light detection is efficiently done by phototransistors having three layer MoS_2. The anodes of polymer based LEDs can be fabricated by chemically exfoliated nanosheets of MoS_2. In this case high work function of nanosheets of MoS_2 injects holes easily whereas the wide band gap of MoO_3 layer is blocked by the electrons in MoO_3/MoS_2 hybrid structure. This enhances the light emission efficiency as the carrier balance is effectively improved. Hence, direct band gap in MoS_2 monolayers make it a suitable and highly efficient material for developing low power optoelectronic devices. In present times many research groups are working on the investigation of optoelectronic devices based on 2D layered MoS_2. Due to their physically and chemically important characteristics, the layered MoS_2 can form hetrostructures with other materials also. This opens up the doors of MoS_2 towards transparent and flexible optoelectronic devices.

3.11 MoS_2-Based Photodetectors

MoS_2 due to its unique crystal phase, easy synthesis, structural ability and tunable photoelectronic features has gained interest in the optoelectronic applications. The 2D MoS_2-based photodetectors have been excellent device in low-power consumption, broadband optical detection, 2D integrated circuits, high speed response. MoS_2 has gained lot of interest for photodetection applications due to its strong light absorption property, good mobility, and also due to the possibility of fabricating single layer. All these features make MoS_2 an excellent candidate for next-generation optoelectronics. The MoS_2- based photodetectors are of the simplest forms in which the MoS_2 nanosheets are grown on insulating substrates (for e.g., Al_2O_3, SiO_2, Si_3N_4). The pristine MoS_2 acts as active medium to produce charge carriers when light of suitable wavelength falls on the photodectors. Physical vapour deposition methods are adopted to fabricate metal contacts on MoS_2 for external biasing. The direct contacts made on MoS_2 act as source and drain, and contact on the insulating substrate act as gate. Proper design of the metal contacts in form of inter-digitated finger contacts can control the response as well as the device speed. These types of photodetectors find its application as field effect transistor.

Mechanically exfoliated monolayer of MoS_2 was deposited on a Si/SiO_2 substrate by Gonzalez Marin et al. (2019) to fabricate a phototransistor with large visible range photoresponsivity of $R = 10^3$ A/W (Fig. 10). However, a significantly faster responsivity device ($R = 0.57$ A/W) was fabricated by Tsai et al. (2013). In this, wet-synthesis approach was adopted for obtaining few-layered structure of MoS_2 to be deposited on the Si/SiO_2 substrate. Photolithography was utilized to fabricate Au electrodes in an interdigitated manner to get a Schottky photodiode with a metal–semiconductor–metal (MSM) sequence (Fig. 11). The interdigitated manner of Au electrode deposition plays significant role in minimizing the carrier collection time which enhances the responsivity.

Another way of fabricating pristine MoS_2 photodetector is by depositing mechanically exfoliated MoS_2 multilayer structure on Al_2O_3 substrate (Fig. 12) (Choi et al., 2012). Visible region responsivity of 0.12 A/W lowers the infrared region due to weak absorption in the IR spectral region. This type of MoS_2 based photodetector exhibited good response alongwith low noise, low dark current (10^{-11} A) and good detectivity of 10^{-11} J in the visible region.

Hence, high values of responsivity of pristine MoS_2 photodetectors can be obtained with optimum bias voltage which attributes to large photogain and proper geometry of detector (interdigitated metal electrodes) as discussed earlier. Most of the MoS_2 based photodetectors show good response in the 400–700 nm of visible range. However, fewer MoS_2 based photodetectors show their response near infrared (NIR) and ultraviolet (NUV) spectrum.

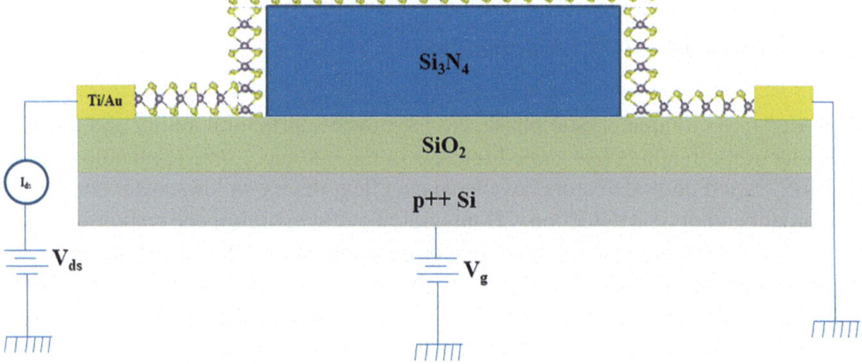

Fig. 10 Schematic representation of a phototransistor

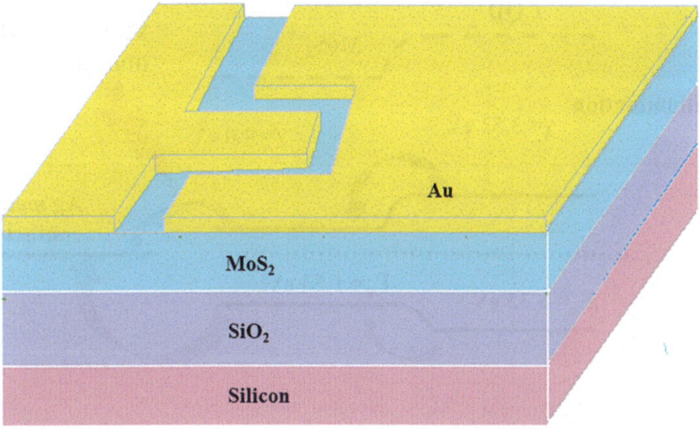

Fig. 11 Schematic representation of Au electrodes in an inter-digitated manner with a metal–semiconductor–metal (MSM) sequence

Fig. 12 Schematic representation of MoS_2 – Au phototransistor deposited on Al_2O_3 substrate

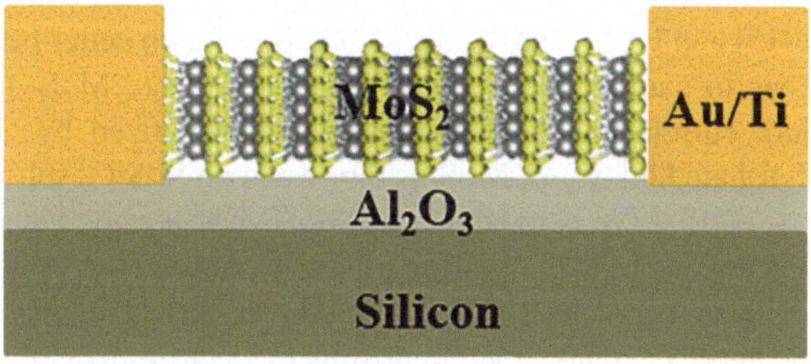

3.12 *MoS_2–Quantum Dots Based Photodetectors*

To overcome the limitation of response of MoS_2 based photodetectors in near to mid infrared (NIR-MIR) and NUV spectrum, quantum dots (QDs) can be combined with the MoS_2 layer to enhance performance. In MoS_2- QD based photodetectors, MoS_2 thin layers are grown hydrothermally on paper substrate (Sahatiya et al., 2018; Gomathi et al., 2017). Higher band gap of QDs in comparison to MoS_2 makes this type of photodetector suitable for optical detection in the UV region (Fig. 13). The van der Waal hetrostructure of MoS_2- QD is formed when MoS_2 induces photo-excited holes successfully and they get trapped subsequently due to hetrojuction small energy barrier. The longer carrier lifetime of QDs will lead to recirculation of photoexcited carriers at the interface. However, MoS_2 layer will re-absorb the light emitted from these QDs with further generation of additional photocarriers. Such device shows sensitivity of 8.4 mA/W to a wavelength of 365 nm in NUV region with 3% enhancement of quantum efficiency as compared to other photodetectors. However, this device is not applicable for operations in high frequency range due to 0.57 s response time. The device performance is also limited due to high vales of dark current (10^{-6} A).

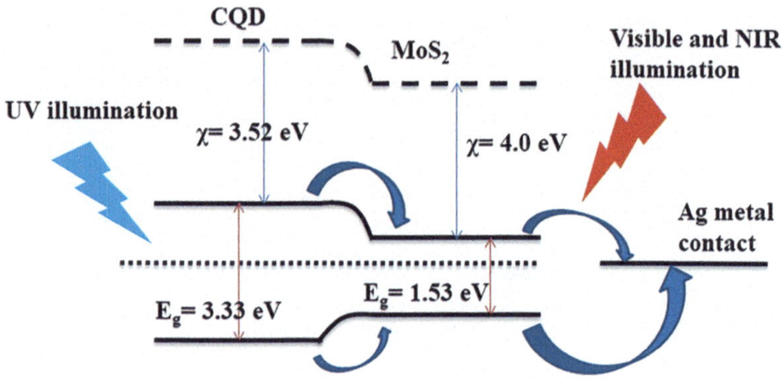

Fig. 13 Electronic band alignment of MoS_2 and QDs photodetector

3.13 MoS_2–Nanoparticles Based Photodetectors

Photodetectors sensitized with nanoparticles CdSe (Ra et al., 2016), PbS (Kufer et al., 2015) and HgTe (Huo et al., 2017), show much better performance. Though, these are toxic in nature but the nanoparticles sensitized MoS_2 photodetectors show reduced dark current. A MoS_2 phototransistor sensitized by CdSe is shown in (Fig. 14). Exfoliated bilayer of MoS_2 is fabricated on Si/SiO_2 substrate. Highest responsivity of 2.5×10^5 A/W is exhibited by MoS2-based photodetector in the NUV region (for 405 nm) with small response time of 60 ms due to formation of n–n heterojunction by CdSe nanocrystals and the MoS_2 bilayer. In such device when UV light is incident on the interface region, there is drifting of electrons to MoS_2 region from CdSe nanocrystals, whereas holes stay intact in the CdSe nanoparticles. With switching off the light, there is recombination of the accumulated electrons in the valley and a n-n barrier is formed which prevents the carrier transfer between the regions (Fig. 15).

3.14 MoS_2–Graphene Based Photodetectors

High carrier mobility of 2D graphene (about 200,000 $cm^2/(V\ s)$ makes it a suitable candidate for MoS_2 based photodetectors (Bolotin et al., 2008). The zero band gap energy and semimetallic behaviour limits the optoelectronic applications of graphene (Mak & Shan, 2016) and absent forbidden band gap can lead to injection of large currents into the material that limits the photodetector's sensitivity due to large dark current. This enhances the possibility to exploit the versatility of 2D graphene in form of contact material upon the MoS_2 layer (Li et al., 2017) and the high mobility characteristic of graphene can be combined with the semiconducting properties of MoS_2. For realization of MoS_2–graphene based photodetectors, mechanically exfoliated few layers of MoS_2 is deposited on a graphene sheet. Pd and Ti metal contacts are fabricated to collect the charges (Li et al., 2015) (Fig. 16).

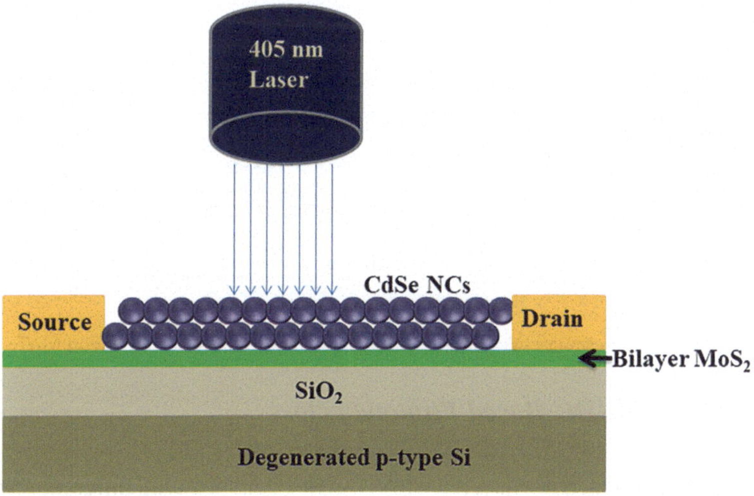

Fig. 14 Schematic representation of MoS_2 phototransistor sensitized by CdSe nanoparticles

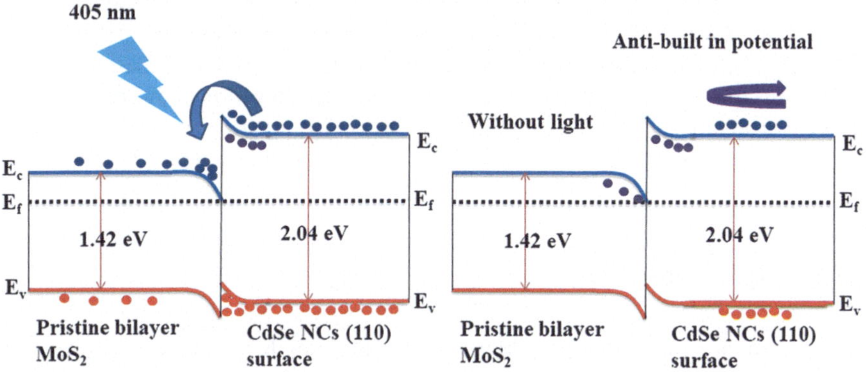

Fig. 15 Energy band gap allignment of MoS_2 and CdSe nanoparticles in presence and absence of light

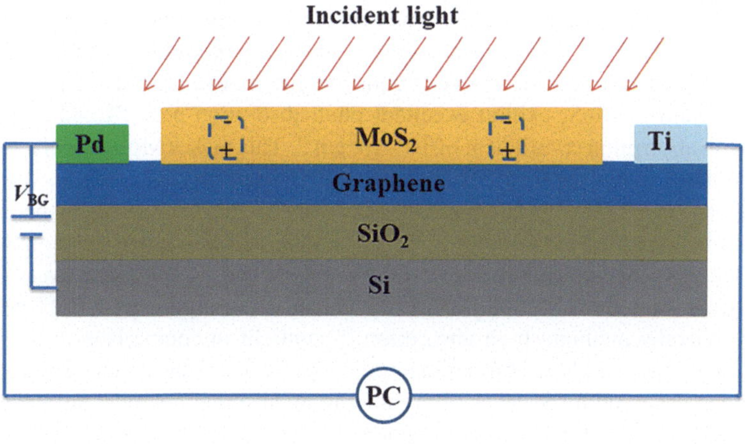

Fig. 16 Schematic diagram of MoS_2-graphene heterostructure phototransistor

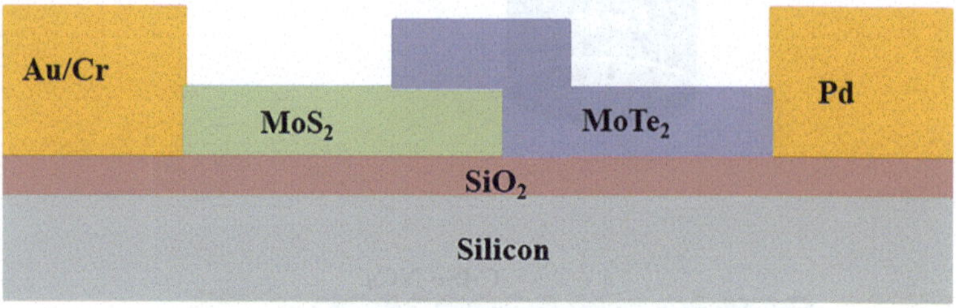

Fig. 17 Schematic representation of hetrojunction photodetector between MoS_2 and $MoTe_2$

3.15 MoS_2–TMDC Based Photodetectors

Mechanically exfoliated few layered MoS_2 can be combined with different TMDCs (like $MoTe_2$) to form heterostructure photodetectors (Chen et al., 2018) Fig. 17. The smaller band gap of $MoTe_2$ varying from $MoTe_2$ shows transition from bulk to monolayer structure. In such photodetecting devices, the electronic band structure of MoS_2 and $MoTe_2$ develops an electric field at the interface of the hetrostructures. This built-in electric field enhances the separation of photogenerated charge carriers and widens the spectral response range of the photodetector formed by combination of heterostructure semiconductor materials. Due to exploitation of band alignment of type II, electrons are injected into MoS_2 and holes are injected into $MoTe_2$.

3.16 MoS_2 Solar Cells

The dependency of band gap of MoS_2 on its layer numbers makes it an efficient material for absorbing solar light for utilization in photovoltaic applications. Light absorption ability and its efficient emission by MoS_2 strongly depends upon the structure of electronic band gap of MoS_2. Due to direct band gap of MoS_2 monolayer, there is easy absorption of photons by the MoS_2 monolayer and they can be easily overcome the band gap and emitted for higher energy states. Whereas in case of bulk or multi-layer MoS_2, this absorption and emission of photons require extra phonon for absorption or emission to conserve the momentum of the crystal. 2H crystal system of MoS_2 exhibit excellent photoabsorption near IR and in visible range with absorption co-efficient of 10^5–10^6 cm^{-1}. This makes MoS_2 largely acceptable for photo voltaic applications in optoelectronics (Singh et al., 2017a). Variation in band gap (1.0–2.0 eV) in comparison to the solar spectrum gives MoS_2 the opportunity to be exploited as absorber material in solar cell like other semiconductors CdTe, GaAs, Si having energy gap of 1.5 eV, 1.4 eV and 1.1 eV respectively (Tang et al., 2017). 1 nm thick MoS_2 layers exhibit enhanced absorption by 5–10% per unit volume, thereby attaining high absorption of sunlight in comparison to the other commonly utilized light absorbing materials like GaAs, Si layers etc. The number of photons absorbed by Si layer of 50 nm can be absorbed by even monolayer of MoS_2 and exhibit high value of photocurrent (2.0–4.5 mA/cm^2) (Bernardi et al., 2013).

3.17 MoS_2 in Organic Solar Cells

Remarkable properties of organic solar cells like facile synthesis, flexibility, light-weight etc. enhances the applicability of these devices in various fields. Typically, an organic solar cell consists of a photoactive material having electron acceptor and electron donor properties as well. In organic solar cells, the hole transport layer consists of Poly (3,4-ethylenedioxythiophene) and poly(styrene sulfonate) represented as PEDOT: PSS. MoS_2 can be efficiently utilized as the electron transport layer (ETL), the hole transport layer (HTL), as an intermediate layer between the buffer layer as well in organic solar cells (Balis et al., 2016; Kakavelakis et al., 2019; Gu et al., 2013). MoS_2 can be also be utilized as anode buffer layer for preventing the leakage current in organic solar cells (Li et al., 2013). MoS_2 can be combined with PEDOT: PSS as nanocomposite to be utilized as hole extraction layer for increasing the efficiency by 3.74% (Ramasamy et al., 2019). Profound work function is an important characteristic of hole transport material since low work function is needed in organic solar cells for electron extraction. Tunable band gap of MoS_2 makes it an efficient material for hole withdrawal and electron extraction layer as well (Yun et al., 2013). For electron extraction layer n-type doped MoS_2 layer is required whereas p-type doping in MoS_2 can be utilized as hole withdrawing layer. *p*-type doping in hole extraction layer of MoS_2 enhances the efficiency of photon conversion by 2.8–3.4% . MoS_2 interfacial layer combined with silver nanowires (AgNW-MoS_2) can be utilized as transparent electrode in organic solar cells (Hu et al., 2015) (Fig. 18a). This type of electrode exhibit transparency of 93.1% at 550 nm with lower sheet resistance value (9.8 Ω/sq). The MoS_2 interfacial layer in the electrode of AgNW-MoS_2 provides improved resistance against moisture and oxidation and also enhances environmental reliability in comparison to pristine AgNW. P type -MoS_2 nanosheets/Ag anode utilized for hole extraction and n type-MoS_2/AgNWs-MoS_2 nanosheets cathode utilized for electron extraction in organic solar cells (PBDTTT-C-T: PC70BM based) enhances the efficiency by 8.72% (Fig. 18a). MoS_2 combined with MoO_3 or MoS_2 combined with Au nanoparticles in organic solar cells can also be utilized as hole extraction layers (Qin et al., 2014; Yang et al., 2014). Chuang et al. (2015) designed organic solar cells with Au nanoparticles decorated MoS_2 nanosheets which can act as an electron extraction layer with efficiency of 4.91%.

3.18 MoS_2 in Perovskite Solar Cells

Pervoskite materials have generic formula ABX_3 where A is $CH_3NH_3^+$, B is Pb and X an anion can be occupied by any halogen atom (Cl, Br, I) in organometallic halide perovskites solar cells (Tsai et al., 2014). In this type solar cells also there are hole and electron extraction layers for charge separation and collection. These layers are inserted inbetween the electrode and the active layer. The generally used Spiro-OMeTAD (2,2',7,7'-tetrakis(N,N-di-p- methoxyphenyl-amine)9,9'-spirobifluoren) hole extraction layer in perovskite solar cells has energy levels at 5.03 eV and valence band of $CH_3NH_3PBI_3$ maxima has energy 5.04 eV. The conduction of holes from $CH_3NH_3PBI_3$ active layer by Spiro-OMeTAD to the anode is not quite cost-effective. Hence, there is a need of other materials that can make this hole transport mechanism in pervoskite solar cells quite efficient. Among many like graphene oxide (GO) (Yang et al., 2017), reduced graphene oxide (rGO) (Sahoo et al., 2019; Palma et al., 2016), poly(3,4-ethylene dioxythiophene) polystyrene sulfonate (PEDOT:PSS) (Wang et al., 2018b), copper(I) iodide (CuI) (Uthayaraj et al., 2019) etc., the hole transport characteristic of MoS_2 can be utilized due to its effective work function (4.95 eV). Such MoS_2-based perovskite solar cells show improved efficiency of 9.53% (Singh et al., 2017a; Tang et al., 2017; Bernardi et al., 2013).

The MoS_2 and GO perovskite solar cell has been fabricated with cell configuration of ITO/PEDOT: PSS/GO/perovskite/PCBM/MoS_2/Ag (Yang et al., 2017) (Fig. 18b). With $MAPbI_3$ as an active layer, efficiency of 19.14% can be achieved. The efficiency of photon conversion of TMDC is quite high due to lower value of series resistance, well aligned valence band of perovskite material and MoS_2 and higher shunt resistance. MoS_2 can be incorporated as interlayer between hole transport layer and perovskite active material. Incorporation of MoS_2 can enhance the perovskite solar cells efficiency by 6% in comparison to without MoS_2 perovskite solar cells. MoS_2/TiO_2 composite utilized as electron extraction layer can improve the efficiency of perovskite solar cells and enhance the solar cell absorption range from 350 to 800 nm (Narayanasamy & Arul, 2019). In this case, the power conversion efficiency (PCE) enhances from 3.74% to 4.43% in comparison to pure TiO_2 -perovskite solar cells.

3.19 MoS_2 Silicon-Based Heterojunction Solar Cells

Fascinating optoelectronic characteristics like light emission and solar energy conversion of heterostructure TMDCs can be combined to fabricate another type of solar cell- MoS_2 silicon based hetrojunction solar cells (Pradhan et al., 2017; Wirth-Lima et al., 2019; Ma et al., 2018). Hetrojunction solar cell of MoS_2/Silicon can be fabricated by direct growth of MoS_2 on silicon substrate by sol-gel method (Mohamed Ismail et al., 2023) (Fig. 18c). Photovoltaic performance of ITO/MoS_2/p-Si/Ag-based solar cell was enhanced due to reduction in interfacial defects by the in situ growth of MoS_2/Silicon hetrojunction interface. This leads to high V_{OC} and decreased recombinations which consequently increase the J_{SC}. WSe_2/MoS_2 fabricated hetrojunction solar cell exhibited 0.2% of power conversion efficiency (Furchi et al., 2014). N-type MoS_2 monolayer can be fabricated on p-type silicon substrate to form Al/MoS_2/p-Si/Cr-Ag (Fig. 18c) based hetrojunction solar cell to exhibit improved PCE of 5.23% and J_{SC} of 22.36 mA/cm^2 in comparison to Al/p-Si based solar cells. MoS_2 can be deposited on buffered substrate of p-type silicon to form Pd/MoS_2/(3–5 nm) SiO_2/Si based heterojunction solar cell (Hao et al., 2015). There is three fold (4.5%) increase in the PCE due to this SiO_2 buffered layer as compared to only 1.4% increase in PCE for normal MoS_2/Si heterojunction solar cell. This increase in PCE is due to reduction in electron-hole recombinations due to the buffer layer. However, the PCE increased to 15.8% for Graphene/MoS_2/Si based

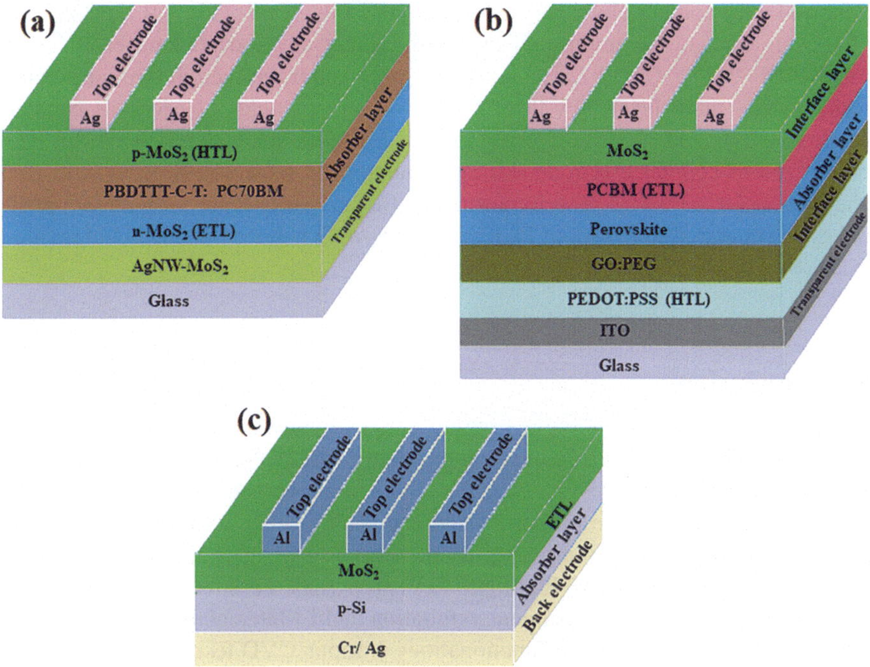

Fig. 18 (**a**) Schematic representation of glass/AgNW-MoS_2/n-MoS_2/PBDTTT-C-T: PC70BM/p-MoS_2/Ag based organic solar cell (Hu et al., 2015). (**b**) ITO/PEDOT:PSS/GO/ perovskite/PCBM/MoS_2/Ag based perovskite solar cell (Wang et al., 2018a). (**c**) MoS_2/p-Si based heterojunction solar cells (Tsai et al., 2014)

heterojunction solar cells (Ma et al., 2018; Jiao et al., 2015). The MoS_2 layer fabricated inbetween silicon/graphene substrate acts as passivation and blocks electron which improves the PCE significantly (Tsuboi et al., 2015). PCE improves by enhancing graphene layers. 9 nm thick MoS_2 inserted between graphene tri-layer and n-type Si substrate increases PCE (11.1%). Another way of forming hetrojunction solar cell is fabricating hetrostructures MoS_2/WSe_2 (Furchi et al., 2014). Bulk hetrojunction solar cells may face limitation in the absorption range (≈100–200 nm). To improve absorption ability of solar cells and its performance came the tandem solar cells. In this type, various single junction solar cells are vertically piled upon each other (Bin Mohd Yusoff et al., 2015; Saraswat et al., 2019) so that thermalization effect is reduced due to the stacking sequence which absorbs specific part of sunlight separately.

3.20 Counter Electrode MoS_2 in Dye-Sensitized Solar Cells

Dye-sensitized solar cells (DSSCs) is another solar cell type. In these solar cells, there is a counter electrode that transfers electrons for the reduction process through an external circuit. This photoelectrochemical process undergoing in DSSCs are affected by the counter electrode electrocatalytic effect (Dou et al., 2012). MoS_2 can effectively replace the traditional Pt based counter electrode (Jeong et al., n.d.; Singh et al., 2017b; Cheng & Hsieh, 2015). MoS_2 can be utilized in three structural forms: MoS_2 multi-layer, few-layer MoS_2 and nanoparticles of MoS_2 as counter electrodes (Lei et al., 2014). MoS_2 deposited alongwith flakes of graphene can also be utilized as counter electrode. MoS_2 stimulates the reduction process with fast charge transfer. 6.07% of power conversion efficiency was achieved with graphene flakes as compared to 6.04% of PCE with MoS_2/rGO nanocomposite counter electrode (Singh et al., 2017b).

4 Conclusions

2D layered TMDCs exhibit fascinating chemistry that explores in depth investigation of varied opportunities for their fundamental phenomena. Diversity of MoS_2 lies in its versatile tunable properties and abundant availability thereby making it available for wider applications range from energy storage to catalysis. In this chapter, we have attempted to highlight key properties of single or few layers MoS_2TMDCs. Large scale chemical exfoliation and Li intercalation explores controlled structures with diverse functionalities. Various CVD techniques have been discussed for fabrication of uniform single and few layered sheets. Mono-layer TMDCs surface chemistry prove to be significant for electrocatalytic hydrogen evolution, energy storage and photochemical water-splitting. Besides these many key challenges need to be resolved like cycling stability of TMDC nanosheets and electrical conductance etc. must be improved for energy storage purposes. Active sites density needs to be enhanced for the purpose of catalysis which makes the basal plane of TMDC entire nanosheets to be catalytically active inspite of only the edges. Thus, well designed and controlled heterostructures have been synthesized for their application in electronic devices. Different possibilities of designing and fabricating varied compounds and their integration into hetro-lattices laterally and vertically have to be achieved for tailoring properties for specific applications which has open up substantial interactions between theorists and experimentalists.

References

Allen, M. J., Tung, V. C., & Kaner, R. B. (2010). Honeycomb carbon: A review of graphene. *Chemical Reviews, 110*, 132–145. https://doi.org/10.1021/cr900070d

Azhagurajan, M., Kajita, T., Itoh, T., Kim, Y. G., & Itaya, K. (2016). In situ visualization of lithium ion intercalation into MoS_2 single crystals using differential optical microscopy with atomic layer resolution. *Journal of the American Chemical Society, 138*(10), 3355–3361. https://doi.org/10.1021/jacs.5b11849

Baby, M., & Kumar, K. R. (2021). Synthesis of transition metal dichalcogenides based hybrid nanostructures by thermal decomposition method. *Journal of Physics: Conference Series, 2070*(1), 012095. https://doi.org/10.1088/1742-6596/2070/1/012095

Bae, S., Kim, H., Lee, Y., Xu, X., Park, J. S., Zhang, Y., et al. (2010). Roll-to-roll production of 30-inch graphene films for transparent electrodes. *Nature Nanotechnology, 5*, 574–578. https://doi.org/10.1038/nnano.2010.132

Balis, N., Stratakis, E., & Kymakis, E. (2016). Graphene and transition metal dichalcogenide nanosheets as charge transport layers for solution processed solar cells. *Biochemical Pharmacology, 19*(10), 580–594. https://doi.org/10.1016/j.mattod.2016.03.018

Bernardi, M., Palummo, M., & Grossman, J. C. (2013). Extraordinary sunlight absorption and one nanometer thick photovoltaics using two-dimensional monolayer materials. *Nano Letters, 13*(8), 3664–3670. https://doi.org/10.1021/nl401544y

Bertolazzi, S., Brivio, J., & Kis, A. (2011). Stretching and breaking of ultrathin MoS_2. *ACS Nano, 5*, 9703–9709. https://doi.org/10.1021/nn203879f

Bin Mohd Yusoff, A. R., Kim, D., Schneider, F. K., da Silva, W. J., & Jang, J. (2015). Au-doped single layer graphene nanoribbons for a record-high efficiency ITO-free tandem polymer solar cell. *Energy & Environmental Science, 8*, 1523–1537. https://doi.org/10.1039/C5EE00749F

Bolotin, K. I., Sikes, K. J., Jiang, Z., Klima, M., Fudenberg, G., Hone, J., Kim, P., & Stormer, H. L. (2008). Ultrahigh electron mobility in suspended graphene. *Solid State Communications, 146*(9–10), 351–355. https://doi.org/10.1016/j.ssc.2008.02.024

Buscema, M., Steele, G. A., Van der Zant, H. S. J., et al. (2014). The effect of the substrate on the Raman and photoluminescence emission of single-layer MoS_2. *Nano Research, 7*, 561–571. https://doi.org/10.1007/s12274-014-0424-0

Butler, S. Z., Hollen, S. M., Cao, L., Cui, Y., Gupta, J. A., Gutierrez, H. R., et al. (2013). Progress, challenges, and opportunities in two-dimensional materials beyond graphene. *ACS Nano, 7*(4), 2898–2926. https://doi.org/10.1021/nn400280c

Castellanos-Gomez, A., Poot, M., Steele, G. A., van der Zant, H. S. J., Agraït, N., & Rubio-Bollinger, G. (2012). Elastic properties of freely suspended MoS2 nanosheets. *Advanced Materials, 24*(6), 772–775. https://doi.org/10.1002/adma.201103965

Castro Neto, A. H. (2001). Charge density wave, superconductivity, and anomalous metallic behavior in 2D transition metal Dichalcogenides. *Physical Review Letters, 86*(19), 4382–4385. https://doi.org/10.1103/PhysRevLett.86.4382

Castro Neto, A. H., Guinea, F., Peres, N. M. R., Novoselov, K. S., & Geim, A. K. (2009). The electronic properties of graphene. *Reviews of Modern Physics, 81*(1), 109–162. https://doi.org/10.1103/RevModPhys.81.109

Chaudhary, N., Khanuja, M., Abid, & Islam, S. S. (2018). Hydrothermal synthesis of MoS2 nanosheets for multiple wavelength optical sensing applications. *Sensors and Actuators A: Physical, 277*, 190–198. https://doi.org/10.1016/J.SNA.2018.05.008

Chen, Y., Wang, X., Wu, G., Wang, Z., Fang, H., Lin, T., Sun, S., Shen, H., Hu, W., Wang, J., et al. (2018). High-performance photovoltaic detector based on MoTe2/MoS_2 Van der Waals Heterostructure. *Small, 14*(9), 1703293. https://doi.org/10.1002/smll.201703293

Cheng, C. K., & Hsieh, C. K. (2015). Electrochemical deposition of molybdenum sulfide thin films on conductive plastic substrates as platinum-free flexible counter electrodes for dye-sensitized solar cells. *Thin Solid Films, 584*, 52–60. https://doi.org/10.1016/j.tsf.2015.02.025

Chhowalla, M., Shin, H. S., Eda, G., Li, L. J., Loh, K. P., & Zhang, H. (2014). The chemistry of two-dimensional layered transition metal dichalcogenide nanosheets. *Nature Chemistry, 5*(4), 263–275. https://doi.org/10.1038/nchem.1589

Choi, W., Cho, M. Y., Konar, A., Lee, J. H., Cha, G. B., Hong, S. C., Kim, S., Kim, J., Jena, D., Joo, J., et al. (2012). High-detectivity multilayer MoS2 phototransistors with spectral response from ultraviolet to infrared. *Advanced Materials, 24*(43), 5832–5836. https://doi.org/10.1002/adma.201201909

Chuang, M. K., Yang, S. S., & Chen, F. C. (2015). Metal nanoparticle-decorated two-dimensional molybdenum sulfide for plasmonic-enhanced polymer photovoltaic devices. *Materials (Basel), 8*(8), 5414–5425. https://doi.org/10.3390/ma8085252

Dines, M. (1975). Lithium intercalation via n-butyllithium of the layered transition metal dichalcogenides. *Materials Research Bulletin, 10*, 287–291. https://doi.org/10.1016/0025-5408(75)90115-4

Ding, S., Zhang, D., Chen, J. S., & Lou, X. W. (2012). Facile synthesis of hierarchical MoS_2 microspheres composed of few-layered nanosheets and their lithium storage properties. *Nanoscale, 4*, 95–98. https://doi.org/10.1039/C1NR11552A

Dou, Y. Y., Li, G. R., Song, J., & Gao, X. P. (2012). Nickel phosphide-embedded graphene as counter electrode for dye-sensitized solar cells. *Physical Chemistry Chemical Physics, 14*(4), 1339–1342. https://doi.org/10.1039/c2cp23775j

Duraisamy, S., Ganguly, A., Sharma, P. K., Benson, J., Davis, J., & Papakonstantinou, P. (2021). One-step hydrothermal synthesis of phase-engineered MoS_2/MoO_3 electrocatalysts for hydrogen evolution reaction. *ACS Applied Nano Materials, 4*(3), 2642–2656. https://doi.org/10.1021/acsanm.0c03274

Eda, G., Yamaguchi, H., Voiry, D., Fujita, T., Chen, M., & Chhowalla, M. (2011). Photoluminescence from chemically exfoliated MoS_2. *Nano Letters, 11*(12), 5111–5116. https://doi.org/10.1021/nl201874w

Eda, G., Fujita, T., Yamaguchi, H., Voiry, D., Chen, M., & Chhowalla, M. (2012). Coherent atomic and electronic heterostructures of single-layer MoS_2. *ACS Nano, 6*, 7311–7317. https://doi.org/10.1021/nn302422x

Feng, C., Ma, J., Li, H., Zeng, R., Guo, Z., & Liu, H. (2009). Synthesis of molybdenum disulfide (MoS2) for lithium ion battery applications. *Materials Research Bulletin, 44*(9), 1811–1815. https://doi.org/10.1016/j.materresbull.2009.05.018

Frindt, R. F., & Yoffe, A. D. (1963). Physical properties of layer structures: Optical properties and photoconductivity of thin crystals of molybdenum disulphide. *Proceedings of the Royal Society of London A, 273*, 69–83. https://doi.org/10.1098/rspa.1963.0075

Furchi, M. M., Pospischil, A., Libisch, F., Burgdörfer, J., & Mueller, T. (2014). Photovoltaic effect in an electrically tunable van der Waals heterojunction. *Nano Letters, 14*(8), 4785–4791. https://doi.org/10.1021/nl501962c

Ganal, P., Olberding, W., & Butz, T. (1993). Soft chemistry induced host metal coordination change from octahedral to trigonal prismatic in 1T-TaS_2. *Solid State Ionics, 59*, 313–319. https://doi.org/10.1016/0167-2738(93)90067-D

Ganatra, R., & Zhang, Q. (2014). Few-layer MoS_2: A promising layered semiconductor. *ACS Nano, 8*(5), 4074–4099. https://doi.org/10.1021/nn405938z

Garah, M. E., Bertolazzi, S., Ippolito, S., Eredia, M., Janica, I., Melinte, G., Ersen, O., Marletta, G., Ciesielski, A., & Samorì, P. (2018). MoS_2 nanosheets via electrochemical lithium-ion intercalation under ambient conditions. *FlatChem Chemistry, 9*, 33–39. https://doi.org/10.1016/J.FLATC.2018.06.001

Gomathi, P. T., Sahatiya, P., & Badhulika, S. (2017). Large-area, flexible broadband photodetector based on ZnS–MoS_2 hybrid on paper substrate. *Advanced Functional Materials, 27*(15), 1701611. https://doi.org/10.1002/adfm.201701611

Gonzalez Marin, J. F., Unuchek, D., Watanabe, K., Taniguchi, T., & Kis, A. (2019). MoS_2 photodetectors integrated with photonic circuits. *npj 2D Materials and Applications, 3*(1), 14. https://doi.org/10.1038/s41699-019-0096-4

Gu, X., Cui, W., Li, H., Wu, Z., Zeng, Z., Lee, S. T., Zhang, H., & Sun, B. (2013). A solution-processed hole extraction layer made from ultrathin MoS_2 nanosheets for efficient organic solar cells. *Advanced Energy Materials, 3*(10), 1262–1268. https://doi.org/10.1002/aenm201300549

Guan, R., Duan, J., Yuan, A., Wang, Z., Yang, S., Han, L., Zhang, B., Li, D., & Luo, B. (2021). Chemical vapor deposition of clean and pure MoS_2 crystals by the inhibition of MoO_{3-x} intermediates. *CrystEngComm, 23*(1), 146–152. https://doi.org/10.1039/D0CE01354D

Gupta, A., Arunachalam, V., & Vasudevan, S. (2015). Water dispersible, positively and negatively charged MoS2 Nanosheets: Surface chemistry and the role of surfactant binding. *The Journal of Physical Chemistry Letters, 6*(4), 739–744. https://doi.org/10.1021/acs.jpclett.5b00158

Hao, L. Z., Gao, W., Liu, Y. J., Han, Z. D., Xue, Q. Z., Guo, W. Y., Zhu, J., & Li, Y. R. (2015). High-performance n-MoS_2/i-SiO2/p-Si heterojunction solar cells. *Nanoscale, 7*, 8304–8308. https://doi.org/10.1039/C5NR01275A

Heising, J., & Kanatzidis, M. G. (1999). Exfoliated and restacked MoS_2 and WS_2: Ionic or neutral species? Encapsulation and ordering of hard electropositive cations. *Journal of the American Chemical Society, 121*(50), 11720–11732. https://doi.org/10.1021/ja991644d

Hong, S., Sheng, C., Krishnamoorthy, A., Rajak, P., Tiwari, S., Nomura, K., Misawa, M., Shimojo, F., Kalia, R. K., Nakano, A., & Vashishta, P. (2018). Chemical vapor deposition synthesis of MoS_2 layers from the direct sulfidation of MoO_3 surfaces using reactive molecular dynamics simulations. *The Journal of Physical Chemistry C, 122*(13), 7494–7503. https://doi.org/10.1021/acs.jpcc.7b12035

Hu, X., Chen, L., Tan, L., Zhang, Y., Hu, L., Xie, B., & Chen, Y. (2015). Versatile MoS_2 nanosheets in ITO-free and semi-transparent polymer power-generating glass. *Scientific Reports, 5*(1), 1–13. https://doi.org/10.1038/srep12161

Huang, Y., Pan, Y. H., Yang, R., Bao, L. H., Meng, L., Luo, H. L., et al. (2020). Universal mechanical exfoliation of large-area 2D crystals. *Nature Communications, 11*, 2453. https://doi.org/10.1038/s41467-020-16733-4

Hui, F., Pan, C., Shi, Y., Ji, Y., Grustan-Gutierrez, E., & Lanza, M. (2016). On the use of two dimensional hexagonal boron nitride as dielectric. *Microelectronic Engineering, 163*, 119–133. https://doi.org/10.1016/j.mee.2016.06.015

Huo, N., Gupta, S., & Konstantatos, G. (2017). MoS_2–HgTe quantum dot hybrid photodetectors beyond 2 μm. *Advanced Materials, 29*(17), 1606576. https://doi.org/10.1002/adma.201606576

Hyun, C. M., Choi, J. H., Lee, S. W., Park, J. H., Lee, K. T., & Ahn, J. H. (2018). Synthesis mechanism of MoS_2 layered crystals by chemical vapor deposition using MoO3 and sulfur powders. *Journal of Alloys and Compounds, 765*. https://doi.org/10.1016/j.jallcom.2018.06.183

Jariwala, D., Sangwan, V. K., Lauhon, L. J., Marks, T. J., & Hersam, M. C. (2014). Emerging device applications for semiconducting two-dimensional transition metal dichalcogenides. *ACS Nano, 8*(2), 1102–1120. https://doi.org/10.1021/nn500064s

Jeong, H., Jae-Yup, K., Koo, B., Son, H., Kim, D., & Ko, M. (n.d.). Rapid sintering of MoS_2 counter electrode using near-infrared pulsed laser for use in highly efficient dye-sensitized solar cells. *Journal of Power Sources, 330*, 104–110. https://doi.org/10.1016/j.jpowsour.2016.09.002

Jiao, K., Duan, C., Wu, X., Chen, J., Wang, Y., & Chen, Y. (2015). The role of MoS_2 as an interfacial layer in graphene/silicon solar cells. *Physical Chemistry Chemical Physics, 17*, 8182–8186. https://doi.org/10.1039/C5CP00321K

Jin, Z., Shin, S., Kwon, D., Han, S. J., & Min, Y. S. (2014). Novel chemical route for atomic layer deposition of MoS_2 thin film on SiO_2/Si substrate. *Nanoscale, 6*, 14453–14458. https://doi.org/10.1039/C4NR04816D

Kakavelakis, G., Gouda, L., Tischler, Y., Kaliakatsos, I., & Petridis, K. (2019). *2D transition metal dichalcogenides for solution-processed organic and perovskite solar cells, two dimensional transition metal dichalcogenides* (pp. 203–239). Springer. https://doi.org/10.1007/978-981-13-9045-6_7

Kim, V., Kim, J. Y., & Kwon, M. K. (2023). Synthesis of MoS_2 using chemical vapor deposition and conventional hydrothermal methods: Applications to respiration sensing. *Applied Sciences, 13*(10), 6084. https://doi.org/10.3390/app13106084

Korn, M., Andrade, S., Borges, S., Sousa, S., & Oliveira, F. (2003). Reagent generation assisted by ultrasonic irradiation. *Journal of the Brazilian Chemical Society, 14*. https://doi.org/10.1590/S0103-50532003000200011

Kufer, D., Nikitskiy, I., Lasanta, T., Navickaite, G., Koppens, F. H. L., & Konstantatos, G. (2015). Hybrid 2D-0D MoS2-PbS quantum dot photodetectors. *Advanced Materials, 27*(1), 176–180. https://doi.org/10.1002/adma.201402471

Kumar, N. A., Dar, M. A., Gul, R., & Baek, J. B. (2015). Graphene and molybdenum disulfide hybrids: Synthesis and applications. *Materials Today, 18*, 286–298. https://doi.org/10.1016/j.mattod.2015.01.016

Lee, C., Yan, H., Brus, L. E., Heinz, T. F., Hone, J., & Ryu, S. (2010). Anomalous lattice vibrations of single-and few-layer MoS_2. *ACS Nano, 4*(5), 2695–2700. https://doi.org/10.1021/nn1003937

Lee, Y. H., Zhang, X. Q., Zhang, W., Chang, M. T., Lin, C. T., Chang, K. D., et al. (2012). Synthesis of large-area MoS_2 atomic layers with chemical vapor deposition. *Advanced Materials, 24*, 2320–2325. https://doi.org/10.1002/adma.201104798

Lei, B., Li, G. R., & Gao, X. P. (2014). Morphology dependence of molybdenum disulfide transparent counter electrode in dye-sensitized solar cells. *Journal of Materials Chemistry A, 2*, 3919–3925. https://doi.org/10.1039/C3TA14313A

Li, X., & Zhu, H. (2015). Two-dimensional MoS_2: Properties, preparation, and applications. *Journal of Metriomics*, 33–44. https://doi.org/10.1016/j.jmat.2015.03.003

Li, Y., Wang, H., Xie, L., Liang, Y., Hong, G., & Dai, H. (2011). MoS_2 nanoparticles grown on graphene: An advanced catalyst for the hydrogen evolution reaction. *Journal of the American Chemical Society, 133*, 7296–7299. https://doi.org/10.1021/ja201269b

Li, X., Zhang, W., Wu, Y., Min, C., & Fang, J. (2013). Solution-processed MoSx as an efficient anode buffer layer in organic solar cells. *ACS Applied Materials & Interfaces, 5*(18), 8823–8827. https://doi.org/10.1021/am402105d

Li, X., Zang, X., Li, X., Zhu, M., Chen, Q., Wang, K., et al. (2014a). Hybrid heterojunction and solid-state photoelectrochemical solar cells. *Advanced Energy Materials, 4*, 1400224. https://doi.org/10.1016/j.jmat.2017.11.004

Li, X., Xie, D., Park, H., Zeng Helen, T., Wang, K., Wei, J., et al. (2014b). Anomalous behaviors of graphene transparent conductors in graphene-silicon heterojunction solar cells. *Advanced Energy Materials, 3*, 1029–1034.

Li, H., Wu, J., Yin, Z., & Zhang, H. (2014c). Preparation and applications of mechanically exfoliated single-layer and multilayer MoS_2 and WSe_2 nanosheets. *Accounts of Chemical Research, 47*(4), 1067–1075. https://doi.org/10.1021/ar4002312

Li, X., Wu, J., Mao, N., Zhang, J., Lei, Z., Liu, Z., & Xu, H. (2015). A self-powered graphene-MoS_2 hybrid phototransistor with fast response rate and high on-off ratio. *Carbon N. Y., 92*, 126–132. https://doi.org/10.1016/j.carbon.2015.03.064

Li, M., Wang, D., Li, J., Pan, Z., Ma, H., Jiang, Y., & Tian, Z. (2016). Facile hydrothermal synthesis of MoS_2 nano-sheets with controllable structures and enhanced catalytic performance for anthracene hydrogenation. *RSC Advances, 6*, 71534–71542. https://doi.org/10.1039/C6RA16084K

Li, X., Tao, L., Chen, Z., Fang, H., Li, X., Wang, X., Xu, J. B., & Zhu, H. (2017). Graphene and related two-dimensional materials: Structure-property relationships for electronics and optoelectronics. *Applied Physics Reviews, 4*(2), 021306. https://doi.org/10.1063/1.4983646

Lin, Y. C., Zhang, W., Huang, J. K., Liu, K. K., Lee, Y. H., Liang, C. T., et al. (2012). Wafer-scale MoS_2 thin layers prepared by MoO_3 sulfurization. *Nanoscale, 4*, 6637–6641. https://doi.org/10.1039/C2NR31833D

Liu, K. K., Zhang, W., Lee, Y. H., Lin, Y. C., Chang, M. T., Su, C. Y., Chang, C. S., Li, H., Shi, Y., Zhang, H., Lai, C. S., & Li, L. J. (2012). Growth of large-area and highly crystalline MoS_2 thin layers on insulating substrates. *Nano Letters, 12*(3), 1538–1544. https://doi.org/10.1021/nl2043612

Lorenz, T., Teich, D., Joswig, J. O., & Seifert, G. (2012). Theoretical study of the mechanical behavior of individual TiS2 and. MoS2 nanotubes. *Journal of Physical Chemistry C, 116*(21), 11714–11721. https://doi.org/10.1021/jp300709w

Luo, L., Shi, M., Zhao, S., Tan, W., Lin, X., Wang, H., & Jiang, F. (2019). Hydrothermal synthesis of MoS_2 with controllable morphologies and its adsorption properties for bisphenol A. *Journal of Saudi Chemical Society, 23*(6), 762–773. https://doi.org/10.1016/j.jscs.2019.01.005

Lv, R., Robinson, J. A., Schaak, R. E., Sun, D., Sun, Y., Mallouk, T. E., et al. (2015). Transition metal dichalcogenides and beyond: Synthesis, properties, and applications of single- and few-layer nanosheets. *Accounts of Chemical Research, 48*, 56–64. https://doi.org/10.1021/acs.accounts.5b00076

Ma, J., Bai, H., Zhao, W., Yuan, Y., & Zhang, K. (2018). High efficiency graphene/MoS_2/Si Schottky barrier solar cells using layer-controlled MoS_2 films. *Solar Energy, 160*, 76–84. https://doi.org/10.1016/j.solener.2017.11.066

Mak, K. F., & Shan, J. (2016). Photonics and optoelectronics of 2D semiconductor transition metal dichalcogenides. *Nature Photonics, 10*(4), 216–226. https://doi.org/10.1038/nphoton.2015.282

Mak, K. F., Lee, C., Hone, J., Shan, J., & Heinz, T. F. (2010). Atomically thin MoS_2: A new direct-gap semiconductor. *Physical Review Letters, 105*(13), 136805. https://doi.org/10.1103/PhysRevLett.105.136805

Meyer, J. C., Geim, A. G., Katnelson, M. I., Novoselov, K. S., & Roth, S. (2006). The structure of suspended graphene sheets. *Nature, 446*, 60–63. https://doi.org/10.1038/nature05545

Miao, H., Hu, X., Sun, Q., Hao, Y., Wu, H., Zhang, D., Bai, J., Liu, E., Fan, J., & Hou, X. (2016). Hydrothermal synthesis of MoS_2 nanosheets films: Microstructure and formation mechanism research. *Materials Letters, 166*, 121–124. https://doi.org/10.1016/j.matlet.2015.12.010

Mohamed Ismail, K. B., Arun Kumar, M., Mahalingam, S., Kim, J., & Atchudan, R. (2023). Recent advances in molybdenum disulfide and its nanocomposites for energy applications: Challenges and development. *Materials (Basel), 16*(12), 4471. https://doi.org/10.3390/ma16124471

Molina-Sánchez, A., Hummer, K., & Wirtz, L. (2015). Vibrational and optical properties of MoS_2: From monolayer to bulk. *Surface Science Reports, 70*(4), 554–586. https://doi.org/10.1016/j.surfrep.2015.10.001

Mukherjee, S., Maiti, R., Midya, A., Das, S., & Ray, S. K. (2015). Tunable direct bandgap optical transitions in MoS_2 nanocrystals for photonic devices. *ACS Photonics, 2*, 760–768. https://doi.org/10.1021/acsphotonics.5b00111

Muralikrishna, S., Manjunath, K., Samrat, D., Reddy, Ramakrishnappa, T., & Nagaraju, D. H. (2015). Hydrothermal synthesis of 2D MoS_2 nanosheets for electrocatalytic hydrogen evolution reaction. *RSC Advances, 5*, 89389–89396. https://doi.org/10.1039/C5RA18855E

Narayanasamy, S., & Arul, V. D. N. (2019). *Two dimensional transition metal dichalcogenides: Synthesis, properties and applications*. Springer. https://doi.org/10.1007/978-981-13-9045-6

Nguyen, T. P., Sohn, W., Oh, J. H., Jang, H. W., & Kim, S. Y. (2016). Size-dependent properties of two-dimensional MoS_2 and WS_2. *The Journal of Physical Chemistry C, 120*, 10078–10085. https://doi.org/10.1021/acs.jpcc.6b01838

Nicolosi, V., Chhowalla, M., Kanatzidis, M., Strano, M. S., & Coleman, J. N. (2013). Liquid exfoliation of layered materials. *Science, 340*, 1226419. https://doi.org/10.1126/science.1226

Niu, L., Coleman, J. N., Zhang, H., Shin, H., Chhowalla, M., & Zheng, Z. (2016). Production of two-dimensional nanomaterials via liquid based direct exfoliation. *Small, 12*, 272–293. https://doi.org/10.1002/smll.201502207

Novoselov, K. S., Geim, A. K., Morozov, S. V., Jiang, D., Zhang, Y., Dubonos, S. V., et al. (2004). Electric field in atomically thin carbon films. *Science, 30*, 6666–6669.

Ottaviano, L., Palleschi, S., Perrozzi, F., D'Olimpio, G., Priante, F., Donarelli, M., Benassi, P., Nardone, M., Gonchigsuren, M., Gombosuren, M., et al. (2017). Mechanical exfoliation and layer number identification of MoS_2 revisited. *2D Materials, 4*(4), 045013. https://doi.org/10.1088/2053-1583/aa8764

Palma, A. L., Cinà, L., Pescetelli, S., Agresti, A., Raggio, M., Paolesse, R., Bonaccorso, F., & Di Carlo, A. (2016). Reduced graphene oxide as efficient and stable hole transporting material in mesoscopic perovskite solar cells. *Nano Energy, 22*, 349–360. https://doi.org/10.1016/nanoen.2016.02.027

Pradhan, N. R., Talapatra, S., Terrones, M., Ajayan, P. M., & Balicas, L. (2017). Optoelectronic properties of heterostructures: The most recent developments based on graphene and transition-metal dichalcogenides. *IEEE Nanotechnology Magazine, 11*, 18–32. https://doi.org/10.1109/MNANO.2017.2676185

Py, M. A., & Haering, R. R. (1983). Structural destabilization induced by lithium intercalation in MoS_2 and related compounds. *Canadian Journal of Physics, 61*(1), 76–84. https://doi.org/10.1139/p83-013

Qin, P., Fang, G., Ke, W., Cheng, F., Zheng, Q., Wan, J., Lei, H., & Zhao, X. (2014). In situ growth of double-layer MoO3/MoS_2 film from MoS_2 for hole-transport layers in organic solar cell. *Journal of Materials Chemistry A, 2*, 2742–2756. https://doi.org/10.1039/C3TA13579A

Ra, H. S., Kwak, D. H., & Lee, J. S. (2016). A hybrid MoS_2 nanosheet-CdSe nanocrystal phototransistor with a fast photoresponse. *Nanoscale, 8*(39), 17223–17230. https://doi.org/10.1039/C6NR05393A

Ramakrishna Matte, H. S. S., Gomathi, A., Manna, A. K., Late, D. J., Datta, R., Pati, S. K., & Rao, C. N. R. (2010). MoS_2 and WS_2 analogues of graphene. *Angewandte Chemie (International Ed. in English), 49*(24), 4059–4062. https://doi.org/10.1002/anie.201000009

Ramasamy, M. S., Ryu, K. Y., Lim, J. W., Bibi, A., Kwon, H., Lee, J. E., Kim, D. H., & Kim, K. (2019). Solution-processed PEDOT: PSS/MoS_2 nanocomposites as efficient hole-transporting layers for organic solar cells. *Nanomaterials, 9*(9), 1328. https://doi.org/10.3390/nano9091328

Rani, A., Singh, K., Patel, A. S., Ckakraborti, A., Kumar, S., Ghosh, K., & Sharma, P. (2020). Visible light driven photocatalysis of organic dyes using SnO_2 decorated MoS_2 nanocomposites. *Chemical Physics Letters, 738*, 136874. https://doi.org/10.1016/j.cplett.2019.136874

Rani, A., Singh, K., Patel, A. S., Chakraborti, A., & Sharma, P. (2021). Enhanced photocatalytic activity of plasmonic Au nanoparticles incorporated MoS_2 nanosheets for degradation of organic dyes. *Journal of Materials Science: Materials in Electronics, 32*, 6168. https://link.springer.com/article/10.1007/s10854-021-05334-6

Rani, A., Singh, K., & Sharma, P. (2022). Investigation of visible light photocatalytic degradation of organic dyes by MoS_2 nanosheets synthesized by different routes. *Bulletin of Materials Science, 45*, 63. https://doi.org/10.1007/s12034-022-02655-y

Rani, A., Singh, K., Patel, A. S., & Sharma, P. (2023). Factors affecting photocatalytic degradation of methyl red by MoS_2 nanostructures prepared by hydrothermal technique. *Bulletin of Materials Science, 46*, 94. https://doi.org/10.1007/s12034-023-02929-z

Sahatiya, P., Jones, S. S., & Badhulika, S. (2018). 2D MoS_2–carbon quantum dot hybrid based large area, flexible UV–vis–NIR photodetector on paper substrate. *Applied Materials Today, 10*, 106–114. https://doi.org/10.1016/j.apmt.2017.12.013

Sahoo, S., Tiwari, S. K., & Nayak, G. C. (2019). *Surface engineering of graphene*. Springer. ISBN: 978-3-030-30206-1.

Saraswat, K. C., Park, J. H., Islam, R., Nazif, K. N., Elder, E., & Associate, L. (2019). *Low cost silicon (Si)/transition metal Dichalcogenides (TMDs) tandem solar cells with goal of >30% efficiency* (Vol. S18–187, pp. 1–2). Stanford Univ.

Savjani, N., Lewis, E. A., Pattrick, R. A. D., Haigh, S. J., & O'Brien, P. (2014). MoS_2 nanosheet production by the direct exfoliation of molybdenite minerals from several type-localities. *RSC Advances, 4*, 35609–35613. https://doi.org/10.1039/C4RA03982C

Seravalli, L., & Bosi, M. (2021). A review on chemical vapour deposition of two-dimensional MoS_2 flakes. *Materials (Basel), 14*(24), 7590. https://doi.org/10.3390/ma14247590

Singh, E., Kim, K. S., Yeom, G. Y., & Nalwa, H. S. (2017a). Atomically thin-layered molybdenum disulfide (MoS_2) for bulk-heterojunction solar cells. *ACS Applied Materials & Interfaces, 9*(4), 3223–3245. https://doi.org/10.1021/acsami.6b13582

Singh, E., Kim, K. S., Yeom, G. Y., & Nalwa, H. S. (2017b). Two-dimensional transition metal dichalcogenide-based counter electrodes for dye-sensitized solar cells. *RSC Advances, 7*, 28234–28290. https://doi.org/10.1039/C7RA03599C

Soldano, C., Mahmood, A., & Dujardin, E. (2010). Production, properties and potential of graphene. *Carbon, 48*, 2127–2150. https://doi.org/10.1016/j.carbon.2010.01.058

Splendiani, A., Sun, L., Zhang, Y., Li, T., Kim, J., Chim, C. Y., et al. (2010). Emerging photoluminescence in monolayer MoS_2. *Nano Letters, 10*(4), 1271–1275. https://doi.org/10.1021/nl903868w

Stephenson, T., Li, Z., Olsen, B., & Mitlin, D. (2014). Lithium ion battery applications of molybdenum disulfide (MoS_2) nanocomposites. *Energy & Environmental Science, 7*, 209–231. https://doi.org/10.1039/C3EE42591F

Tang, Q., & Zhou, Z. (2013). Graphene-analogous low-dimensional materials. *Progress in Materials Science, 58*, 1244–1315. https://doi.org/10.1016/J.PMATSCI.2013.04.003

Tang, W., Rassay, S. S., & Ravindra, N. M. (2017). Electronic & optical properties of transition-metal dichalcogenides. *Madridge Journal of Nanotechnology & Nanoscience, 2*, 58–64. https://doi.org/10.18689/mjnn.2017-110

Toh, R. J., Sofer, Z., Luxa, J., Sedmidubský, D., & Pumera, M. (2017). 3R phase of MoS_2 and WS_2 outperforms the corresponding 2H phase for hydrogen evolution. *Chemical Communications, 53*, 3054–3305. https://doi.org/10.1039/C6CC09952A

Tsai, D. S., Liu, K. K., Lien, D. H., Tsai, M. L., Kang, C. F., Lin, C. A., Li, L. J., & He, J. H. (2013). Few-layer MoS_2 with high broadband photogain and fast optical switching for use in harsh environments. *ACS Nano, 7*(5), 3905–3911. https://doi.org/10.1021/nn305301b

Tsai, M. L., Su, S. H., Chang, J. K., Tsai, D. S., Chen, C. H., Wu, C. I., Li, L. J., Chen, L. J., & He, J. H. (2014). Monolayer MoS_2 heterojunction solar cells. *ACS Nano, 8*(8), 8317–8322. https://doi.org/10.1021/nn502776h

Tsuboi, Y., Wang, F., Kozawa, D., Funahashi, K., Mouri, S., Miyauchi, Y., Takenobu, T., & Matsuda, K. (2015). Enhanced photovoltaic performances of graphene/Si solar cells by insertion of a MoS_2 thin film. *Nanoscale, 7*, 14476–14482. https://doi.org/10.1039/C5NR03046C

Uthayaraj, S., Karunarathne, D. G. B. C., & Kumara, G. R. A. (2019). Powder pressed cuprous iodide (CuI) as A hole transporting material for perovskite solar cells. *Materials (Basel), 12*(13), 2037. https://doi.org/10.3390/ma12132037

Wang, X., Feng, H., Wu, Y., & Jiao, L. (2013). Controlled synthesis of highly crystalline MoS_2 flakes by chemical vapour deposition. *Journal of the American Chemical Society, 135*(14), 5304–5307. https://doi.org/10.1021/ja4013485

Wang, D., Pan, Z., Wu, Z., Wang, Z., & Liu, Z. (2014). Hydrothermal synthesis of MoS2 nanoflowers as highly efficient hydrogen evolution reaction catalysts. *Journal of Power Sources, 264*, 229–234. https://doi.org/10.1016/j.jpowsour.2014.04.066

Wang, Y., Wang, S., Chen, X., Li, Z., Wang, J., Li, T., & Deng, X. (2018a). Largely enhanced VOC and stability in perovskite solar cells with modified energy match by coupled 2D interlayers. *Journal of Materials Chemistry A, 6*(11), 4860–4867. https://doi.org/10.1039/C7TA11295E

Wang, C., Zhang, C., Tong, S., Xia, H., Wang, L., Xie, H., Gao, Y., & Yang, J. (2018b). Energy level and thickness control on PEDOT:PSS layer for efficient planar heterojunction perovskite cells. *Journal of Physics D: Applied Physics, 51*(2), 025110. https://doi.org/10.1088/1361-6463/aa9d30

Wi, S., Kim, H., Chen, M., Nam, H., Guo, L. J., Meyhofer, E., et al. (2014). Enhancement of photovoltaic response in multilayer MoS_2 induced by plasma doping. *ACS Nano, 8*(5), 5270–5281. https://doi.org/10.1021/nn5013429

Wilson, J. A., & Yoffe, A. D. (1969). The transition metal dichalcogenides discussion and interpretation of the observed optical, electrical and structural properties. *Advances in Physics, 18*(73), 193–335. https://doi.org/10.1080/00018736900101307

Wirth-Lima, A. J., Alves-Sousa, P. P., & Bezerra-Fraga, W. (2019). Graphene/silicon and 2D-MoS_2/silicon solar cells: A review. *Applied Physics A: Materials Science & Processing, 125*, 241. https://doi.org/10.1007/S00339-019-2540-X

Wu, W., Wang, L., Li, Y., Zhang, F., Lin, L., Niu, S., et al. (2014). Piezoelectricity of single-atomic-layer MoS_2 for energy conversion and piezotronics. *Nature, 514*(7523), 470–474. https://doi.org/10.1038/nature13792

Wu, Z., Li, F., Li, X., Yang, Y., Huang, X., & Li, H. (2022). Direct synthesis of MoS_2 nanosheets in reduced graphene oxide nanoscroll for enhanced photodetection. *Nanomaterials, 12*(9), 1581. https://doi.org/10.3390/nano12091581

Xiong, K., Kim, H., Marstell, R. J., Göritz, A., Wipf, C., Li, L., Park, J. H., Luo, X., Wietstruck, M., & Madjar, A. (2018). CMOS-compatible batch processing of monolayer MoS_2 MOSFETs. *Journal of Physics D: Applied Physics, 51*(15). https://doi.org/10.1088/1361-6463/aab4ba

Yang, X., & Li, B. (2020). Monolayer MoS_2 for nanoscale photonics. *Nano, 9*(7), 1557–1577. https://doi.org/10.1515/nanoph-2019-0533

Yang, X., Liu, W., Xiong, M., Zhang, Y., Liang, T., Yang, J., Xu, M., Ye, J., & Chen, H. (2014). Au nanoparticles on ultrathin MoS_2 sheets for plasmonic organic solar cells. *Journal of Materials Chemistry A, 2*, 14798–14806. https://doi.org/10.1039/C4TA03178D

Yang, Q.-D., Li, J., Cheng, Y., Li, H.-W., Guan, Z., Yu, B., & Tsang, S.-W. (2017). Graphene oxide as an efficient hole-transporting material for high-performance perovskite solar cells with enhanced stability. *Journal of Materials Chemistry A, 5*, 9852–9858. https://doi.org/10.1039/C7TA01752A

Yao, Y., Lin, Z., Li, Z., Song, X., Moon, K. S., & Wong, C. P. (2012). Large-scale production of two-dimensional nanosheets. *Journal of Materials Chemistry, 22*, 13494–13499. https://doi.org/10.1039/C2JM30587A

Yi, M., & Shen, Z. A. (2015). A review on mechanical exfoliation for the scalable production of graphene. *Journal of Materials Chemistry A, 3*, 11700–11715. https://doi.org/10.1039/C5TA00252D

Yue, J., Jian, J., Dong, P., Luo, L., Chang, F., et al. (2019). Growth of single-layer MoS_2 by chemical vapor deposition on sapphire substrat. *IOP Conference Series: Materials Science and Engineering, 592*(1), 012044. https://doi.org/10.1088/1757-899X/592/1/012044

Yun, J.-M., Noh, Y.-J., Yeo, J.-S., Go, Y.-J., Na, S.-I., Jeong, H.-G., Kim, J., Lee, S., Kim, S.-S., Koo, H. Y., et al. (2013). Efficient work-function engineering of solution-processed MoS_2 thin-films for novel hole and electron transport layers leading to high-performance polymer solar cells. *Journal of Materials Chemistry C, 1*, 3777–3783. https://doi.org/10.1039/C3TC30504J

Zang, X., Chen, Q., Li, P., He, Y., Li, X., Zhu, M., et al. (2014). Highly flexible and adaptable, all-solid-state supercapacitors based on graphene woven-fabric film electrodes. *Small, 10*, 2583–2588. https://doi.org/10.1002/smll.201303738

Zeng, Z., Zhang, X., Bustillo, K., Niu, K., Gammer, C., Xu, J., & Zheng, H. (2015). In situ study of lithiation and delithiation of MoS_2 nanosheets using electrochemical liquid cell transmission electron microscopy. *Nano Letters, 15*(8), 5214–5220. https://doi.org/10.1021/acs.nanolett.5b02483

Zhan, Y., Liu, Z., Najmaei, S., Ajayan, P. M., & Lou, J. (2012). Large-area vapor-phase growth and characterization of MoS(2) atomic layers on a SiO(2) substrate. *Small, 8*(7), 966–971. https://doi.org/10.1002/smll.201102654

Zhang, H., & Li, Z. (2020). Stable edge structures and electronic states in zigzag 1T'-dichalcogenide nanoribbons. *Journal of Physics: Condensed Matter, 32*(36), 365303. https://doi.org/10.1088/1361-648X/ab9051

Zhang, W., Wang, Y., Zhang, D., Yu, S., Zhu, W., Zheng, F., Wang, S., & Wang, J. (2015). A one-step approach to the large-scale synthesis of functionalized MoS_2 nanosheets by ionic liquid assisted grinding. *Nanoscale, 7*, 10210–10217. https://doi.org/10.1039/C5NR02253C

Zhang, X., Wang, C., Xue, M., Lin, B. C., Ye, X., & Lei, W. N. (2016). Hydrothermal synthesis and characterization of ultrathin MoS_2 nanosheets. *Chalcogenide Letters, 13*, 27–34.

Zhang, X., Ma, G., & Wang, J. (2019). Hydrothermal synthesis of two-dimensional MoS2 and its applications. *Tungsten, 1*. https://doi.org/10.1007/s42864-019-00014-9

Zhang, Y., Chen, X., Zhang, H., Hu, S., Zhao, G., Zhang, M., Qin, W., Wang, Z., Huang, X., & Wang, J. (2021). Facile exfoliation for high-quality molybdenum disulfide nanoflakes and relevant field-effect transistors developed with thermal treatment. *Frontiers in Chemistry: Nanoscience, 9*. https://doi.org/10.3389/fchem.2021.650901

Photocatalytic Hydrogen Production of Perovskite Based Nanocomposites by Green Laser Irradiation Techniques

Mohamed Jaffer Sadiq Mohamed and Mohammed Ashraf Gondal

1 Introduction

Recent technological advances have raised the need for energy in modern life (Abbasi et al., 2022; Dawood et al., 2020; Zhang et al., 2022a, b). Because of the scarcity of fossil fuels, there has been raised concerns about the efficiency of energy-gathering systems. Furthermore, it is vital to recognize that the use of fossil-fuels has a substantial detrimental influence on the environment, particularly in terms of global warming. The observed behaviour is mostly caused by CO_2 emissions into the Earth's atmosphere (Akrami et al., 2022; Kumar et al., 2018; Wang et al., 2020). Globally, the shift from non-renewable-energy to renewable-energy sources is being prioritized. Most of the current renewable energy research is focused on designing and improving energy-collecting devices in order to improve their efficiency (Lee et al., 2021; Li et al., 2022; Yue et al., 2021).

Photocatalytic hydrogen (H_2) production technology has the potential to address the world's significant energy and environmental challenges (Chen et al., 2022; Ma et al., 2018; Ollis et al., 1991). As previously research, the process converts solar energy into chemical energy, making it an appealing choice (Li et al., 2009; Saeed et al., 2022). Significant research is being undertaken on several semiconductor materials, such as transition metal oxides, sulphides, and selenides, to investigate their potential as photocatalysts for effectively producing hydrogen (Abdulrazzak et al., 2016; Gultom et al., 2019; Ismael, 2021; Liu et al., 2019, 2022; Ramesh Reddy et al., 2020; Singh & Kansal, 2022; Sun et al., 2018; Tasleem & Tahir, 2020a, b; Wu et al., 2016; Zhang et al., 2022a, b). Most recent academic research has

M. J. S. Mohamed (✉) · M. A. Gondal
Laser Research Group, Department of Physics & Interdisciplinary Research Center for Hydrogen Technologies and Carbon Management, King Fahd University of Petroleum and Minerals (KFUPM), Dhahran, Saudi Arabia

S. A. Bansal et al. (eds.), *Emerging Applications of Novel Nanoparticles*, Lecture Notes in Nanoscale Science and Technology 37, https://doi.org/10.1007/978-3-031-57843-4_12

focused on ABO_3 inorganic perovskite materials. In this context, the sign A denotes an alkaline-earth metal, the sign B represents a transition-metal, and the sign O represents oxygen. The possible potential photocatalytic properties of these materials have piqued the interest of researchers (Adnan et al., 2018; Deshmukh et al., 2021; Du et al., 2022; Pei et al., 2015; Tasleem & Tahir, 2020a). Strontium titanate ($SrTiO_3$) is classified as an inorganic perovskite material with an n-type bandgap (Deshmukh et al., 2021; Kiss et al., 2017; Kumar et al., 2018; Manchala et al., 2022; Zhang et al., 2020). The poor photostability, and tall electron-hole recombination of $SrTiO_3$ restrict its catalytic effectiveness. To enhance the crystalline structure of $SrTiO_3$, metals or metal chalcogenides may be added. The combination of these components affects both the catalytic efficacy and the material's durability. The desired outcome is achieved by changing the electrical configuration and introducing oxygen-vacancies, which increases the efficiency of segregating photo-generated pairs, and also increases a greater number of catalytic sites (Jang et al., 2021; Kanagaraj & Thiripuranthagan, 2017; Mohamed et al., 2022; Pan et al., 2021; Qin et al., 2021; Rabuffetti et al., 2008; Yang et al., 2016; Yoon et al., 2018).

Laser irradiation technique is a versatile and well-known technology for producing nanoparticles (Fazio et al., 2020; Johny et al., 2022; Lee et al., 2022). Despite the availability of traditional technologies such as chemical-vapour deposition, sol-gel, hydrothermal, thermal-evaporation, solvothermal, and spray pyrolysis, this technology has piqued the interest of many researchers (Jaffer & Samson, 2014; Jamkhande et al., 2019; Liang et al., 2019; Shukla & Srivastava, 2016). The laser irradiation approach enables the synthesis of nanoparticles with a wide set of characteristics that are not bound by any limitations (Hadadi et al., 2023; Ilyas et al., 2017; Mohamed et al., 2022; Naik et al., 2021; Naser et al., 2022). The synthesis approaches are constrained by several factors, including the composition of the precursor solutions, the formation of by-products, the specific temperature and pressure requirements, the duration of the reaction, the purity of the resulting product, and the potential toxicity associated with the process (Fazio et al., 2020; Hassan et al., 2021; Naser et al., 2022). The laser irradiation technique is a very efficient top-down strategy for producing high-purity nanomaterials and composites. The laser irradiation technique is well-known for its environmental sustainability since it operates independently of established temperature and pressure conditions (Fazio et al., 2020; Gondal et al., 2016).

According to the best of our knowledge, this is the first time. The present work effectively synthesised $SrTiO_3$ (ST)@CdSe (CS) composites by using laser irradiation technology, which is recognized for its direct and environmentally friendly approach. Furthermore, the physio-chemical characteristics of the prepared nanocomposite materials were thoroughly examined and discussed. The photo catalytic H_2 activity of the synthesized nanocomposite was assessed in depth using extensive photocatalytic H_2 production experiments.

2 Materials and Methods

2.1 *Materials*

$SrTiO_3$ (strontium-titanate), CdSe (cadmium-selenide), KOH (potassium-hydroxide), 5% Nafion, and CH_3OH (methanol) were procured from Sigma and used without additional processing or treatment.

2.2 *Synthesis Procedure*

STCS-x (x = 1.0–4.0%) nanocomposites were prepared using the direct laser irradiation approach. In brief, to prepare STCS3, 30 mg of CS and 100 mg of ST were mixed in 25 mL of deionized water to form a dispersed resolution. The mixture was then ultrasonically homogenized for 30 min. During laser irradiation process, Nd-YAG laser with a 532 nm wavelength and an 8 ns pulse duration was utilized to irradiate the ST and CS specimens. The use of a focusing lens boosted the strength of the laser beam. Each sample was constantly agitated for 30 min while being exposed to the laser light. The solution, which included ST and CS nanoparticles, was dried for 2 h at 120 °C. Moreover, STCS with varied CS wt.% were created using the same manner. The synthesized STCS-x contains x through 0, 1.0, 2.0, 3.0, and 4.0 wt.% of CS are named as ST, CS, STCS1, STCS2, STCS3, and STCS4, respectively.

2.3 *Measurements of Photocatalytic Hydrogen*

The photocatalytic technique used to produce H_2 was examined. In the experiment, a custom-made quartz reactor containing 0.03 g of a synthesized materials was dispersed in 30 mL of CH_3OH (10%) and it act as a hole scavenger. Then, the mixed solution was ultrasonically dispersed to ensure complete dispersion. A high-purity N_2-gas was injected into the dispersed resolution for around 30 min to remove any dissolved oxygen. A 300-W Xenon-lamp (PLS-SXE300) was used in the study as an ultraviolet (UV) light source. The light irradiated the quartz reactor for an extended period of time, enabling the reaction to take place at ambient temperature. The rate of H_2 production was measured hourly using a GC (gas-chromatograph). A gas injector was used for manual injection. A magnetic stirrer was used to keep the fluid circulating throughout the experiment to prevent particle accumulation in the reactor. It is important to highlight that each sample was subjected to three consecutive trials to assess the reliability of the data acquired. In the same way, the photocatalytic stability and recyclability of the prepared catalysts were also studied.

2.4 Characterization

The crystalline structure of the prepared materials was carried out using an XRD (X-ray diffraction, Rigaku) at a Cu-Kα X-ray radiation source with a wavelength (0.15406 nm). The materials were analysed using a Raman spectrometer (HORIBA Scientific LabRAM) with a He-Ne laser as the light source at a 532 nm excitation wavelength and a 10% power level. The morphology of the materials was investigated using SEM (scanning electron microscopy, Zeiss) equipped with EDX (Energy-dispersive X-ray spectroscopy), and TEM (transmission electron microscopy, Titan-FEI-Morgagni-268). The elemental composition and oxidation state of the prepared materials was analysed using an XPS (X-ray photoelectron spectroscopy, Kratos DLD) at a monochromatic Al-Kα X-ray radiation source. The optical characteristics were analysed using an UV-vis spectrophotometer (JASCO V-750) throughout a 200–800 nm wavelength range. A JASCO FP-8500 with an Xe (xenon) light source was used for photoluminescence (PL) spectroscopy.

3 Results and Discussion

3.1 Synthesis Mechanism Process

The STCS nanocomposite was prepared using a cutting-edge, eco-friendly laser irradiation technique (Fig. 1). The experiment used a 532 nm neodymium-doped-yttrium-aluminium-garnet (ND: YAG) laser-operating at the second-harmonic wavelength. Using the laser irradiation technique, a laser beam irradiated a water dispersion containing ST and CS nanoparticles for around 30 min. The experimental conditions were optimized using previous literature (Gondal et al., 2011; Hassan et al., 2021; Ilyas et al., 2017; Mohamed et al., 2022). When the laser beam comes into contact with a solid material, it is successfully captivated, quickly turning absorbed energy into heat energy. The metallic specimen experiences a phase shift, moving from solid form to liquid form, because of the heat energy in its immediate surroundings. Solid laser pulses increase irradiance, resulting in increased energy transfer to the surface of the sample. The increased energy inside the focus zone may produce localized vaporization of a specific piece of the solid sample surface. The application of laser-induced temperature causes the solid surface of an object to evaporate. The laser-beam ionizes the vapour-gas, resulting in production of plasma. The plasma bubble's distinguishing characteristics, such as high temperature and pressure, are the result of focused irradiance at a critical region on the samples surface. The laser-formed plasma bubbles quickly grow to its maximum size. Because of the large gravity transformation amid the bubbles and the surrounding liquids, the bubble cools and disintegrates. When particles in a plasma interact with molecules in a liquid medium, nanomaterials are formed. Nanomaterials are atoms produced from a solid sample and its surroundings. This phenomenon occurs because of the

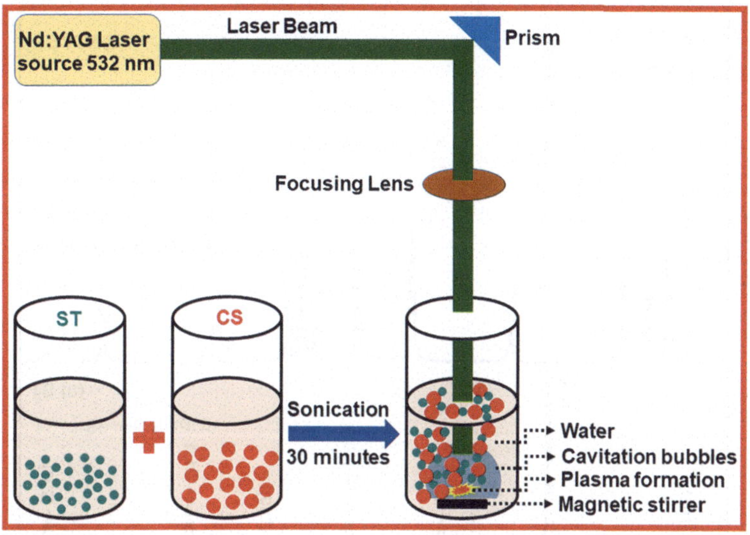

Fig. 1 Depicts the nanocomposite preparation process for laser irradiation technique

particular characteristics of the plume. The existence of a discernible change in the colour of the solution following laser ablation, especially when the minerals ST and CS are mixed, indicates a possible compositional arrangement. By minimizing intermediary chemicals or catalysts, the laser irradiation technique reduces the need for post-synthesis purification. As a result, the production method yields a high-purity nanocomposite material.

3.2 XRD Analysis

XRD was used to investigate the structural characteristics of the ST, CS, and STCS3 composites, as presented in Fig. 2. The diffraction peaks detected at 23.0°, 32.6°, 40.2°, 46.7°, 52.6°, 58.0°, and 68.0° correspond to the (100), (110), (111), (200), (210), (211), and (220) planes of ST (ICDD No. 35–0734), respectively (Luo et al., 2019; Tan et al., 2020). The diffraction peaks found at 24.1°, 25.6°, 27.3°, 35.3°, 42.3°, 46.0°, 49.0°, 49.9°, 50.5°, 56.0°, 64.1°, 66.6°, and 68.1° match to the (100), (002), (101), (102), (110), (103), (200), (112), (201), (202), (203), (210), and (211) crystallographic planes of CS (ICDD No. 08–0459), respectively (Bang et al., 2012; Suresh & Arunseshan, 2014). After CS modification, the diffraction peaks in ST within STCS3 samples remain unchanged. Based on these information, it is feasible to conclude that the presence of CS in STCS3 samples has no influence on the crystal structure of ST.

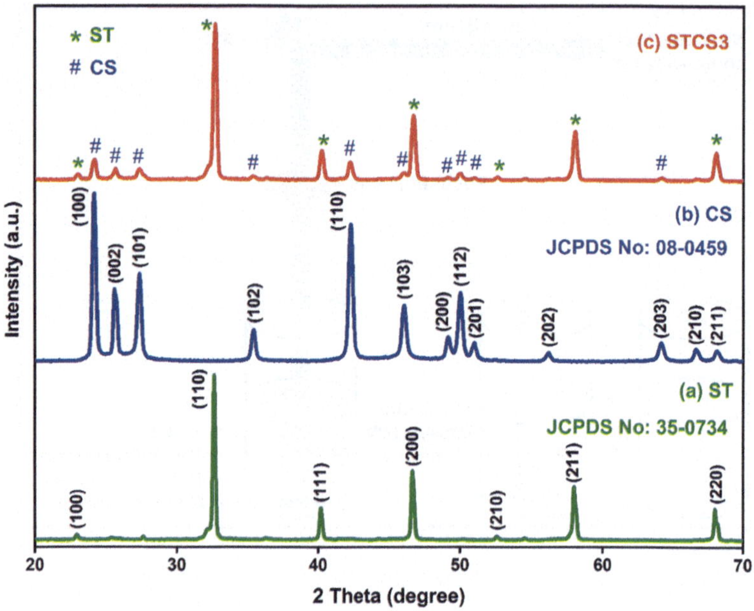

Fig. 2 Depicts the XRD spectra of ST, CS, and STCS3

3.3 *Raman Analysis*

The Raman analysis technique is well-known as a potential tool for analysing the crystallographic and electrical characteristics of diverse materials. This method is also well-known for its high sensitivity and lack of destructiveness, and it may be used to acquire insight into structural data. The Raman spectra of CS, ST, and the composite STCS3 are shown in Fig. 3. There are multiple frequency bands in the STCS3 systems, which are 82.2, 184.6, 251.6, 298.4, 359.3, and 640.1 cm^{-1}, respectively. The peaks in the ST sample indicate a high degree of agreement, which might be attributable to a variety of characteristics. This result suggests that the sample closely matches the predicated traits. The observed broad range of spectra may be ascribed to the ST second-order Raman active mode (Nilsen & Skinner, 1968; Rabuffetti et al., 2008). The STCS3 system shows a unique longitudinal optical (LO) mode with a frequency of 206.4 cm^{-1}. The existence of the CS component inside the system is related to the mode under consideration. The first overtone (2 LO) is responsible for a peak at 414.4 cm^{-1} in the CS sample (Akilarasan et al., 2021; Islam et al., 2014).

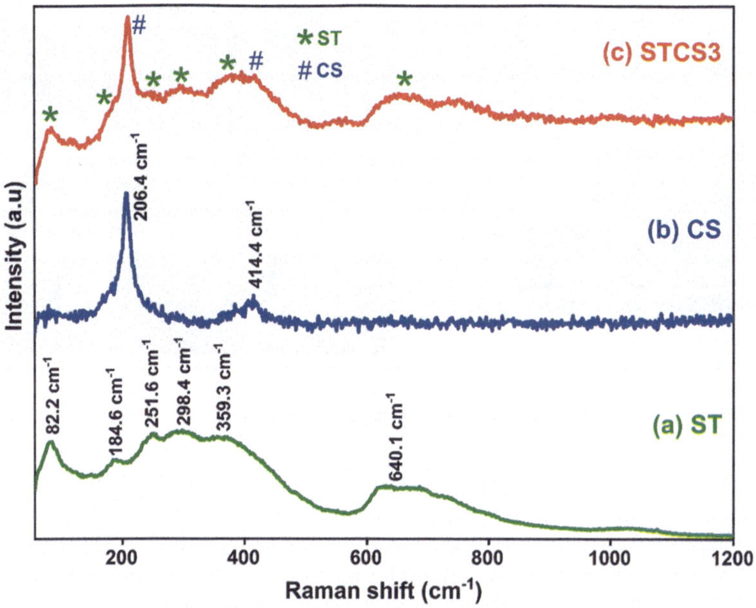

Fig. 3 Depicts the Raman spectra of ST, CS, and STCS3

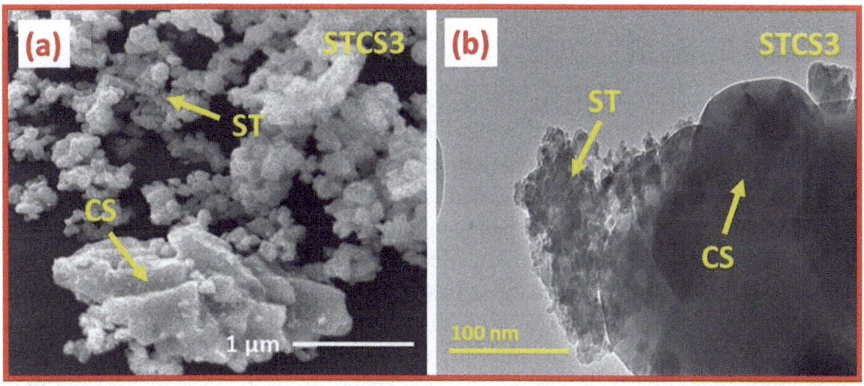

Fig. 4 Depicts the SEM and TEM image of STCS3

3.4 *Morphological Study*

Figure 4a shows a SEM image of STCS3. In this study, spherical ST particles and polygonal CS particles displayed significant aggregation and interconnectivity. TEM was also used to investigate the morphology of the STCS3 nanocomposites. Figure 4b shows a TEM image of STCS3 that has a structure and conforms to SEM micrographs. A detailed EDX and elemental mapping study was carried out to

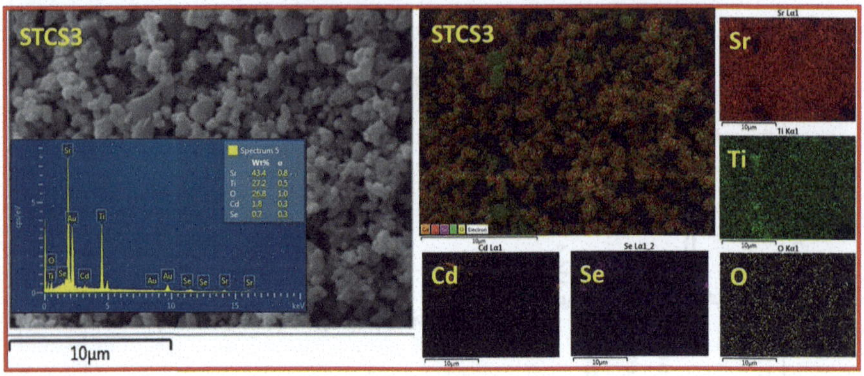

Fig. 5 Depicts the EDX spectra and elemental mapping of the STCS3

analyze the primary components of the STCS3 sample, as shown in Fig. 5. The investigation of this analysis looked at the distribution of the elements strontium (Sr), titanium (Ti), oxygen (O), cadmium (Cd), and selenium (Se). The appearance of Au peaks in the EDX spectrum might be attributed to the gold-coating during the sample analysis. There were no further peaks detected, indicating that the nanocomposite components were free from impurities.

3.5 XPS Analysis

The elemental composition and oxidation state of a material's surface was determined using XPS. The experiment sought to prove the existence of CS on ST by standardizing the binding energies of the characteristic elements using the C 1 s reference-peak at 284.4 eV. The chemical composition of STCS3 as shown by survey spectra is depicted in Fig. 6a. The obtained peaks in the STCS3 sample corresponding to Sr, Ti, O, Cd, and Se co-exist to the results of the EDX analysis. Figure 6b shows a spectrum of Sr 3d, with binding energies of 134.4 eV for Sr $3d_{3/2}$ and 132.7 eV for Sr $3d_{5/2}$, indicating the existence of the Sr^{2+} state (Kanagaraj & Thiripuranthagan, 2017). Figure 6c shows a spectrum of Ti 2p, which reveal two distinct peaks with binding energies of 463.9 eV and 458.1 eV, which are associated with the Ti $2p_{1/2}$ and Ti $2p_{3/2}$ states. The difference in energy of 5.8 eV amid the Ti $2p_{1/2}$ and Ti $2p_{3/2}$ suggests that the titanium is in the Ti^{4+} oxidation state (Yang et al., 2016). Figure 6d shows a spectrum of Cd 3d, which reveal two distinct peaks at 411.5 eV and 404.7 eV, which match to the energy levels of Cd $3d_{3/2}$ and Cd $3d_{5/2}$ in Cd^{2+} within the CS combination (Du et al., 2022). Figure 6e shows a spectrum of Se with two distinct peaks at 54.1 eV and 53.5 eV. The detected peaks in the CS complex suggest a selenium-metal bonding configuration, with the Se^{4+} state moiety's Se $3d_{5/2}$ and Se $3d_{3/2}$ states (Liu et al., 2019). Figure 6f shows the core level O 1 s spectra, with a significant signal at 529.3 eV indicating the presence of lattice

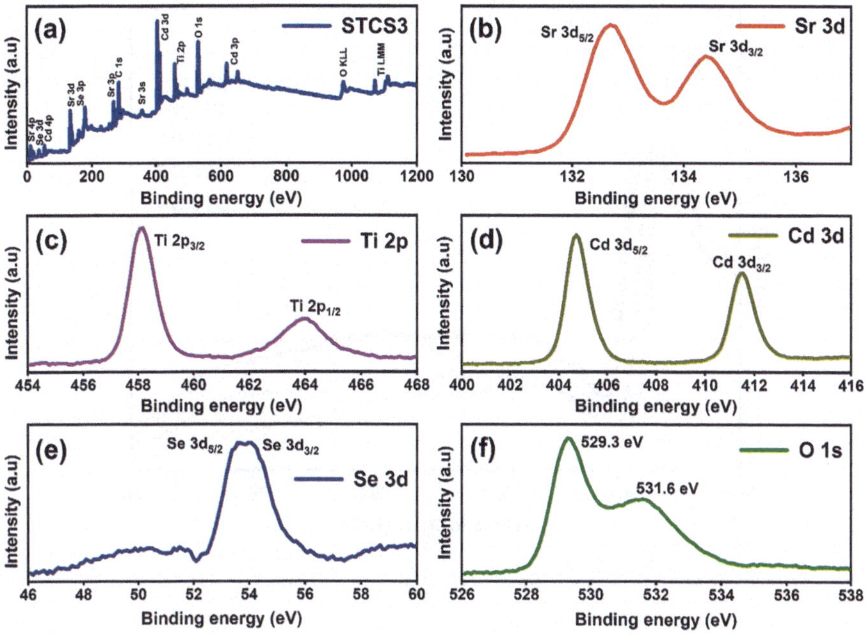

Fig. 6 Depicts the STCS3 XPS survey spectra, and high-resolution spectra of Sr 3d, Ti 2p, Cd 3d, Se 3d, and O 1 s

oxygen within the STCS3 composite. A high peak at 531.6 eV must also be considered into account as evidence for the presence of chemisorbed oxygen due to oxygen vacancies (Kumar et al., 2018).

3.6 *Optical Analysis*

The optical characteristic of the prepared materials was used to investigate the UV-Vis absorbance spectrum ranging from 300 to 800 nm. Figure 7 reveals a considerable increase in absorption intensity in the visible region. Furthermore, molecules with co-substitution exhibit a little redshift in the absorption edge.

The bandgap is a fundamental property of semiconductor materials. The energy bandgaps of the specimens in Fig. 8 are measured using Tauc's Relation (Mohamed et al., 2018; Sadiq et al., 2017b), which may be expressed mathematically as $(\alpha h\nu)^2 = A(Eg - h\nu)^n$. The symbols in the equation represent various parameters. The symbol 'α' indicates the Absorption-constant, 'hv' the photon-energy, 'Eg' the bandgap, 'A' the scaling-factor, and 'n' a variable factor depending on the semiconductor material. The exponent 'n' is indirect bandgap semiconductors is 1/2, whereas it is 2 in direct bandgap semiconductors. The bandgap of ST, a direct bandgap semiconductor is determined using a linear extrapolation approach. In this

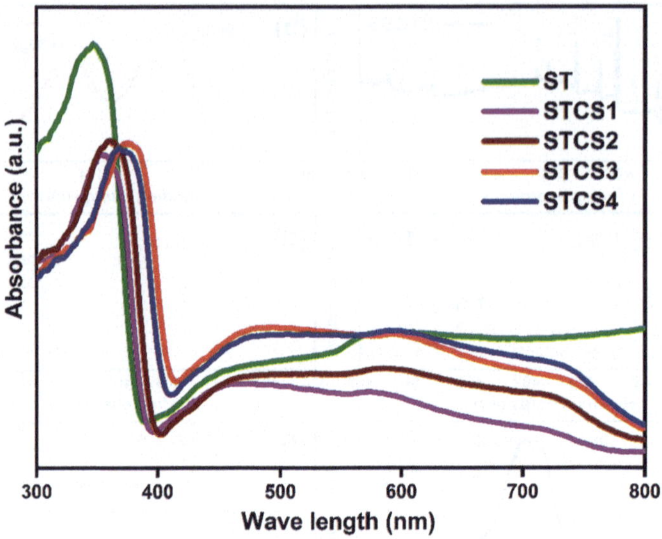

Fig. 7 Depicts the UV-Vis DRS spectra of bare ST and various ratios on STCS

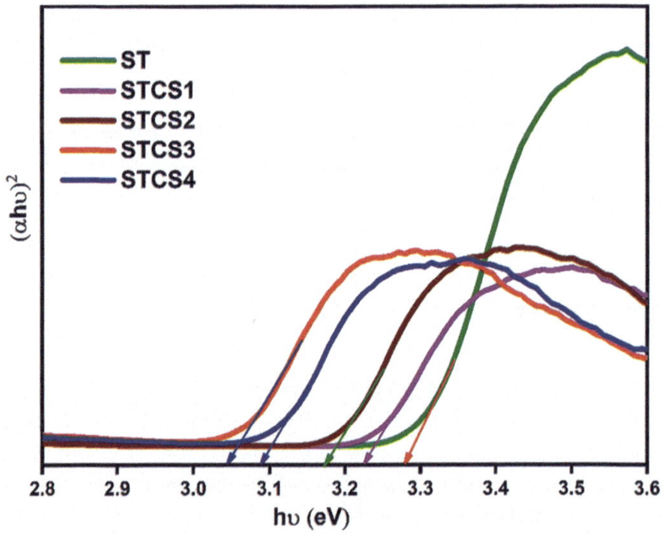

Fig. 8 Depicts the Tauc's plot of bare ST and various ratios on STCS

method, the plot between αhv and $(hv)^2$ is extended until it intersects the x-axis. The direct bandgap values for different compositions are ST (3.27 eV), STCS1 (3.22 eV), STCS2 (3.17 eV), STCS3 (3.01 eV), and STCS4 (3.09 eV), respectively. The bandgap value rapidly drops as the concentration of CS approaches 3%. However, the actual behaviour of the bandgap at high CS concentrations, is uncertain. The reduction in bandgap energy might be due to microstrain near the absorption limits.

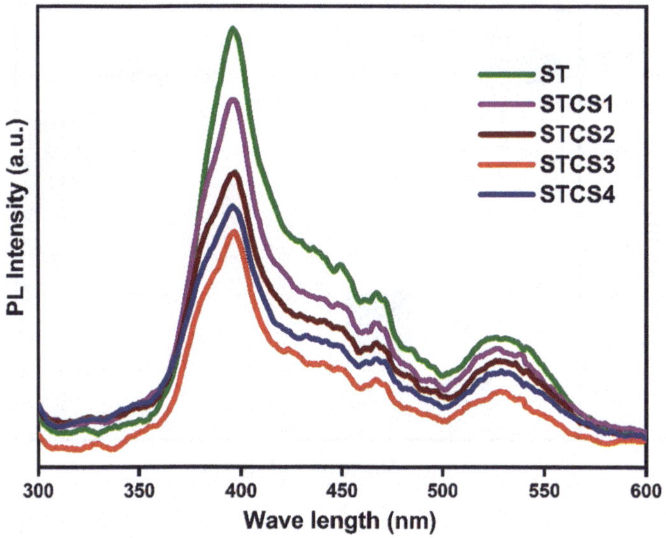

Fig. 9 Depicts the PL spectra of bare ST and various ratios on STCS

PL is a physical phenomenon that involves the emission of light from a material after photon absorption (Sadiq et al., 2016, 2017a). The emission peaks in the obtained samples had a wavelength of 396 nm. As seen in Fig. 9, the peak intensity nearly corresponds to the recombination period of photo generated e^{-}-h^{+} pairs. It has been proven that the intensity of PL and the concentration of the composite material have a negative connection. The STCS3 sample showed the lesser PL intensity when compared to the other samples. The observed reduction in emission peak intensity in STCS3 implies a robust electrical interaction and amended charge-carrier separation at the ST-CS interface.

3.7 Photocatalytic Hydrogen Activity

In the study, the effect of CS on ST in terms of H_2 production was investigated using a 10% CH_3OH solution and a 250 W Xenon lamp. The findings reveal that the presence of CS on ST significantly reduces the generation of H_2. According to observed data, adding CS to ST improves H_2 production efficiency. The hydrogen production rate of STCS3 was 3214.29 μmol g^{-1} h^{-1}. The quantity of CS on the ST substrate is inversely related to the activity of the hydrogen production. The observed behaviour might be explained by CS on the active surfaces of ST, which may produce recombination centres rather than increasing the number of active sites for H_2 production. Figure 10 shows that STCS3 produces much more H_2 than ST and CS. STCS3 has an H_2 that is 6.0 times that of ST and 2.6 times that of CS. STCS3 nanocomposite has a substantially greater H_2 production activity than pure CS and ST

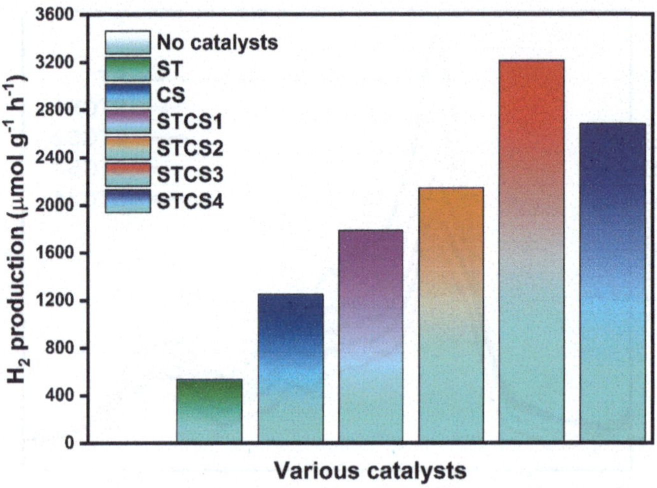

Fig. 10 Depicts the hydrogen production histogram of bare ST, CS, and various ratios on STCS

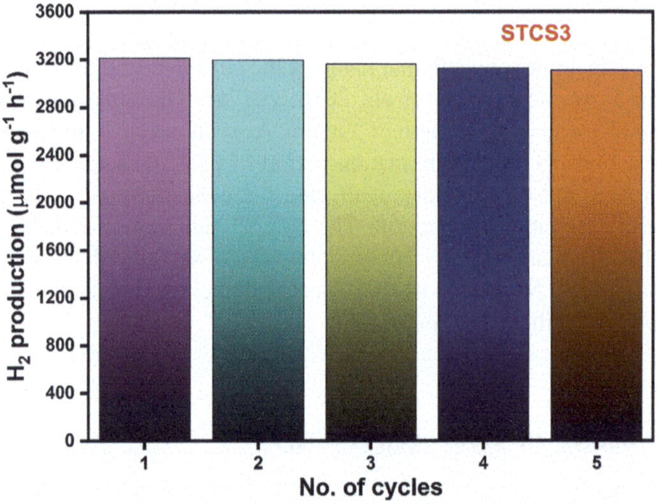

Fig. 11 Depicts the hydrogen production stability performance histogram of STCS3

nanocomposite. The H_2 production rates for various materials are as follows: ST (535.70 μmol g^{-1} h^{-1}), CS (1250.0 μmol g^{-1} h^{-1}), STCS1 (1785.71 μmol g^{-1} h^{-1}), STCS2 (2142.86 μmol g^{-1} h^{-1}), STCS3 (3214.29 μmol g^{-1} h^{-1}), STCS4 (2678.57 μmol g^{-1} h^{-1}), respectively.

These study focuses on the stability of photocatalytic H_2 production. Figure 11 shows that STCS3 produces the most H_2 during the course of the five cycles. Despite this, H_2 production falls with each consecutive cycle, as seen by the figures falling from 3214.29 to 3107.14 μmol g^{-1} h^{-1}. The decrease in H_2 production might be attributed to catalyst loss during each washing and drying process. The findings reveal that the composite material STCS3 performed well throughout the experiment. It was also very stable throughout the photocatalysis process.

3.8 Mechanism for Hydrogen

According to the suggested photocatalytic method, the STCS heterojunction nanocomposite produces more H_2 than pure ST and CS materials. The observed phenomenon is caused by the interaction of two semiconductors, which results in the formation of a solid-solid interfacial layer. The above-mentioned finding is supported by significant research and empirical data. The direct Z-scheme, a proposed photocatalytic process, is shown in Fig. 12. The previous strategy outlines the kinetics of photogenerated carriers as well as the subsequent reactions inside the system (Liu et al., 2019; Sun et al., 2018). Understanding the basic ideas behind the photocatalytic hydrogen-generation process is aided by semiconductor energy band theory. Semiconductors have a forbidden energy band, which corresponds to the range of energy levels in which electrons cannot exist. When semiconductors create pairs of electrons and holes when driven by photon-energy more than the forbidden bandgap size. When irradiated to UV-light, the narrow band gap in ST and CS enables for effective electron-excitation from the valence-band (V_B) to the conduction-band (C_B). The previous approach simplifies the generation of electron-hole pairs.

To investigate photogenerated charge-carrier separation in binary heterojunction devices, the classic-transfer mechanism and the direct Z-scheme mechanism are often employed (Qi et al., 2017; Xu et al., 2018). The existing mechanism was

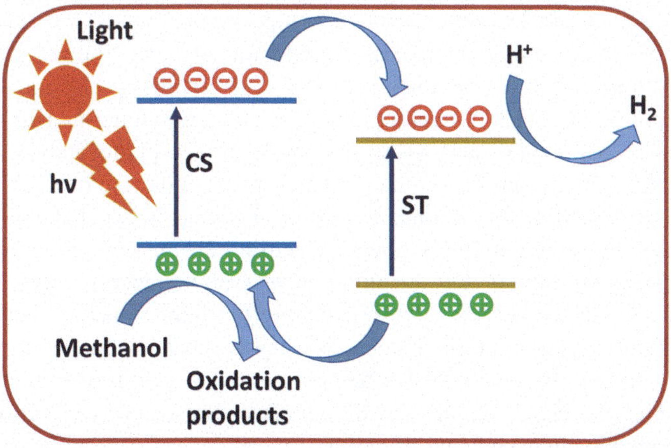

Fig. 12 Depicts the proposed photocatalytic H_2 production mechanism using CH_3OH as the sacrificial agent over STCS

expected to be used by the system. The capacity of photogenerated electrons to participate in reduction processes may be diminished when they are transported from the CS C_B to the ST C_B, which might be explained by a change in potential energy ranging from −1.12 to 0.23 eV. Because of the comparably unfavourable standard reduction potential of H_2O/H_2 (−0.42 eV), the extra electrons in the ST C_B have limited efficiency in accelerating the reduction of H^+ to H_2.

As a result, our photocatalytic method was chosen over the usual transfer method. A Z-scheme mechanism is anticipated to transfer charge inside the STCS system. When the C_B of ST and the V_B of CS were exposed to light, they generated electrons and holes, respectively. These charge carriers migrated independently towards the solid-solid hetero-interface due to electrostatic interactions inside the material. The barrier between electrons and holes is then recombined. Recombination dynamics enhanced the spatial-separation of photo-generated charge-carriers in both ST and CS materials. An overabundance of electrons in the CS C_B aided the H_2 production process, resulting in the desired product. The use of CH_3OH (methanol) as a sacrificial-agent aid in the capture and removal of excess holes from the ST and CS V_B, and above-mentioned method was beneficial in increasing the photocatalytic effect.

4 Conclusions

In summary and conclusion, a heterostructure photocatalyst of STCS was successfully synthesized using the laser irradiation technique, which is recognized for its ease of use and environmental friendliness. Several characterization approaches were used to produce a strong bond between ST and CS nanoparticles. When exposed to UV light, the STCS samples consistently demonstrated improved efficiency in photocatalytic H_2 production. The composite material STCS3 has the most exceptional H_2 evolution rate, with a measured value of 3214.29 μmol g^{-1} h^{-1}. The obtained result is 6.0 times more than that of ST (535.70 μmol g^{-1} h^{-1}) and 2.6 times greater than that of CS (1250.0 μmol g^{-1} h^{-1}). It's also worth noting that the STCS heterojunction nanocomposite displayed outstanding photocatalytic stability throughout the reaction phase. The increased H_2 production rate in STCS composites may be explained by the significant interfacial contact between the two component minerals. This interaction encourages electron transmission and integration in ST and hole integration in CS, resulting in less photogenerated charge carrier recombination in both materials. By broadening the light-response spectrum, the STCS Z-scheme heterostructure photocatalysis system has the potential to improve solar energy usage efficiency. It also has the potential to improve photocatalysis efficiency. The photocatalysts described above are well-known for their effectiveness, low cost, and environmental friendliness.

References

Abbasi, K. R., Shahbaz, M., Zhang, J., Irfan, M., & Alvarado, R. (2022). Analyze the environmental sustainability factors of China: The role of fossil fuel energy and renewable energy. *Renewable Energy, 187*, 390–402.

Abdulrazzak, F. H., Hussein, F. H., Alkaim, A. F., Ivanova, I., Emeline, A. V., & Bahnemann, D. W. (2016). Sonochemical/hydration–dehydration synthesis of Pt–TiO_2 NPs/decorated carbon nanotubes with enhanced photocatalytic hydrogen production activity. *Photochemical & Photobiological Sciences, 15*(11), 1347–1357.

Adnan, M. A. B., Arifin, K., Minggu, L. J., & Kassim, M. B. (2018). Titanate-based perovskites for photochemical and photoelectrochemical water splitting applications: A review. *International Journal of Hydrogen Energy, 43*(52), 23209–23220.

Akilarasan, M., Tamilalagan, E., Chen, S. M., Maheshwaran, S., Chen, T. W., Al-Mohaimeed, A. M., Al-Onazi, W. A., & Elshikh, M. S. (2021). An eco-friendly low-temperature synthetic approach towards micro-pebble-structured GO@$SrTiO_3$ nanocomposites for the detection of 2,4,6-trichlorophenol in environmental samples. *Microchimica Acta, 188*(3), 72.

Akrami, S., Murakami, Y., Watanabe, M., Ishihara, T., Arita, M., Fuji, M., & Edalati, K. (2022). Defective high-entropy oxide photocatalyst with high activity for CO_2 conversion. *Applied Catalysis B: Environmental, 303*, 120896.

Bang, J. U., Lee, S. J., Jang, J. S., Choi, W., & Song, H. (2012). Geometric effect of single or double metal-tipped CdSe nanorods on photocatalytic H_2 generation. *The Journal of Physical Chemistry Letters, 3*(24), 3781–3785.

Chen, L., Ren, J. T., & Yuan, Z. Y. (2022). Design strategies of phosphorus-containing catalysts for photocatalytic, photoelectrochemical and electrocatalytic water splitting. *Green Chemistry, 24*(2), 713–747.

Dawood, F., Anda, M., & Shafiullah, G. M. (2020). Hydrogen production for energy: An overview. *International Journal of Hydrogen Energy, 45*(7), 3847–3869.

Deshmukh, V. V., Ravikumar, C. R., Kumar, M. R. A., Ghotekar, S., Kumar, A. N., Jahagirdar, A. A., & Murthy, H. C. A. (2021). Structure, morphology and electrochemical properties of $SrTiO_3$ perovskite: Photocatalytic and supercapacitor applications. *Environmental Chemistry and Ecotoxicology, 3*, 241–248.

Du, C., Nie, S., Zhang, C., Wang, T., Wang, S., Zhang, J., Yu, C., Lu, Z., Dong, S., Feng, J., Liu, H., & Sun, J. (2022). Dual-functional Z-scheme CdSe/Se/BiOBr photocatalyst: Generation of hydrogen peroxide and efficient degradation of ciprofloxacin. *Journal of Colloid and Interface Science, 606*, 1715–1728.

Fazio, E., Gökce, B., De Giacomo, A., Meneghetti, M., Compagnini, G., Tommasini, M., Waag, F., Lucotti, A., Zanchi, C. G., & Ossi, P. M. (2020). Nanoparticles engineering by pulsed laser ablation in liquids: Concepts and applications. *Nanomaterials, 10*(11), 2317.

Gondal, M. A., Chang, X., Ali, M. A., Yamani, Z. H., Zhou, Q., & Ji, G. (2011). Adsorption and degradation performance of Rhodamine B over BiOBr under monochromatic 532 nm pulsed laser exposure. *Applied Catalysis A: General, 397*(1–2), 192–200.

Gondal, M. A., Ilyas, A. M., & Baig, U. (2016). Pulsed laser ablation in liquid synthesis of ZnO/TiO_2 nanocomposite catalyst with enhanced photovoltaic and photocatalytic performance. *Ceramics International, 42*(11), 13151–13160.

Gultom, N. S., Abdullah, H., Kuo, D. H., & Ke, W. C. (2019). Oriented p–n heterojunction Ag_2O/Zn(O,S) nanodiodes on mesoporous SiO_2 for photocatalytic hydrogen production. *ACS Applied Energy Materials, 2*(5), 3228–3236.

Hadadi, N. A., Baig, U., Gondal, M. A., Mohamed, M. J. S., & Dastageer, M. A. (2023). Pulsed laser induced synthesis of graphitic carbon nitride-cadmium selenide nanocomposite for photo-catalytic degradation of organic dyes, and electro-catalytic hydrogen evolution reaction. *Colloids and Surfaces A: Physicochemical and Engineering Aspects, 658*, 130711.

Hassan, M., Gondal, M. A., Cevik, E., Dastageer, M. A., Baig, U., Moqbel, R. A., Qahtan, T. F., Bozkurt, A., & al Abass, N. (2021). Laser assisted anchoring of cadmium sulfide nanospheres into tungsten oxide nanosheets for enhanced photocatalytic and electrochemical energy storage applications. *Colloids and Surfaces A: Physicochemical and Engineering Aspects, 617*, 126318.

Ilyas, A., Gondal, M. A., Yamani, Z. H., & Baig, U. (2017). Facile synthesis of titanium dioxide-cadmium sulfide nanocomposite using pulsed laser ablation in liquid and its performance in photovoltaic and photocatalytic applications. *International Journal of Energy Research, 41*(10), 1422–1435.

Islam, S. K., Sohel, M. A., & Lombardi, J. R. (2014). Coupled exciton and charge-transfer resonances in the Raman enhancement of phonon modes of CdSe quantum dots (QDs). *The Journal of Physical Chemistry C, 118*(33), 19415–19421.

Ismael, M. (2021). Latest progress on the key operating parameters affecting the photocatalytic activity of TiO_2-based photocatalysts for hydrogen fuel production: A comprehensive review. *Fuel, 303*, 121207.

Jaffer, S. M. M., & Samson, N. A. (2014). Reflux condensation synthesis and characterization of CO_3O_4 nanoparticles for photocatalytic applications. *Iranian Journal of Catalysis, 4*(4), 219–226.

Jamkhande, P. G., Ghule, N. W., Bamer, A. H., & Kalaskar, M. G. (2019). Metal nanoparticles synthesis: An overview on methods of preparation, advantages and disadvantages, and applications. *Journal of Drug Delivery Science and Technology, 53*, 101174.

Jang, H. J., Park, S. J., Yang, J. H., Hong, S.-M., Rhee, C. K., Kim, D., & Sohn, Y. (2021). Photocatalytic and photoelectrocatalytic properties of Eu(III)-doped perovskite $SrTiO_3$ nanoparticles with dopant level approaches. *Materials Science in Semiconductor Processing, 132*, 105919.

Johny, J., Li, Y., Kamp, M., Prymak, O., Liang, S. X., Krekeler, T., Ritter, M., Kienle, L., Rehbock, C., Barcikowski, S., & Reichenberger, S. (2022). Laser-generated high entropy metallic glass nanoparticles as bifunctional electrocatalysts. *Nano Research, 15*(6), 4807–4819.

Kanagaraj, T., & Thiripuranthagan, S. (2017). Photocatalytic activities of novel $SrTiO_3$—BiOBr heterojunction catalysts towards the degradation of reactive dyes. *Applied Catalysis B: Environmental, 207*, 218–232.

Kiss, B., Manning, T. D., Hesp, D., Didier, C., Taylor, A., Pickup, D. M., Chadwick, A. V., Allison, H. E., Dhanak, V. R., Claridge, J. B., Darwent, J. R., & Rosseinsky, M. J. (2017). Nano-structured rhodium doped $SrTiO_3$–visible light activated photocatalyst for water decontamination. *Applied Catalysis B: Environmental, 206*, 547–555.

Kumar, A., Sharma, G., Naushad, M., Al-Muhtaseb, A. H., Guo, C., Iglesias-Juez, A., & Stadler, F. J. (2018). High-performance photocatalytic hydrogen production and degradation of levofloxacin by wide spectrum-responsive Ag/Fe_3O_4 bridged $SrTiO_3/g\text{-}C_3N_4$ plasmonic nanojunctions: Joint effect of Ag and Fe_3O_4. *ACS Applied Materials & Interfaces, 10*(47), 40474–40490.

Lee, S. J., Theerthagiri, J., Nithyadharseni, P., Arunachalam, P., Balaji, D., Kumar, A. M., Madhavan, J., Mittal, V., & Choi, M. Y. (2021). Heteroatom-doped graphene-based materials for sustainable energy applications: A review. *Renewable and Sustainable Energy Reviews, 143*, 110849.

Lee, S. J., Theerthagiri, J., & Choi, M. Y. (2022). Time-resolved dynamics of laser-induced cavitation bubbles during production of Ni nanoparticles via pulsed laser ablation in different solvents and their electrocatalytic activity for determination of toxic nitroaromatics. *Chemical Engineering Journal, 427*, 130970.

Li, G. S., Zhang, D. Q., & Yu, J. C. (2009). A new visible-light photocatalyst: CdS quantum dots embedded mesoporous TiO_2. *Environmental Science & Technology, 43*(18), 7079–7085.

Li, Z., Wang, S., Wu, J., & Zhou, W. (2022). Recent progress in defective TiO_2 photocatalysts for energy and environmental applications. *Renewable and Sustainable Energy Reviews, 156*, 111980.

Liang, Q., Liu, X., Zeng, G., Liu, Z., Tang, L., Shao, B., Zeng, Z., Zhang, W., Liu, Y., & Cheng, M. (2019). Surfactant-assisted synthesis of photocatalysts: Mechanism, synthesis, recent advances and environmental application. *Chemical Engineering Journal, 372*, 429–451.

Liu, Y., Dai, F., Zhao, R., Huai, X., Han, J., & Wang, L. (2019). Aqueous synthesis of core/shell/shell CdSe/CdS/ZnS quantum dots for photocatalytic hydrogen generation. *Journal of Materials Science, 54*(11), 8571–8580.

Liu, Y., Lin, Y., Liu, Z., Tang, J., Chen, L., Liu, X., Tian, Y., Fang, D., & Wang, J. (2022). Self-generating $CoFe_2O_4$ as electric channel in Z-scheme CoO(111)/$CoFe_2O_4$/Fe_2O_3 photocatalyst for synchronous photocatalytic degradation and hydrogen production. *Materials Science in Semiconductor Processing, 140*, 106382.

Luo, Y., Deng, B., Pu, Y., Liu, A., Wang, J., Ma, K., Gao, F., Gao, B., Zou, W., & Dong, L. (2019). Interfacial coupling effects in g-C_3N_4/$SrTiO_3$ nanocomposites with enhanced H_2 evolution under visible light irradiation. *Applied Catalysis B: Environmental, 247*, 1–9.

Ma, R., Dong, L., Li, B., Su, T., Luo, X., Qin, Z., & Ji, H. (2018). g-C_3N_4/$BiYO_3$ composite for photocatalytic hydrogen evolution. *ChemistrySelect, 3*(21), 5891–5899.

Manchala, S., Gandamalla, A., Rao, V. N., Venkatakrishnan, S. M., & Shanker, V. (2022). Solar-light responsive efficient H_2 evolution using a novel ternary hierarchical $SrTiO_3$/CdS/carbon nanospheres photocatalytic system. *Journal of Nanostructure in Chemistry, 12*(2), 179–191.

Mohamed, M. J. S., Shenoy, S., & Bhat, D. K. (2018). Novel NRGO-$CoWO_4$-Fe_2O_3 nanocomposite as an efficient catalyst for dye degradation and reduction of 4-nitrophenol. *Materials Chemistry and Physics, 208*, 112–122.

Mohamed, M. J. S., Gondal, M. A., Hassan, M., Khan, A. Z., Surrati, A. M., & Almessiere, M. A. (2022). Exceptional co-catalysts free $SrTiO_3$ perovskite coupled CdSe nanohybrid catalyst by green pulsed laser ablation for electrochemical hydrogen evolution reaction. *Chemical Engineering Journal Advances, 11*, 100344.

Naik, S. S., Lee, S. J., Theerthagiri, J., Yu, Y., & Choi, M. Y. (2021). Rapid and highly selective electrochemical sensor based on ZnS/Au-decorated f-multi-walled carbon nanotube nanocomposites produced via pulsed laser technique for detection of toxic nitro compounds. *Journal of Hazardous Materials, 418*, 126269.

Naser, H., Hassan, Z., Mohammad, S. M., Shanshool, H. M., & Al-Hazeem, N. Z. (2022). Parameters influencing the absorbance of gold-silver alloy nanomaterials using the pulsed laser ablation in liquid (PLAL) approach: A review. *Brazilian Journal of Physics, 52*(3), 100.

Nilsen, W. G., & Skinner, J. G. (1968). Raman spectrum of strontium titanate. *The Journal of Chemical Physics, 48*(5), 2240–2248.

Ollis, D. F., Pelizzetti, E., & Serpone, N. (1991). Photocatalyzed destruction of water contaminants. *Environmental Science & Technology, 25*(9), 1522–1529.

Pan, J., Chen, Z., Wang, P., Wang, P., Yu, Q., Zhao, W., Wang, J., Zhu, M., Zheng, Y., & Li, C. (2021). The overall water splitting of CdS/Ti^{3+}-$SrTiO_3$ core–shell heterojunction via OER enhancement of MnOx nanoparticles. *Chemical Engineering Journal, 424*, 130357.

Pei, Y., Du, W., Li, Y., Shen, W., Wang, Y., Yang, S., & Han, S. (2015). The effect of carbon–polyaniline hybrid coating on high-temperature electrochemical performance of perovskite-type oxide $LaFeO_3$ for MH–Ni batteries. *Physical Chemistry Chemical Physics, 17*(27), 18185–18192.

Qi, K., Cheng, B., Yu, J., & Ho, W. (2017). A review on TiO_2-based Z-scheme photocatalysts. *Chinese Journal of Catalysis, 38*(12), 1936–1955. https://doi.org/10.1016/S1872-2067(17)62962-0

Qin, Y., Fang, F., Xie, Z., Lin, H., Zhang, K., Yu, X., & Chang, K. (2021). La,Al-Codoped $SrTiO_3$ as a photocatalyst in overall water splitting: Significant surface engineering effects on defect engineering. *ACS Catalysis, 11*(18), 11429–11439.

Rabuffetti, F. A., Kim, H.-S., Enterkin, J. A., Wang, Y., Lanier, C. H., Marks, L. D., Poeppelmeier, K. R., & Stair, P. C. (2008). Synthesis-dependent first-order Raman scattering in $SrTiO_3$ nanocubes at room temperature. *Chemistry of Materials, 20*(17), 5628–5635.

Ramesh Reddy, N., Bhargav, U., Mamatha Kumari, M., Cheralathan, K. K., & Sakar, M. (2020). Review on the interface engineering in the carbonaceous titania for the improved photocatalytic hydrogen production. *International Journal of Hydrogen Energy, 45*(13), 7584–7615.

Sadiq, M. M. J., Shenoy, U. S., & Bhat, D. K. (2016). Novel RGO–$ZnWO_4$–Fe_3O_4 nanocomposite as high performance visible light photocatalyst. *RSC Advances, 6*(66), 61821–61829.

Sadiq, M. M. J., Shenoy, U. S., & Bhat, D. K. (2017a). Enhanced photocatalytic performance of N-doped RGO-$FeWO_4/Fe_3O_4$ ternary nanocomposite in environmental applications. *Materials Today Chemistry, 4*, 133–141.

Sadiq, M. M. J., Shenoy, U. S., & Bhat, D. K. (2017b). $NiWO_4$-ZnO-NRGO ternary nanocomposite as an efficient photocatalyst for degradation of methylene blue and reduction of 4-nitro phenol. *Journal of Physics and Chemistry of Solids, 109*, 124–133.

Saeed, M., Muneer, M., Haq, A. ul, & Akram, N. (2022). Photocatalysis: An effective tool for photodegradation of dyes—A review. *Environmental Science and Pollution Research, 29*(1), 293–311.

Shukla, K., & Srivastava, V. C. (2016). Diethyl carbonate: Critical review of synthesis routes, catalysts used and engineering aspects. *RSC Advances, 6*(39), 32624–32645.

Singh, S., & Kansal, S. K. (2022). Recent progress in red phosphorus-based photocatalysts for photocatalytic water remediation and hydrogen production. *Applied Materials Today, 26*, 101345.

Sun, B. J., Zhou, W., Li, H. Z., Ren, L. P., Qiao, P. Z., Li, W., & Fu, H. G. (2018). Synthesis of particulate hierarchical tandem heterojunctions toward optimized photocatalytic hydrogen production. *Advanced Materials, 30*, 1804282.

Suresh, S., & Arunseshan, C. (2014). Dielectric properties of cadmium selenide (CdSe) nanoparticles synthesized by solvothermal method. *Applied Nanoscience, 4*(2), 179–184.

Tan, C. E., Lee, J. T., Su, E. C., & Wey, M. Y. (2020). Facile approach for Z-scheme type Pt/g-$C_3N_4/SrTiO_3$ heterojunction semiconductor synthesis via low-temperature process for simultaneous dyes degradation and hydrogen production. *International Journal of Hydrogen Energy, 45*(24), 13330–13339.

Tasleem, S., & Tahir, M. (2020a). Current trends in strategies to improve photocatalytic performance of perovskites materials for solar to hydrogen production. *Renewable and Sustainable Energy Reviews, 132*, 110073.

Tasleem, S., & Tahir, M. (2020b). Recent progress in structural development and band engineering of perovskites materials for photocatalytic solar hydrogen production: A review. *International Journal of Hydrogen Energy, 45*(38), 19078–19111.

Wang, Z., Choi, J., Xu, M., Hao, X., Zhang, H., Jiang, Z., Zuo, M., Kim, J., Zhou, W., & Meng, X. (2020). Optimizing electron densities of Ni-N-C complexes by hybrid coordination for efficient electrocatalytic CO_2 reduction. *ChemSusChem, 13*(5), 929–937.

Wu, Q., Huang, F., Zhao, M., Xu, J., Zhou, J., & Wang, Y. (2016). Ultra-small yellow defective TiO_2 nanoparticles for co-catalyst free photocatalytic hydrogen production. *Nano Energy, 24*, 63–71.

Xu, Q., Zhang, L., Yu, J., Wageh, S., Al-Ghamdi, A. A., & Jaroniec, M. (2018). Direct Z-scheme photocatalysts: Principles, synthesis, and applications. *Materials Today, 21*(10), 1042–1063.

Yang, D., Sun, Y., Tong, Z., Nan, Y., & Jiang, Z. (2016). Fabrication of bimodal-pore $SrTiO_3$ microspheres with excellent photocatalytic performance for Cr(VI) reduction under simulated sunlight. *Journal of Hazardous Materials, 312*, 45–54.

Yoon, H. J., Kim, S. K., Huang, W., & Sohn, Y. (2018). Comparable electrocatalytic performances of carbon- and Rh-loaded $SrTiO_3$ nanoparticles. *Chinese Chemical Letters, 29*(6), 800–804.

Yue, M., Lambert, H., Pahon, E., Roche, R., Jemei, S., & Hissel, D. (2021). Hydrogen energy systems: A critical review of technologies, applications, trends and challenges. *Renewable and Sustainable Energy Reviews, 146*, 111180.

Zhang, L., Yin, J., Wei, K., Li, B., Jiao, T., Chen, Y., Zhou, J., & Peng, Q. (2020). Fabrication of hierarchical $SrTiO_3@MoS_2$ heterostructure nanofibers as efficient and low-cost electrocatalysts for hydrogen-evolution reactions. *Nanotechnology, 31*(20), 205604.

Zhang, Y. P., Han, W., Yang, Y., Zhang, H. Y., Wang, Y., Wang, L., Sun, X. J., & Zhang, F. M. (2022a). S-scheme heterojunction of black TiO_2 and covalent-organic framework for enhanced photocatalytic hydrogen evolution. *Chemical Engineering Journal, 446*, 137213.

Zhang, S., Ocłoń, P., Klemeš, J. J., Michorczyk, P., Pielichowska, K., & Pielichowski, K. (2022b). Renewable energy systems for building heating, cooling and electricity production with thermal energy storage. *Renewable and Sustainable Energy Reviews, 165*, 112560.

Tuning the Physical Properties of Perovskite Multiferroic Nanoparticles for Green Energy Applications

Manish Kumar, Sandeep Kumar Pundir, Subhash Sharma, O. Raymond Herrera, J. M. Siqueiros, and Manish Kumar

1 Introduction

To channelize the modern economy for further advancement, energy turned up to be the main constituent. It becomes bedrock in the realization of the basic human needs. Thus, it is securing a better future with smooth transportation, food security and good health (Zhu et al., 2016; Ramohlola et al., 2018). In the nineteenth century, the unification of electricity and magnetism lead to the development of one common discipline that subsequently was summarized in Maxwell equations. But

M. Kumar
Department of Physics, University of Lucknow, Lucknow, Uttar Pradesh, India

BSNV PG College, University of Lucknow, Lucknow, Uttar Pradesh, India

Experimental Research Laboratory, Department of Physics, ARSD College, University of Delhi, New Delhi, India

S. K. Pundir
Department of Physics, University of Lucknow, Lucknow, Uttar Pradesh, India

S. Sharma
CONAHCyT-Centro de Nanociencias y Nanotecnologia, Universidad Nacional Autónoma de México, Ensenada, Baja California, México

Centro de Nanociencias y Nanotecnología, Universidad Nacional Autónoma de México, Ensenada, Baja California, México

O. Raymond Herrera · J. M. Siqueiros
Centro de Nanociencias y Nanotecnología, Universidad Nacional Autónoma de México, Ensenada, Baja California, México

M. Kumar (✉)
Experimental Research Laboratory, Department of Physics, ARSD College, University of Delhi, New Delhi, India
e-mail: mkumar2@arsd.du.ac.in

S. A. Bansal et al. (eds.), *Emerging Applications of Novel Nanoparticles*, Lecture Notes in Nanoscale Science and Technology 37, https://doi.org/10.1007/978-3-031-57843-4_13

in solids, electric and magnetic ordering are often considered separately and in general, with the valid logic that the charge effects are due to the electric charges of electrons and ions, whereas electron spins govern magnetic properties.

Based on experimental evidence, Dzyaloshinskii's predictions (Dzyaloshinskii, 1959) and Astrov observations (Astrov, 1960), the existing research progress is largely centered on multiferroic (MF) perovskite materials that simultaneously exhibit both ferroelectric and magnetic behavior (Spaldin & Ramesh, 2019; Kosub et al., 2017; Martin & Ramesh, 2012). In the same way magnetoelectric (ME) materials can control the polarization by applying a magnetic field and the control of magnetization by an electric field. Such class of materials have drawn significant attention and provided a vast field of opportunity for researchers in the field of materials. Thus, research and development of multifunctional devices such as multistate memory devices along with devices for energy applications based on the ME coupling effect is highly anticipated (Vaz et al., 2010; Bibes & Barthelemy, 2008; Lu et al., 2011).

Not only the magnetization appear stimulated by an electric field, or the inverse effect of electric polarization generated by the application of magnetic field in solids (i.e., existence of responses to strong cross-coupling) but, there are systems in which two types of ordering can coexist in the absence of external electric and magnetic fields. One can be, for instance ferromagnetism, that is, the spontaneous ordering of orbital and spin magnetic moments, other one can be ferroelectricity, the spontaneous ordering of electric dipole moments can exist simultaneously in one material. These materials are called multiferroics as shown in Fig. 1 (Sun et al., 2021). Multiferroics often include also a third type of ordering: spontaneous deformation, which leads to ferroelasticity. Boracites were probably the first known multiferroics (Ascher et al., 1966), and soon several others were either found in nature, or synthesized artificially (Smolenskiĭ & Chupis, 1982). Figure 1a illustrates well the association of properties between the multiferroic/ferromagnetic/ferroelectric/

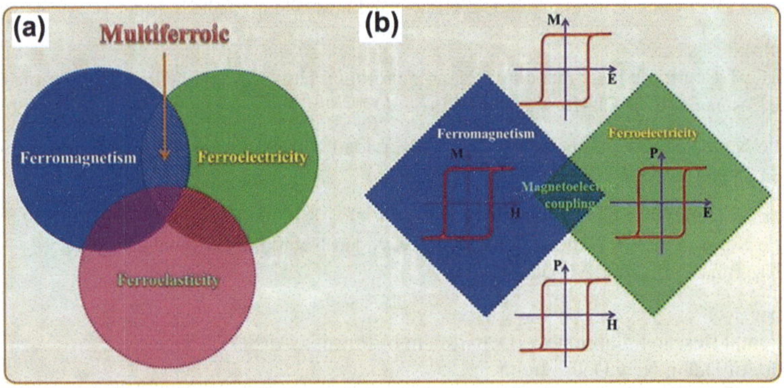

Fig. 1 (**a**) Schematic representation of multiferroic, ferromagnetism, ferroelectricity and ferroelasticity in materials, (**b**) Schematic connection in the middle of magnetoelectric coupling, ferromagnetism and ferroelectricity. (Adapted from the Ref. (Sun et al., 2021))

ferroelastic materials (Sun et al., 2021). The concept range and the respective hysteresis curve involved in magnetoelectric coupling/ferromagnetism/ferroelectricity are clearly displayed in the abstract association in Fig. 1b (Sun et al., 2021). From the practical aspect point of view, $TbMnO_3$ (TMO) and $BiFeO_3$ (BFO) are generally regarded promising multiferroic materials among the observed single-phase multiferroic-magnetoelectric (MF/ME) materials (Lee et al., 2006a, b; Park et al., 2007).

The microscopic origin of magnetism in magnets is attributed to the presence of unpaired localized electrons in partially filled d or f shells of localized of transition-metal or rare-earth ions providing a corresponding localized magnetic moment. Exchange interaction between the localized moments leads to magnetic ordering. However, the circumstances leading the presence of ferroelectricity are completely different. There are different types of multiferroics because the origins of ferroelectricity are different and numerous. Depending on the origin there are two types of multiferroics. Type-I multiferroics, are those materials where ferroelectricity and magnetism have different origin and appear largely independent of one another, although there may be some coupling between them (Khomskii II). In these materials, ferroelectricity typically appears at higher temperatures than magnetism, and the spontaneous polarization is often rather large (of order 10–100 $\mu C/cm^2$). Then there are type-II multiferroics which are relatively recently discovered materials (Kimura et al., 2003; Hur et al., 2004). In this case, ferroelectricity is due to magnetism evincing a strong coupling between both orderings. However, the polarization in these materials is usually much smaller (10^{-2} $\mu C/cm^2$). There are currently investigations on MF composites in the form of multilayers and self-organized nano structures displaying magnetic and ferroelectric behavior (Schmid, 2008; Schmidt, 2011).

To meet the future global energy demand, it is necessary to harness direct radiation in a wide range of the solar spectrum as it is recognized as the clean source of energy and this is achieved with the help of photovoltaic technology. MFs provide a possibility of green energy from new types of multifunctional devices due to their efficient ferroelectric polarization-driven carrier separation and above-bandgap generated photo-voltages. To reduce their bandgap towards the visible region while simultaneously retaining one of the ferroic order parameter turned out to be one of the biggest challenges so far. To address these issues, herein an elemental composition engineering of multiferroic perovskite materials (like in $BiFeO_3$) is suggested by substitutional doping via. suitable transition metal cations, as a means to tune the characteristics of the transition metal–oxygen bond. The aim of this chapter is to discuss and summarize the different effects of doping that can favorably modify the properties of multiferroic perovskites for suitable energy application devices. A brief description of comprehensive modification procedures will enable an understanding of alternative ways of tuning the properties of perovskites. Additionally, a brief special attention is given to the effect of doping on the electric and magnetic properties of MF perovskites and to the use of MF materials in green energy applications.

2 Multiferroic Perovskite Materials

Much attention has been given to multiferroic perovskite materials by scientists due to their various physical properties and extensive applications such as superconductors, metals, semiconductors, insulators, ferromagnets, anti-ferromagnets, ferroelectrics, multiferroics, dielectrics, and piezoelectrics. German mineralogist Gustav Rose discovered calcium titanate, also known as calcium titanium oxide ($CaTiO_3$) in 1839 and was called perovskite, after the Russian mineralogist, Lev A. Perovski (1792–1856) (Dragan et al., 2020; Schmidt, 2011). Since then, all the compounds with as the calcium titanate structure are referred to as perovskite materials. Goldschmidt tolerance factor (*t*) is, a dimensionless number given by Eq. (1) allows to predict the stability of different types of perovskite structure with the chemical formula

$$t = \frac{r_A + r_x}{\sqrt{2}((r_B + r_x)} \tag{1}$$

Where r_A, r_B, and r_X are the radii of cation *A*, cation *B*, and *X* anion, respectively. The value of *t* for the ideal perovskite is 1 but can vary between 0.813 and 1.107 for a stable structure; otherwise the cubic or cubic-like crystal structure will be stressed, warped, and finally destroyed. Figure 2a displays a schematic figure of the generic ABX_3 perovskite crystal structure for a hybrid organic–inorganic metal halide perovskite (Liu & Yang, 2017). Perovskite structure is a network of corner-sharing BX_6 octahedra surrounding a larger *A*-site cation ($r_A > r_B$), where the anion *X*, is usually a chalcogen or halogen and the cations, *A* and *B* can span over a periodic table.

$TbMnO_3$ and $BiFeO_3$ perovskites are generally considered very promising multiferroic materials among the single–phase multiferroic/magnetoelectric materials.

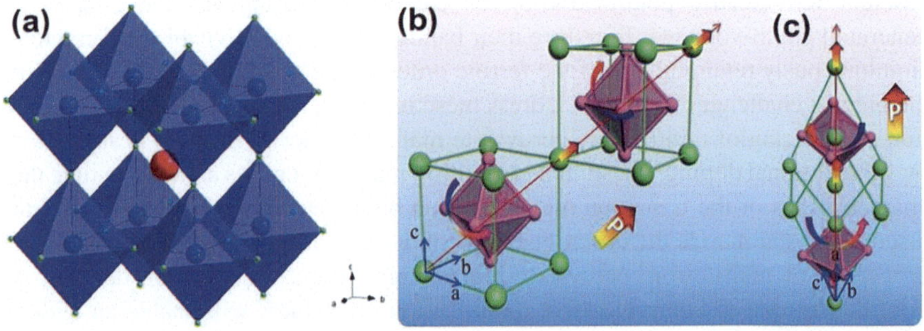

Fig. 2 (**a**) Ideal cubic perovkite ABO_3 structure in the octahedral configuration, where red, blue and green colors represent the A^{2+}, B^{4+} and Oxygen ions respectively. (Adapted from the Ref. (Liu & Yang, 2017)). (**b**) Room temperature $BiFeO_3$ pseudocubic unit cell crystal structure with one unit cell and (**c**) rhombohedral crystal structure of $BiFeO_3$ unit cell. (Adapted from the Ref. (Sun et al., 2021))

In particular, $BiFeO_3$ is a typical ABO_3 type multiferroic perovskite material as shown in Fig. 2b (Sun et al., 2021), has been extensively explored with regard to magnetoelectric coupling, photoelectric and photovoltaic effects (Sarkar et al., 2015; Moubah et al., 2012). Consequently, to substantiate the devices BFO turns out to be a potential candidate on account of its multiple properties. The room temperature crystal structure of BFO in its one and two unit cells representations are shown in Fig. 2b, c (Sun et al., 2021). The distortion of the ferroelectric phase corresponds to the displacement of the cations and the tilt of the octahedron is known as the distortion rotation of the antiferromagnetic state. There is a reversal of rotation directions of the adjacent oxygen octahedral along the crystal direction which reflects the reversing of the antiferroelastic displacement and deformation and causes this point a deviation of the residual magnetization moment from the b-axis direction (Park et al., 2014). An essential role in the stability of the ferroelectric phase is played by the variation in the bond length. The changeable structure of BFO becomes the source of the many special physical properties exhibited by this compound.

BFO belongs to the *R3c* space group and has a distinctive tripartite structure at room temperature. Lone pairs of electrons in Bi ions are responsible for the ferroelectricity in the BFO perovskite along with the presence of G-type antiferromagnetic ordering. It is worth noting the emergence of many opportunities for researchers to study other materials with perovskite structure with multiferroic properties (Shang et al., 2013; Nechache et al., 2011; Fernàndez-Posada et al., 2016). For instance, $TbMnO_3$ a multiferroic material possessing with an orthogonal perovskite structure along with excellent optical properties (Martin & Ramesh, 2012). Hence, $TbMnO_3$ a multifunctional material sums up various properties along with ferroelectricity, ferromagnetism and optical properties.

3 Physics of Multiferroic Perovskite Materials

3.1 The Beginning of Ferroelectric and Magnetic Order Parameters in Perovskites

Soft mode theory is well explaining the emergence of ferroelectricity with the descent in temperature in some perovskite materials (Cochran, 1959). The mode of the lowest-frequency polar-phonon turns softer and at critical temperature T_C the frequency drops to zero while the material's structure transforms form ideal cubic to a lower symmetry structure. Thus the breaking of centro-symmetry and deformation of the perovskite due to such structural instability is generally described by second-order Jahn-Teller Effect (Halasyamani & Poeppelmeier, 1998) such as in the classic ferroelectric $BaTiO_3$, where Ti^{4+} shifts along the C_4, C_2 and C_3 axis, as shown in Fig. 3 (Liu & Yang, 2017) and leading to stabilization in tetragonal, orthorhombic or rhombohedral structure with different spontaneous polarizations in different

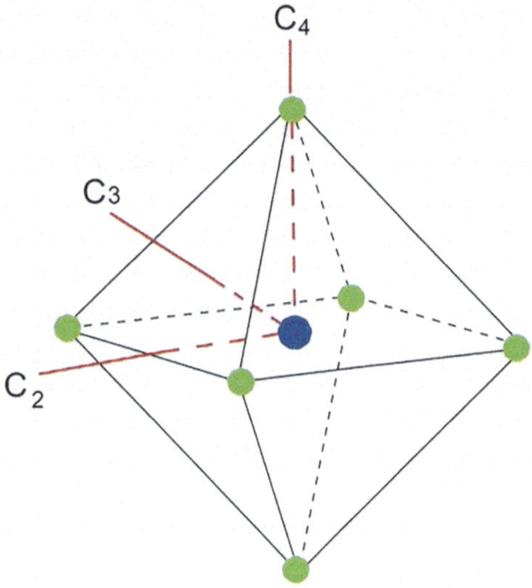

Fig. 3 B-site cation (blue ball) is shifting in octahedral crystal structure of an ideal cubic perovskite. The green balls and red lines symbolize the oxygen ions and high symmetry axes along which B-site cation travels. (Adapted from the Ref. (Liu & Yang, 2017))

temperature ranges (Liu & Yang, 2017). There are empty d electron states (d^0) in large a number of ferroelectric perovskite oxides in which the B site is occupied with transition metal ions as in $BaTiO_3$. However, the Pb^{2+} and Bi^{3+}based oxides in which ferroelectricity is driven by the A-site occupation do not necessarily contain empty d electron states (d^0). Also, the ferroelectric state which is favored by long-range coulomb interaction in both B-site and A-site driven ferroelectrics are also responsible for reducing the B-O or A-O bond lengths (Cohen, 1992).

Quantum-mechanical exchange interactions, including double exchange interaction and super-exchange interaction are in the origin of magnetism in perovskites, a completely different phenomenology from that of ferroelectricity. Ferromagnetism which is a highly desirable magnetic ordering for materials with multiferroic properties is induced by the double exchange interactions and finds use in multi-state memory devices and energy applications (Zener, 1951a, b). Unfortunately, the condition for ferromagnetic ordering in perovskite materials implies the existence of electrons available for conductivity and that leads to the destruction of ferroelectricity with the screening of long-range coulomb interactions even if such interactions exist. Hence, a different set of conditions are required to give place to ferroelectricity in magnetic perovskites materials (Scott, 2008; Loidl et al., 2008). As a matter of fact, perovskite compounds with double exchange interaction can very seldom be considered multiferroic materials.

With respect to double interaction, either ferromagnetism or antiferromagnetism (AFM) can be induced in perovskite-type oxides due to the super-exchange interaction (Kramers, 1934; Anderson, 1950, 1959; Goodenough, 1955; Kanamori, 1960). However, these kinds of compounds largely show insulating behavior thus providing the path to combine ferroelectricity and magnetism together. The presence of partly filled d shell (d^n) in the B site cation for magnetic perovskite oxides which further splits in an oxygen octahedral environment into two energy states of two

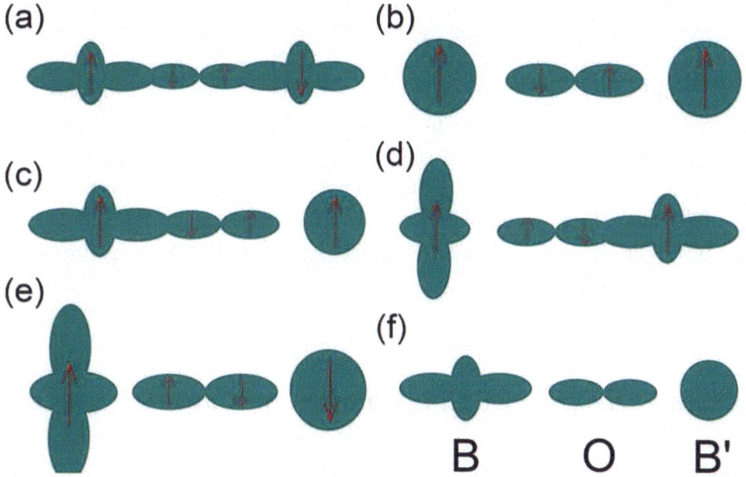

Fig. 4 (**a**) Strong AFM order, (**b**) weak AFM order, (**c**) weak FM order, (**d**) FM order, (**e**) AFM order induced via superexhange interaction. (**f**) Signifies the legends of electronic orbits of B, B′ and O ions. (Adapted from the Ref. (Liu & Yang, 2017))

high energy e_g orbitals ($d_{x^2-y^2}$, $d_{3z^2-r^2}$) and three low energy t_{2g} orbitals (d_{xy}, d_{yz}, d_{zx}) complete the picture (Tokura & Nagaosa, 2000; Dzyaloshinsky, 1958; Moriya, 1960). Moreover, two neighboring magnetic cations coupled through non-magnetic O^{2-} as shown in Fig. 4a–d (Liu & Yang, 2017), mainly decide the different magnetic configurations that can be parallel or orthogonal.

Such electronic structure of B cations reflects that the magnetic ordering of materials is determined by the d electrons configurations. For example, the G-type AFM originated through Fe^{3+}-O^{2-}-Fe^{3+} in BFO, $Pb(Fe_{2/3}W_{1/3})O_3$ (PFW), $PbFe_{0.5}Ta_{0.5}O_3$ (PFT) and $PbFe_{0.5}Nb_{0.5}O_3$ (PFN) where the B site is occupied by Fe^{3+}($3d^5$, $S = 5/2$) with each d orbital filled by a single electron and the orbitals of neighboring Fe^{3+} being parallel with opposite spin directions as shown in Fig. 4e. In case of some double perovskite oxides $A(BB'')O_3$ ferromagnetism produced by the neighboring orbitals of cations which are orthogonal with same magnetic moment directions and ordered B and B′ cations with empty and half-filled e_g orbitals as shown in Fig. 4f.

3.2 *Dzyaloshinsky-Moriya Interaction*

Dzyaloshinsky-Moriya interaction which is generally expressed by Eq. (2) provides that week ferromagnetism can be derived from the antiferromagnetic order. Multiferroic antiferromagnetic oxides like BFO, relaxor ferroelectric PFW, PFN and PFT are not encouraging for further advancement of memory devices, since antiferromagnetism is not considered the main constituent as ferroic order parameter.

$$H_{DM} = D_{BB'}\left(S_B \times S_{B'}\right)D_{BB'} \propto \left(X \times r_{BB'}\right) \tag{2}$$

Here, site B and site B' are connected by the vector $r_{BB'}$ representing the displacement of the ligand ion (in perovskite ion it is O^{2-}) along the BB' line. From, the geometrical analysis, the additional magnetic interaction in a distorted perovskite is provided through AFM based on 180^0 of B-O-B′ bond angle and bending of B-O-B′ bonding. There will be induction of week ferromagnetism with the reducing exchange interaction along with the enhancement of the Dzyaloshinsky-Moriya interaction (Weingart et al., 2012).

3.3 Magnetoelectric Coupling and Mutual Control

For a multiferroic system, the Landau free energy expressed as function of applied magnetic field H and applied electric field E is given by Eq. (3) (Weingart et al., 2012; Fiebig, 2005):

$$\begin{aligned} F(E,H) = F_0 - P_i E_i - M_i H_i - \frac{1}{2}\varepsilon_0 \varepsilon_{ij} E_i E_j - \frac{1}{2}\mu_0 \mu_{ij} H_i H_j \\ -\alpha_{ij} E_i H_j - \frac{1}{2}\beta_{ij} E_i H_j H_k - \frac{1}{2}\gamma_{ijk} H_i E_i E_k \end{aligned} \tag{3}$$

where the linear magnetoelectric coupling is represented by α and the bilinear magnetoelectric coupling by β and γ. P_i and M_i denotes the spontaneous polarization and magnetization in direction i; μ_0 and ε_0 are, respectively, the permeability and permittivity of free space. While ε_{ij} and μ_{ij} are the relative dielectric permittivity and relative magnetic permeability, respectively. α is the linear magnetoelectric coupling coefficients and exist only for the ferromagnetic or antiferromagnetic point groups. There will be induced magnetization (or polarization) due to an electric field (or magnetic field) whenever the coefficients for linear magnetoelectric coupling exist in a multiferroic system. The upper boundary of linear ME coupling was proposed by Brown as Eq. (4) (Eerenstein et al., 2006):

$$\alpha_{ij} \leq \frac{\sqrt{\varepsilon_{ij}\mu_{ij}}}{c_0} \tag{4}$$

where c_0 is the speed of light in vacuum'.

It stands out from the expression that even without ferroic orders, large values of the electric permittivity and magnetic permeability can give rise to linear ME coupling. Ferroelectric and ferromagnetic materials as is well known do possess large permittivity and permeability. Although, it is not a prerequisite for coupling. The linear magnetoelectric coupling coefficient 'α' could be zero for a particular multiferroic material. Thus, multiferroics are not necessarily considered magnetoelectric materials (Eerenstein et al., 2006). An electric or magnetic field can control the direction of the magnetization or the spontaneous polarization, respectively, and often these direction turns out to be the interesting properties in applications in data storage (Brown et al., 1968; Schmid, 1994). Usually, the complex scenarios make such effects achievable.

In view of possible applications, it is desirable to have polarization controlled by an applied magnetic field. Such as in TMO 90^0 rotation of small polarization (0.08 $\mu C/cm^2$) was achieved with the help of a magnetic field at low temperature (Kimura et al., 2003). Mostovoy was the first to explain the flipping of polarization (Cheong & Mostovoy, 2007; Mostovoy, 2006; Kenzelmann & Harris, 2008). Thus, Landau free energy expansion was made to include the tri linear magnetoelectric coupling term (Harris et al., 2005). In view of the structure, flipping of polarization occurred simultaneously with a first-order transition to a magnetic commensurate phase (Aliouane et al., 2006). Neutron scattering spectroscopy revealed the control of spin helicity in the materials with an electric field (Yamasaki et al., 2007).

4 Tuning the Physical Properties of Perovskite Multiferroics

Tuning of the physical properties of multiferroic perovskites to enhance power conversion efficiencies (PCEs), lower processing costs and making stable and environment friendly materials are the most relevant challenges in the current photovoltaic (PV) technologies. An appealing and versatile approach facilitated by all-oxide photovoltaics is suggested to overcome these kinds of challenges. In practice, oxides are increasingly being incorporated as transparent conducting electrodes, passivation components, selective layers, and to an extent as light harvesters in solar cells (Rühle et al., 2012; Pérez-Tomás et al., 2018; Barquinha et al., 2012; Coll et al., 2019). In light of the quick rise in the value of PCE (over to 23.7%) of lead halide perovskites (Park et al., 2016; Bisquert et al., 2017; NREL; Rappe et al., 2017), significant attention has been given to semiconductor ferroelectric oxides possessing perovskite structure as they can be considered for photoactive layers (Coll et al., 2015; Lopez-Varo et al., 2016; Butler et al., 2015).

The distinctive property of having non-centro-symmetric structure of ferroelectric perovskite oxides is mainly responsible for the bulk photovoltaic effect providing a means to separate charge carriers spontaneously (Paillard et al., 2016). In this scenario, extremely large open-circuit voltages beyond the band gap on a material with a single phase can be achieved (Yang et al., 2015; Bhatnagar et al., 2013). In addition to this, the ferroelectric perovskite oxides on account of their spontaneous electric polarization control the direction of the PV current (Lee et al., 2011). However, ferroelectric perovskite oxides usually absorb the incident light in the ultraviolet range (band gap $Eg \approx 3$–4 eV, far from the ideal value of ≈1.4 eV for the maximum PCE) and produce low photo currents under the AM1.5G illumination conditions (Yang et al., 2010; Bhatnagar et al., 2014). The presence of wide band gaps in the ferroelectric perovskites is attributed to the large differences in electronegativity between the transition metal cation and the oxygen (Wang et al., 2015). Therefore, tuning the characteristics of the B–O bond is an effective way to modify the energy bandgap of the material (Arima et al., 1993).

There are attractive approaches to reduce the bandgap while retaining the ferroelectric properties allowing for photovoltaic devices with higher PCE value. In the quest for lead free perovskite oxide materials one of the most attractive is $BiFeO_3$ (BFO) for a possible application as light absorbing layer in ferroelectric based photovoltaic devices. It has a relatively narrow energy band gap of the order of 2.7 eV and ferroelectric properties at room temperature along with high saturated polarization in the pseudocubic [111] direction (90 $\mu C/cm^2$) (Wang et al., 2003). Additionally, it has relatively high refractive index '*n*' (Kumar et al., 2008; Leguy et al., 2016) which makes it suitable for light confinement purposes and enhances its of photo absorption capacity (Pettersson et al., 1999). Regardless of the existence of several approaches to modify the band gap (Wang et al., 2014; Herklotz et al., 2016; Sando et al., 2018; Zhang et al., 2017; Grinberg et al., 2013; Bai et al., 2017; Cao et al., 2014; Nechache et al., 2015), compositional tuning turned out to be one the most successful (Wang et al., 2014; Grinberg et al., 2013). A record PCE of 3.3% in Bi_2FeCrO_6 (8.1% in multilayer architectures) has been achieved (Wang et al., 2014; Chakrabartty et al., 2018) with this approach and a larger one of 4.2% in $BiMnO_3$-based composite films despite showing weaker ferroelectricity than undoped BFO.

Notwithstanding the rapid advancement in ferroelectric oxides with perovskite structure, there are more fundamental issues urgently required to be explored and studied. There are efforts in the direction of realizing the photovoltaic devices through developing new material compositions with reduced band gap and improved electrical polarization via scalable processing and cost-effective methods. In the following section, the effect on the properties of $BiFeO_3$ of the substitution of different elements either in the A, B, and/or oxygen sites is discussed including specific examples for each case.

4.1 *Effect of A-Site Substitution*

The Bi^{3+} ions are replaced with the doping ions in A-site substitution. The A-site substitution can influence the band structure indirectly since electronic levels of ions located at the A-site are situated far-off from the Fermi level. There are effects on the electronic band structure of the perovskite oxide, associated to the A-site substitution, that are well described by adequately adjusting the band width and band filling. When replacing the A-site ion by smaller ionic radius ions, a buckling in the angle of Fe–O–Fe bond takes place along with a smaller value of the tolerance factor that leads a to more insulating character. There are physical properties driven by carrier-doping including high-Tc superconductivity and colossal magnetoresistance that are induced when the trivalent A-site ions in the parent antiferromagnetic insulators are substituted with divalent ions. The novel conducting states of hole-doped $BiFeO_3$ can be produced by replacing the Bi^{3+} by divalent ions like Ca^{2+} and Sr^{2+} if other defects such as oxygen vacancies are not produced in the process. Moreover, there will be vacancies in the valence band of some of the ligand levels provided by the introduction of hole carriers through controlling the band filling on account of charge neutrality Ce^{4+} substitution is well documented to produce electron carriers.

Figure 5 shows the phase diagram of Ca doped BFO films grown on (001) $SrTiO_3$ (Yang et al., 2009). A strong similarity of the phase diagram with the ones of colossal magneto-resistive manganites and high-Tc superconductors where hole doping is responsible for a competition between the ground states which are energetically similar (Yang et al., 2009). The conductive properties of Ca-doped BFO are found to be controlled by the electrons band filling when the electric field is applied to an extent for the realization of formation, erosion and inversion of the p-n junction in the host materials. In the phase diagram, a 'dome-like' feature near a 1/8 Ca concentration represents new phase with pseudo-tetragonal structure where the electric modulation of conduction is found to be largest. Conductive antiferromagnetic measurements revealed the resistive switching character of the material where the reversal of the effect is driven by an applied electric field. Electric conduction with reversible modulation is observed along with ferroelectric state modulation. This is the result of spatial movement of oxygen vacancies which are naturally produced

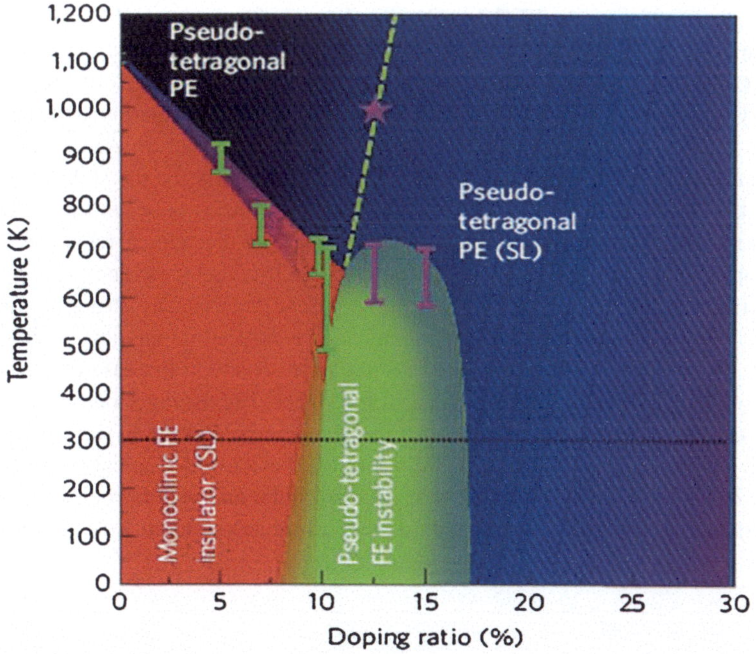

Fig. 5 Ca-doped BFO films phase diagram of $Bi_{1-x}Ca_xFeO_{3-d}$ film grown on (001)$SrTiO_3$ substrates. (Adapted from the Ref. (Yang et al., 2009))

under the applied electric field. Here, the role of donor impurities is well played by the oxygen vacancies to compensate Ca acceptors and maintain a highly stable valence state of Fe^{3+}. Thus, the conductor–insulator transition is well defined by the interchange developed through the band filling control between a mobility edge and the Fermi energy. There is an emergence of new ideas regarding magnetoelectrics and magnetoelectronics via integration of electronic conduction into the electric and magnetic degrees of freedom (Ko et al., 2011; Ramirez et al., 2008, 2009a, b). Moreover, the concept of doping with Ca and Mn could also be applied in the case of domain walls of BFO (Chiu et al., 2011).

There are more possibilities of coexisting of phases with different structures with close energy levels when the materials are modified under the structural limit, for which the chemical substitution method is found to be very useful. As a result of modification of materials under the limit of boundary structure, weak factors like magnetic field, electric field and temperature often exhibit a colossal response. A gigantic piezoelectric effect has been found at the morphotropic phase boundary (MPB) particularly in Pb-based ferroelectrics (Park & Shrout, 1997; Guo et al., 2000; Kutnjak et al., 2006; Fu & Cohen, 2000; Bellaiche et al., 2001). To obtain MPB in ferroelectrics with perovskite structure for improved functional properties there is one well recognized route, i.e., A and/or B site chemical composition tunning. To a great extent, BFO has been acknowledged as a perovskite ferroelectric with eco-friendly Pb free characteristics with extra functionalities at room temperature. Chemical substitution in BFO has been used to produce an MPB. However, there are some issues in BFO like having low electromechanical coefficients in comparison to Pb-based piezoelectrics, large coercive fields and high-leakage currents (Catalan & Scott, 2009; Lebeugle et al., 2007; Fujino et al., 2008; Kan et al., 2010; Cheng et al., 2010). Trivalent ions with twelve coordination number with their corresponding ionic radii are as Bi^{3+}(1.36 Å) > La^{3+}(1.36 Å) > Sm^{3+}(1.28 Å) > Dy^{3+}(1.24 Å) (Cox, 1955). In view of the conditions for the stability of the perovskite structure, there are strong links in ferroelectrics between the crystallographic structure, orientation of polarization and electro mechanical response. Thus, it is not surprising that the changes in the ferroelectric, dielectric and piezoelectric properties induced by the substitution in the parent materials are strongly affected by the ionic radii of the substituting elements. It is noticed remarkably in 2010 the presence of a universal term for effective radius of A-site in the case of thin films made of BFO substituted with rare earth materials. The average ionic size of the A-site cation is strongly responsible for the characteristic of the hysteresis loop of double polarization and the significantly improved electromechanical properties. It is also observed that the structural transition from the undoped ferroelectric rhombohedral phase to an orthorhombic phase takes place independently of the rare earth dopant species.

4.2 Effect of B-Site Substitution

We have a B-site substitution when a transition metal ion takes the place of the Fe^{3+} ions in the $BiFeO_3$ structure. There are physical properties closely regulated by the B-site substitution in the parent materials associated to alterations in the electronic structure close to the Fermi level which resulting from the relation of the d orbital state of the Fe^{3+} ion with the $BiFeO_3$ conduction band. To determine whether ordered structures or random distributions of the dopant ions develop in the parent materials over the substitution process is mainly decided by the radius of the dopant ions. A large mismatch between the radius of the Fe^{3+} and dopant ion is responsible for the induction of an ordered structure in particular, in the heavy doping cases. A good example is the Sr_2FeMoO_6 double perovskite compound where ferromagnetism and metallic conduction are present. The stable ordered structure developed in this compound is seen as a consequence of Mo^{5+} ions having ionic size and charge valence state which are completely different than those of Fe^{3+}at the B-site and since all nearest-neighbor bonds are formed between $Fe^{3+}$and Mo^{5+} ions (Kobayashi et al., 1998). Moreover, there are impurities related with the 3d transition metal located at random positions in the form of point defects and some of which are responsible for the creation of midgap states that will strongly influence the optical and conductive behavior of the material. In principle, these defect levels are determined by the combined effect of the Madelung potential, the number of occupied $d(t_{2g})$ and $d(e_g)$ orbitals and Coulomb potential between the number of positive charges in the nucleus and number of electrons in the ion. The midgap states could be filled partially with control of band filling and to an extent with charges trapped in the electric filed. The electrons trapped at defect sites and are capable to travel to conduction band due to random thermal fluctuation and roam freely before resting in a defect site. As often exhibited, band edge states of transition metal oxides which differ completely form the conventional semiconductors on account of randomness and polaronic self-localization can be further localized adequately into a definite level known as mobility edge. Electronic conduction in such cases can be described by the variable range hopping mechanism (Mott, 1969). The Poole–Frenkel mechanism establishes that the electron can travel by hopping between localized defect sites under the influence of a large electric field even at relatively low temperatures (Frenkel, 1938). There are well acknowledged properties on account of the super-exchange interaction, regarding perceptive magnetism in the insulating compounds. In view of the Pauli Exclusion Principle, even if the hopping of electrons is not forbidden, the on-site Hund rule allows formation of favorable ferromagnetic arrangements between the magnetic moments of the neighboring sites. There is exchange interaction between the Fe^{3+} ions which does not permit half-filled d^5 electrons to hop to the neighboring site without flipping its spin despite all the magnetic moments of the present ions being ordered ferromagnetically. Thus to make antiferromagnetic magnetic momen alignment favorable, the intermediate state with energy $E(d^{n+1} + d^{n-1})$ after electron hopping will become unstable in terms of the on-site Hund rule.

The doping of Cr and Mn is used to regulate magnetism. Cr^{3+} has three d-orbital electrons in that account for its magnetic moment of 3 μB. The only t_{2g} orbitals are filled with three electrons while e_g orbitals are vacant. The interaction between half-filled e_g–empty e_g (d^5–d^3 type) results in the ferromagnetic exchange interaction between Fe^{3+} and Cr^{3+}. Under the limit of small substitution ratio, there are weak antiferromagnetic behaviors shown by neighboring Cr^{3+} ions that arise due to magnetic interaction as result of the interaction between half-filled t_{2g}–half-filled t_{2g} despite the fact that the Cr^{3+} ions hardly interact with each other at the nearest-neighboring sites. To obtain the ferromagnetic state, it is required to avoid the formation of Fe^{3+}–Fe^{3+} bonds and Cr^{3+}–Cr^{3+} bonds and to maximize the number of Fe^{3+}–Cr^{3+} bonds. There is an ideal double perovskite $BiFe_{0.5}Cr_{0.5}O_3$compound in which all bonds are ferromagnetic d^5–d^3 type on account of G-type ordering for Fe^{3+} and Cr^{3+}ions. As an effect of similar chemical valence state and ionic size owing to randomly located B-site ions, the realization of double perovskite compound is somehow not feasible.

The Jahn–Teller active interesting Mn^{3+} ion impurities with a d orbital with four electrons is responsible for the occurrence of distortion in the local structure associated with an anisotropic orbital ($d_{3z^2-r^2}$) (Kugel & Khomskii, 1982). The Mn^{3+} ion surrounded by an oxygen octahedron elongated in the z-axis direction interact antiferromagnetically with the Fe^{3+} ion neighbor along the *z*-axis direction but result in ferromagnetic interaction along the other principal axes. Both Cr^{3+} and Mn^{3+} ions cause the ferromagnetic exchange interactions with neighbor Fe^{3+} ions which result in diluting substitution of the ions and lead to ferromagnetic state stabilization. Random substitution generates distinctive changes in the magnetic state resulting in magnetic frustration effects which cause the suppression of magnetic ordering and decline in the value of Nèel temperature. Under the limit of large magnetic frustration, magnetic responses with slow dynamics with respect to glass behaviors of spin/cluster at low temperature can be expected (Yang et al., 2006). The temperature of magnetic transition can be modified with magnetic impurities substitutions and result in an enhancement of properties related to magnetoelectric/magneto-dielectric effects around this transition temperature (Yang et al., 2005, 2007). In order to prevent magnetic frustration that arises on account of random substitutions, the hetero-epitaxial bi-layers or structures of self-assembled composites like nano-pillars of $BiFeO_3$ and other spinel ferrites produce new magnetic layers with ordered structures and are employed in studying the interfacial exchange coupling/bias with the $BiFeO_3$ (Yu et al., 2010; Zheng et al., 2004).

Chemical substitution is applied to $BiFeO_3$ with intention of modifying electronic and magnetic properties to reduce leakage currents (Lee et al., 2006a, b). There are reports of leakage current reduction by up to three orders of magnitudes when the B-site of BFO is substituted with Ti^{4+} ion. In contrast, its conductivity is increased by two orders of magnitude Ni^{2+} is incorporated into BFO (Qi et al., 2005). Correspondingly, to a great extent a leakage current reduces in the films grown of BFO when Cr or Mn are used as doping agents (Kim et al., 2006).

4.3 Effect of Oxygen-Site Substitution

The loss of oxygen in BFO produces both surface and bulk vacancies. Oxygen vacancies can be produced by controlling the growth parameters when depositing the BFO thin films (Scullin et al., 2008, 2010). To an extent the BFO's chemical and electronic properties are modified by the presence of such vacancies affecting the BFO based device performance. Oxygen vacancies in $Bi_{0.9}Ca_{0.1}FeO_{3-0.05}$ for instance affect the electro chemical properties that further cause the modification in the conductivity of domain walls (Seidel et al., 2012a, b). There are indications of migration of these oxygen vacancies formed in the bulk BFO to the surface. Considering the experimental limitations, it is difficult to measure the oxygen content for very small changes although, there are a few reports of quantification of concentration of O-atom defects (Cuong et al., 2007). The vacancy formation occurred on heating of bulk BFO in vacuum or in an oxygen poor environment. The transport properties including ionic and electronic conductivity and diffusion are profoundly affected with the presence of ionic defects like oxygen vacancies.

Several studies and investigations of $BiFeO_3$ and other ferroic materials resulted in exhibiting the attractive electrical transport properties of the domain walls (Seidel et al., 2011, 2012a, b, 2013; Lubk et al., 2012; Borisevich et al., 2010; Yang et al., 2010; Maksymovych et al., 2011; He et al., 2012; Farokhipoor & Noheda, 2011; Choi et al., 2010; Meier et al., 2012; Catalan et al., 2012; Seidel, 2012). The ferroelectric domain walls in $BiFeO_3$, hexagonal $YMnO_3$ and $Pb(Zr_{0.2}Ti_{0.8})O_3$ consist of variable levels of local electrical conductance and these variable levels have been measured at the domain walls. Such interfaces can cause variation in the bonding between ions, stress induction, and locally broken symmetry. The modification of the electronic structure of these kinds of strongly interconnected materials has been realized and associated to the tuning of bandwidth, level degeneracy and orbital interaction. The carrier densities induced by the charge transfer, are different at the interface with respect to densities in the bulk. Thus, completely different physical properties at the interface resulted in comparison to the corresponding properties of the parent bulk materials (Ogale, 2005; Ohtomo et al., 2002; Dagotto, 2007; Mannhart & Schlom, 2010; Yamada et al., 2004; Zubko et al., 2011; Ferreira et al., 2010; Fan et al., 2011; Frederikse et al., 1964). The physical properties of interfaces and domain boundaries were modified and controlled by doping offering new avenues for research activities and providing ground for characterization of nanoelectronics for prospective nanotechnology advancement (Zubko et al., 2011).

The modified electronic structure occurring in the ferroelectric (multiferroic) arises due to the changing transport behavior caused by doping in the parent materials. Certainly, the different ferroic materials demonstrate different transport behavior on account of the conductivity of the domain walls. In $BiFeO_3$ domain walls are more conductive than the domain itself. In the case of $YMnO_3$ the domain walls shows more conductive or insulating behaviors depending on the orientation of the domains (Farokhipoor & Noheda, 2011; Choi et al., 2010). Close examinations are required on the role of accumulation of defects at the domain walls as in the range of ferroelectric semiconductors. Localized states are observed and these states split at the domain walls and inside but close to the domain walls in the bulk continuum. There is a high dispersion of these non-degenerate states with respect to "heavy-fermion" states at an isolated wall (Idlis et al., 1992). The polarization and space charges coupling at ferroelectric–ferroelastic domain walls result in the formation of double layers of charge (Xiao et al., 2005). There are charged domain walls with energies almost one order of magnitude higher than those of uncharged domain walls (Gureev et al., 2009) observations show the walls decorated with defects like oxygen vacancies. The charge and defect layers are present at the walls and the electrical failure promoted by such walls enables pathways of high conductivity from electrode to electrode. Hence, the electronic structure at the walls and the strain in ferroelectric and ferroelastic oxides are governed with the doping and thus result in effectively modifying the functionalities at the nanoscale (Ferreira et al., 2010; Fan et al., 2011). Varying amounts of vacancies can be introduced with specific conditions of growth, in thin film samples (Frederikse et al., 1964).

5 Green Energy Applications in Perovskite Multiferroics

Perovskite multiferroics are an emerging material class that shows coexisting ferroelectric and magnetic properties in the same crystal structure. These materials have been the center of attraction in recent years due to their potential for various applications, including the field of green energy (Rühle et al., 2012; Pérez-Tomás et al., 2018; Barquinha et al., 2012; Coll et al., 2019; Kumar et al., 2020, 2021; Raj et al., 2021a, b). Here are some potential green energy applications of perovskite multiferroics that could be game changers soon; notwithstanding, it has a wide range of applications. Perovskite multiferroic materials extraordinarily convert mechanical vibrations and temperature variations into electrical energy, making them feasible for self-controlled sensors and energy-effective gadgets. Furthermore, multiferroic materials remarkable properties open doors for creating spintronic gadgets, which depend on the behavior of the electron's spin for information control. These gadgets can outflank conventional hardware regarding energy proficiency, helping several farms and farming with diminished energy consumption. Certain multiferroics display magnetoelectric coupling, permitting their attractive properties to be manipulated by electric fields. This property can be outfitted for magnetic refrigeration, an exceptionally productive and eco-accommodating cooling innovation that can supplant earth-safe refrigerants. Perovskite multiferroics can be applied in sensors for natural observing purposes, including recognizing contaminants and temperature changes. Their reactions to numerous outside elements can prompt more precise and energy-productive detecting gadgets. Multiferroic materials can possibly alter the field of hardware by decreasing power consumption in semiconductors and memory stockpiling components, adding to the improvement of energy-efficient electronic gadgets. The extraordinary properties of multiferroic materials may likewise find applications in cutting-edge energy storage frameworks, by working on the proficiency and life span of batteries and supercapacitors. Multiferroic materials also have the potential to be active materials for solar cell applications using the control of their electronic and magnetic properties; we could find ways of improving the effectiveness and manageability of sun-oriented energy transformation. It is vital to note that perovskite multiferroics are still in the beginning phases of investigation, and there are significant challenges to address before achieving their maximum capacity in environmentally friendly power energy applications. Here, we will show an example of multiferroic thin films as active materials for solar cell applications.

Machado et al. (2019) fabricated a solar cell device using Co-doped BFO thin films on LSMO/STO substrate. Photo-response estimations were done for BFO and Co-doped BFO with $x = 0.1$ and $x = 0.3$. The sample cells are schematically displayed in Fig. 6a (Machado et al., 2019). Here, they characterize the current density (J), where positive carriers stream from the top Pt anode to the LSMO grounded electrode. I–V curves were plotted after pre-poling with different voltages for each composition according to their corresponding ferroelectric coercive fields taken

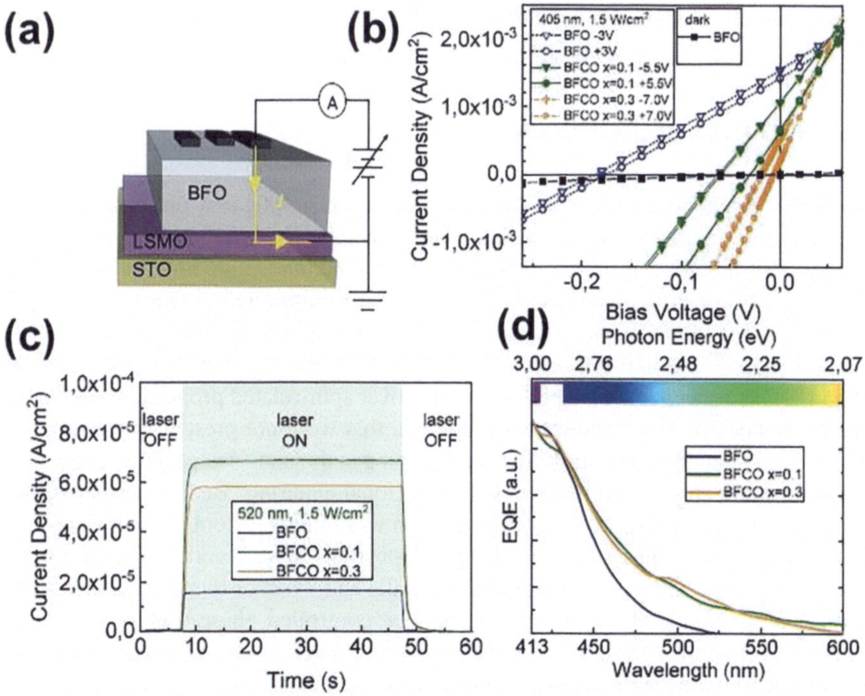

Fig. 6 (**a**) Schematic configuration for measuring photo-response, (**b**) J–V curves for BFO and Co-doped BFO, (**c**) Time dependence of the short-circuit photocurrent for BFO and BFCO $x = 0.1$ and $x = 0.3$ films illuminated under 520 nm with a laser of 1.5 W cm^{-2}. (**d**) External quantum efficiency (EQE) as a function of the wavelength. (Adapted from the Ref. (Machado et al., 2019))

from their P(E) loops, specifically, ±3 V for BFO, ±5 V for BFCO $x = 0.1$, and ±7 V for BFCO $x = 0.3$ (Fig. 6b) (Machado et al., 2019). The direct J–V relation is normal in numerous ferroelectrics perovskite oxides and contrasts with the rectifying behavior of high-performance solar cells (Bhatnagar et al., 2013; Zenkevich et al., 2014; Raj et al., 2022, 2021a, b).

Such a response can relate to the prevailing conduction mechanism examined below. Illuminating a BFO thin film with a 1.5 W cm^{-2} intensity, 405 nm wavelength laser (below BFO bandgap energy) causes an uncommon change in the I–V curve. The authors reported that the short-circuit current and open-circuit voltage are Jsc = 1.56 mA cm^{-2} and Voc = −186 mV, respectively, comparable to those reported in the high-quality PLD BFO/LSMO/STO system (Yang et al., 2016). They used the same methodology with Co-doped BFO films as they did for BFO. Furthermore, they measured the photo-response of the BFO, BFCO $x = 0.1$, and BFCO $x = 0.3$ samples when illuminated with a 520 nm wavelength laser (below the bandgap energy of Co-BFO) and without being pre-polarized as shown in Fig. 6c (Machado et al., 2019). They observed a significant improvement in Jsc as compared to that of undoped BFO. Lastly, the external quantum efficiency (EQE) vs. wavelength (Fig. 6d) (Machado et al., 2019) measurements further confirm that BFCO films show an overall better performance than BFO films between 440–400 and 590 nm (Fig. 6d) (Machado et al., 2019).

6 Summary and Outlook

From the outset, our exploration aimed to uncover intriguing new physical phenomena in modified multiferroic materials, with a specific focus on the case of bismuth ferrite. This pursuit was driven by the desire to create innovative functional material systems. Multiferroic materials, exemplified by bismuth ferrite, represent a unique category due to their capacity to undergo structural and polar order modifications, much like conventional ferroelectrics such as PZT, through the process known as doping. This process enables precise control over spin-related properties, leading to the emergence of novel transport phenomena that were not present in the original multiferroic material. The introduction of doping at the nano scale has consequently ignited significant interest in the field of functional materials. Recent advancements have unveiled properties related to conduction within ferroic domain walls, presenting various opportunities and challenges. Questions arise regarding the limits of charge transport behavior at these domain walls and whether the topological structure of these walls can be designed to induce controlled phase transitions due to correlated electron effects. Additionally, there is much interest surrounding the possibility of initiating an Anderson transition through domain wall doping or straining. Another crucial aspect is the examination of magnetism at these multiferroic domain walls and how the introduction of additional carriers, through doping or electric gating, can impact magnetic interactions, potentially transitioning from super-exchange to double exchange. In summary, multiferroic materials hold promise as photoactive materials, providing a robust platform for the investigation of fundamental physical properties. This exploration includes the intricate relationship between cation substitution and electronic and optical properties. The potential for modifying energy band gaps offers a promising avenue, particularly when combining these modified materials with traditional photovoltaic devices. In the dynamic field of ferroelectric perovskite oxides, it is anticipated that solutions will emerge to address the challenges of creating stable, robust, and efficient all-oxide photovoltaic devices.

Acknowledgments One of the authors (M. Kumar) would like to thank the Experimental Research Laboratory, ARSD College, the University of Delhi, New Delhi, India, for research support. Subhash Sharma acknowledges support from the CONAHCyT Program via the Researchers for Mexico through Project 352-2018. PAPIIT-DGAPA-UNAM Grants IN103323 partially supported this work.

References

Aliouane, N., Argyriou, D. N., Strempfer, J., et al. (2006). Field-induced linear magnetoelastic coupling in multiferroic $TbMnO_3$. *Physical Review B, 73*(2), 020102. https://doi.org/10.1103/PhysRevB.73.020102

Anderson, P. W. (1950). Antiferromagnetism. Theory of superexchange interaction. *Physics Review, 79*(2), 350–356. https://doi.org/10.1103/PhysRev.79.350

Anderson, P. W. (1959). New approach to the theory of super exchange interactions. *Physics Review, 115*(1), 2. https://doi.org/10.1103/PhysRev.115.2

Arima, T., Tokura, Y., & Torrance, J. B. (1993). Variation of optical gapsin perovskite-type 3d transition-metal oxides. *Physical Review B: Condensed Matter and Materials Physics, 48*, 17006–17009. https://doi.org/10.1103/PhysRevB.48.17006

Ascher, E., Rieder, H., Schmid, H., & Stössel, H. (1966). Some properties of ferromagnetoelectric nickel-iodine boracite, $Ni_3B_7O_{13}I$. *Applied Physics, 37*(3), 1404–1405. https://doi.org/10.1063/1.1708493

Astrov, D. N. (1960). *Soviet Physics JETP, 11*, 708.

Bai, Y., Tofel, P., Palosaari, J., Jantunen, H., & Juuti, J. (2017). A game changer: A multifunctional perovskite exhibiting giant ferroelectricity and narrow bandgap with potential application in a truly monolithic multi energy harvester or sensor. *Advanced Materials, 29*, 1700767. https://doi.org/10.1002/adma.201700767

Barquinha, P., Martins, R., Pereira, L., & Fortunato, E. (2012). *Transparent oxide electronics*. Wiley Online Books. ISBN: 978-0-470-68373-6.

Bellaiche, L., Garcia, A., & Vanderbilt, D. (2001). Electric-field induced polarization paths in $Pb(Zr_{1-x}Tix)O_3$ alloys. *Physical Review B: Condensed Matter and Materials Physics, 64*, 060103. https://doi.org/10.1103/PhysRevB.64.060103

Bhatnagar, A., Chaudhuri, A. R., Kim, Y. H., Hesse, D., & Alexe, M. (2013). Role of domain walls in the abnormal photovoltaic effect in $BiFeO_3$. *Nature Communications, 4*, 2835. https://doi.org/10.1038/ncomms3835

Bhatnagar, A., Kim, Y. H., Hesse, D., & Alexe, M. (2014). Persistent photoconductivity in strained epitaxial $BiFeO_3$ thin films. *Nano Letters, 14*, 5224–5228. https://doi.org/10.1021/nl502183j

Bibes, M., & Barthelemy, A. (2008). Multiferroics: Towards a magnetoelectric memory. *Nature Materials, 7*(6), 425–426. https://doi.org/10.1038/nmat2189

Bisquert, J., Qi, Y., Ma, T., & Yan, Y. (2017). Advances and obstacles on perovskite solar cell research from material properties to photovoltaic function. *ACS Energy Letters, 2*, 520–523. https://doi.org/10.1021/acsenergylett.7b00085

Borisevich, A. Y., Ovchinnikov, O., Chang, H. J., Oxley, M., et al. (2010). *ACS Nano, 4*, 6071.

Brown, W. F., Hornreich, R. M., & Shtrikman, S. (1968). Upper bound on the magnetoelectric susceptibility. *Physics Review, 168*(2), 574–577. https://doi.org/10.1103/PhysRev.168.574

Butler, K. T., Frost, J. M., & Walsh, A. (2015). Ferroelectric materials for solar energy conversion: Photoferroics revisited. *Energy and Environmental Science, 8*, 838–848. https://doi.org/10.1039/C4EE03523B

Cao, J., Ji, Y., Tian, C., & Yi, Z. (2014). Synthesis and enhancement of visible light activities of nitrogen-doped $BaTiO_3$. *Alloys and Compounds, 615*, 243–248. https://doi.org/10.1016/j.jallcom.2014.07.008

Catalan, G., & Scott, J. F. (2009). Physics and applications of bismuth ferrite. *Advanced Materials, 21*, 2463–2485. https://doi.org/10.1002/adma.200802849

Catalan, G., Seidel, J., Ramesh, R., & Scott, J. F. (2012). Domain wall nanoelectronics. *Reviews of Modern Physics, 8*, 119–156. https://doi.org/10.1103/RevModPhys.84.119

Chakrabartty, J., Harnagea, C., Celikin, M., Rosei, F., & Nechache, R. (2018). Improved photovoltaic performance from inorganic perovskite oxide thin films with mixed crystal phases. *Nature Photonics, 12*, 271–276. https://doi.org/10.1038/s41566-018-0137-0

Cheng, C. J., Kan, D., Anbusathaiah, V., Takeuchi, I., & Nagarajan, V. (2010). Microstructure-electromechanical property correlations in rare-earth-substituted $BiFeO_3$ epitaxial thin films at morphotropic phase boundaries. *Applied Physics Letters, 97*, 212905. https://doi.org/10.1063/1.3520642

Cheong, S. W., & Mostovoy, M. (2007). Multiferroics: A magnetic twist for ferroelectricity. *Nature Materials, 6*(1), 13–20. https://doi.org/10.1038/nmat1804

Chiu, Y. P., Chen, Y. T., Huang, B. C., Shih, M. C., et al. (2011). *Advanced Materials, 23*, 1530–1534.

Choi, T., Horibe, Y., Yi, H. T., Choi, Y. J., Wu, W., & Cheong, S. W. (2010). Insulating interlocked ferroelectric and structural antiphase domain walls in multiferroic $YMnO_3$. *Nature Materials, 9*(253), 63–66. https://doi.org/10.1038/nmat2632

Cochran, W. (1959). Crystal stability and the theory of ferroelectricity. *Physical Review Letters, 3*(9), 412–414. https://doi.org/10.1103/PhysRevLett.3.412

Cohen, R. E. (1992). Origin of ferroelectricity in perovskite oxides. *Nature, 358*(6382), 136–138. https://doi.org/10.1038/358136a0

Coll, M., Gomez, A., Mas-Marza, E., Almora, O., et al. (2015). Polarization switching and light-enhanced piezoelectricity in lead halide perovskites. *The Journal of Physical Chemistry Letters, 6*, 1408–1413. https://doi.org/10.1021/acs.jpclett.5b00502

Coll, M., Fontcuberta, J., Althammer Gross, M. R., Grollier, J., et al. (2019). Towards oxide electronics: A roadmap. *Applied Surface Science, 482*, 1–93. https://doi.org/10.1016/j.apsusc.2019.03.312

Cox, P. A. (1955). *Transition metal oxides*. Clarendon Press.

Cuong, D. D., Lee, B., Choi, K. M., Ahn, H. S., et al. (2007). Oxygen vacancy clustering and electron localization in oxygen-deficient SrTiO3: LDA+U study. *Physical Review Letters, 98*, 115503. https://doi.org/10.1103/PhysRevLett.98.115503

Dagotto, E. R. (2007). When oxides meet face to face. *Science, 318*, 1076–1077. https://doi.org/10.1126/science.1151094

Dragan, M., Enache, S., Varlam, M., & Petrov, K. (2020). *Perovskite-based materials for energy applications*. IntechOpen. https://doi.org/10.5772/intechopen.91271

Dzyaloshinskii, I. E. (1959). *Soviet Physics JETP, 10*, 628.

Dzyaloshinsky, I. (1958). A thermodynamic theory of "weak" ferromagnetism of antiferromagnetics. *The Journal of Physics and Chemistry of Solids, 4*(4), 241–255. https://doi.org/10.1016/0022-3697(58)90076-3

Eerenstein, W., Mathur, N. D., & Scott, J. F. (2006). Multiferroic and magnetoelectric materials. *Nature, 442*, 759–765. https://doi.org/10.1038/nature05023

Fan, W., Cao, J., Seidel, J., Gu, Y., Yim, J. W., Barrett, C., et al. (2011). Large kinetic asymmetry in the metal-insulator transition nucleated at localized and extended defects. *Physical Review B: Condensed Matter and Materials Physics, 83*, 235102. https://doi.org/10.1103/PhysRevB.83.235102

Farokhipoor, S., & Noheda, B. (2011). Conduction through 71° domain walls in $BiFeO_3$ thin films. *Physical Review Letters, 107*, 127601. https://doi.org/10.1103/PhysRevLett.107.127601

Fernàndez-Posada, C. M., Castro, A., Kiat, J. M., Porcher, F., et al. (2016). A novel perovskite oxide chemically designed to show multiferroic phase boundary with room-temperature magnetoelectricity. *Nature Communications, 7*, 12772. https://doi.org/10.1038/ncomms12772

Ferreira, L. G., Redfern, S. A. T., Artacho, E., Salje, E., & Lee, W. T. (2010). *Physical Review B: Condensed Matter and Materials Physics, 81*, 024109.

Fiebig, M. (2005). Revival of the magnetoelectric effect. *Journal of Physics D: Applied Physics, 38*(8), R123. https://doi.org/10.1088/0022-3727/38/8/R01

Frederikse, H. P. R., Thurber, W. R., & Hosler, W. R. (1964). Electronic transport in strontium titanate. *Physics Review, 134*, A442. https://doi.org/10.1103/PhysRev.134.A442

Frenkel, J. (1938). On pre-breakdown phenomena in insulators and electronic semi-conductors. *Physics Review, 54*, 647. https://doi.org/10.1103/PhysRev.54.647

Fu, H., & Cohen, R. E. (2000). Polarization rotation mechanism for ultrahigh electromechanical response in single-crystal piezoelectrics. *Nature, 403*, 281–283. https://doi.org/10.1038/35002022

Fujino, S., Murakami, M., Anbusathaiah, V., Lim, S. H., Nagarajan, V., et al. (2008). Combinatorial discovery of a lead-free morphotropic phase boundary in a thin-film piezoelectric perovskite. *Applied Physics Letters, 92*, 202904. https://doi.org/10.1063/1.2931706

Goodenough, J. B. (1955). Theory of the role of covalence in the perovskite-type manganites [La, M(II)] MnO_3. *Physics Review, 100*(2), 564–573. https://doi.org/10.1103/PhysRev.100.564

Grinberg, I., West, D. V., Torres, M., Gou, G., et al. (2013). Perovskite oxides for visible-light-absorbing ferroelectric and photovoltaic materials. *Nature, 503*, 509. https://doi.org/10.1038/nature12622

Guo, R., Cross, L. E., Park, S. E., Noheda, B., Cox, D. E., & Shirane, G. (2000). Origin of the high piezoelectric response in $PbZr_{1-x}Ti_xO_3$. *Physical Review Letters, 84*, 5423. https://doi.org/10.1103/PhysRevLett.84.5423

Gureev, M. Y., Tagantsev, A. K., & Setter, N. (2009). *18th IEEE ISAF proceedings* (pp. 37–42).

Halasyamani, P. S., & Poeppelmeier, K. R. (1998). Noncentrosymmetric oxides. *Chemistry of Materials, 10*(10), 2753–2769. https://doi.org/10.1021/cm980140w

Harris, M., Kenzelmann, A. B., Jonas, S., et al. (2005). Magnetic inversion symmetry breaking and ferroelectricity in $TbMnO_3$. *Physical Review Letters, 95*(8), 087206. https://doi.org/10.1103/PhysRevLett.95.087206

He, Q., Yeh, C. H., Yang, J. C., Bhalla, G. S., Liang, C. W., et al. (2012). Magneto transport at domain walls in BiFeO3. *Physical Review Letters, 108*, 067203. https://doi.org/10.1103/PhysRevLett.108.067203

Herklotz, A., Rus, S. F., & Ward, T. Z. (2016). Continuously controlled optical band gap in oxide semiconductor thin films. *Nano Letters, 16*, 1782–1786. https://doi.org/10.1021/acs.nanolett.5b04815

Hur, N., Park, S., Sharma, P. A., Ahn, J. S., Guha, S., & Cheong, S. W. (2004). Electric polarization reversal and memory in a multiferroic material induced by magnetic fields. *Nature, 429*, 392–395. https://doi.org/10.1038/nature02572

Idlis, B. G., Usmanov, M. S., & Parsons, D. (1992). *JETP Letters, 56*, 264–268.

Kan, D., Palova, L., Anbusathaiah, V., Cheng, C. J., Fujino, S., et al. (2010). Universal behavior and electric-field-induced structural transition in rare-earth-substituted $BiFeO_3$. *Advanced Functional Materials, 20*, 1108–1115.

Kanamori, J. (1960). Crystal distortion in magnetic compounds. *Applied Physics, 31*(5), S14–S23. https://doi.org/10.1063/1.1984590

Kenzelmann, M., & Harris, A. B. (2008). Comment on "ferroelectricity in spiral magnet". *Physical Review Letters, 100*(8), 089701. https://doi.org/10.1103/PhysRevLett.100.089701

Khomskii II, D. Physikalisches Institut, Universitätzu Köln, Zülpicher Strasse 77, 50937 Köln, Germany.

Kim, J. K., Kim, S. S., Kim, W. J., Bhalla, A. S., & Guo, R. (2006). Enhanced ferroelectric properties of Cr-doped $BiFeO_3$ thin films grown by chemical solution deposition. *Applied Physics Letters, 88*, 132901. https://doi.org/10.1063/1.2189453

Kimura, T., Goto, T., Shintani, H., et al. (2003). Magnetic control of ferroelectric polarization. *Nature, 426*, 55–58. https://doi.org/10.1038/nature02018

Ko, K. T., Jung, M. H., He, Q., Lee, J. H., Woo, C. S., Chu, K., et al. (2011). Concurrent transition of ferroelectric and magnetic ordering near room temperature. *Nature Communications, 2*, 567. https://doi.org/10.1038/ncomms1576

Kobayashi, K. L., Kimura, T., Sawada, H., Terakura, K., & Tokura, Y. (1998). Room-temperature magnetoresistance in an oxide material with an ordered double-perovskite structure. *Nature, 395*, 677. https://doi.org/10.1038/27167

Kosub, T., Kopte, M., Huhne, R., Appel, P., Shields, B., Maletinsky, P., Hubner, R., Liedke, M. O., Fassbender, J., Schmidt, O. G., & Makarov, D. (2017). Purely antiferromagnetic magnetoelectric random access memory. *Nature Communications, 8*, 13985. https://doi.org/10.1038/ncomms13985

Kramers, H. A. (1934). L'interaction entre les atomesmagn_etog_enesdans un cristalparamagn_etique. *Physica, 1*(1–6), 182–192. https://doi.org/10.1016/S0031-8914(34)90023-9

Kugel, K. I., & Khomskii, D. I. (1982). The Jahn-Teller effect and magnetism: Transition metal compounds. *Soviet Physics Uspekhi, 25*, 231. https://doi.org/10.1070/PU1982v025n04ABEH004537

Kumar, A., Rai, R. C., Podraza, N. J., Denev, S., et al. (2008). Linear and nonlinear optical properties of $BiFeO_3$. *Applied Physics Letters, 92*, 121915. https://doi.org/10.1063/1.2901168

Kumar, M., Raj, A., Kumar, A., & Anshul, A. (2020). Effect of band-gap tuning on lead-free double perovskite heterostructure devices for photovoltaic applications via SCAPS simulation. *Materials Today Communications, 26*, 101851. https://doi.org/10.1016/j.mtcomm.2020.101851

Kumar, M., Raj, A., Kumar, A., & Anshul, A. (2021). Theoretical evidence of high power conversion efficiency in double perovskite solar cell device. *Optical Materials, 111*, 110565. https://doi.org/10.1016/j.optmat.2020.110565

Kutnjak, Z., Petzelt, J., & Blinc, R. (2006). The giant electromechanical response in ferroelectric relaxors as a critical phenomenon. *Nature, 441*, 956–959. https://doi.org/10.1038/nature04854

Lebeugle, D., Colson, D., Forget, A., & Viret, M. (2007). Very large spontaneous electric polarization in $BiFeO_3$ single crystals at room temperature and its evolution under cycling fields. *Applied Physics Letters, 91*, 022907. https://doi.org/10.1063/1.2753390

Lee, J. H., Murugavel, P., Ryu, H., Lee, D., Jo, J. Y., et al. (2006a). Epitaxial stabilization of a new multiferroic hexagonal phase of $TbMnO_3$ thin films. *Advanced Materials, 18*, 3125–3129. https://doi.org/10.1002/adma.200601621

Lee, Y. H., Wu, J. M., & Lai, C. H. (2006b). Influence of La doping in multiferroic properties of $BiFeO_3$ thin films. *Applied Physics Letters, 88*, 042903. https://doi.org/10.1063/1.2167793

Lee, D., Baek, S. H., Kim, T. H., Yoon, J. G., Folkman, C. M., Eom, C. B., & Noh, T. W. (2011). Polarity control of carrier injection at ferroelectric/metal interfaces for electrically switchable diode and photovoltaic effects. *Physical Review B: Condensed Matter and Materials Physics, 84*, 125305. https://doi.org/10.1103/PhysRevB.84.125305

Leguy, A. M. A., Azarhoosh, P., Alonso, M. I., Campoy-Quiles, M., et al. (2016). Experimental and theoretical optical properties of methylammonium lead halide perovskites. *Nanoscale, 8*, 6317–6327. https://doi.org/10.1039/C5NR05435D

Li, W., & Ji, L. J. (2018). Perovskite ferroelectrics go metal free. *Science, 361*, 132. https://doi.org/10.1126/science.aat5729

Liu, H., & Yang, X. (2017). A brief review on perovskite multiferroics. *Ferroelectrics, 507*(1), 69–85. https://doi.org/10.1080/00150193.2017.1283171

Loidl, A., Krohns, S., Hemberger, J., et al. (2008). Bananas go paraelectric. *Journal of Physics: Condensed Matter, 20*(19), 191001. https://doi.org/10.1088/0953-8984/20/19/191001

Lopez-Varo, P., Bertoluzzi, L., Bisquert, J., Alexe, M., et al. (2016). Physical aspects of ferroelectric semiconductors for photovoltaic solar energy conversion. *Physics Reports, 653*, 1–40. https://doi.org/10.1016/j.physrep.2016.07.006

Lu, X., Kim, Y., Goetze, S., Li, X., Dong, S., Werner, P., Alexe, M., & Hesse, D. (2011). Magnetoelectric coupling in ordered arrays of multilayered heteroepitaxial $BaTiO_3/CoFe_2O_4$ nanodots. *Nano Letters, 11*, 3202–3206. https://doi.org/10.1021/nl201443h

Lubk, A., Rossell, M. D., Seidel, J., He, Q., Yang, S. Y., et al. (2012). Evidence of sharp and diffuse domain walls in $BiFeO_3$ by means of unit-cell-wise strain and polarization maps obtained with high resolution scanning transmission electron microscopy. *Physical Review Letters, 109*, 047601. https://doi.org/10.1103/PhysRevLett.109.047601

Machado, P., Scigaj, M., Gazquez, J., Rueda, E., et al. (2019). Band gap tuning of solution-processed ferroelectric perovskite $BiFe_{1-x}Co_xO_3$ thin films. *Chemistry of Materials, 31*, 947–954. https://doi.org/10.1021/acs.chemmater.8b04380

Maksymovych, P., Seidel, J., Chu, Y. H., Baddorf, A., et al. (2011). Dynamic conductivity of ferroelectric domain walls in $BiFeO_3$. *Nano Letters, 11*, 1906–1912. https://doi.org/10.1021/nl104363x

Mannhart, J., & Schlom, D. G. (2010). Oxide interfaces an opportunity for electronics. *Science, 327*, 1607–1611. https://doi.org/10.1126/science.1181862

Martin, L. W., & Ramesh, R. (2012). Multiferroic and magnetoelectric heterostructures. *Acta Materialia, 60*, 2449–2470. https://doi.org/10.1016/j.actamat.2011.12.024

Meier, D., Seidel, J., Cano, A., Delaney, K., Kumagai, Y., et al. (2012). Anisotropic conductance at improper ferroelectric domain walls. *Nature Materials, 11*, 284–288. https://doi.org/10.1038/nmat3249

Moriya, T. (1960). Anisotropic superexchange interaction and weak ferromagnetism. *Physics Review, 120*(1), 91–98. https://doi.org/10.1103/PhysRev.120.91

Mostovoy, M. (2006). Ferroelectricity in spiral magnets. *Physical Review Letters, 96*(6), 067601. https://doi.org/10.1103/PhysRevLett.96.067601

Mott, N. F. (1969). Conduction in non-crystalline materials. *Philosophical Magazine, 19*, 835. https://doi.org/10.1080/14786436908216338

Moubah, R., Rousseau, O., Colson, D., Artemenko, A., Maglione, M., & Viret, M. (2012). Photoelectric effects in single domain $BiFeO_3$ crystals. *Advanced Functional Materials, 22*, 4814–4818. https://doi.org/10.1002/adfm.201201150

National Renewable Energy Laboratory (NREL). https://www.nrel.gov/pv/assets/pdfs/pv-efficiency-chart.20181221.pdf. Accessed 14 Jan 2019.

Nechache, R., Cojocaru, C. V., Harnagea, C., Nauenheim, C., et al. (2011). Epitaxial patterning of Bi_2FeCrO_6 double perovskite nanostructures: Multiferroic at room temperature. *Advanced Materials, 23*, 1724–1729. https://doi.org/10.1002/adma.201004405

Nechache, R., Harnagea, C., Li, S., Cardenas, L., Huang, W., Chakrabartty, J., & Rosei, F. (2015). Bandgap tuning of multiferroic oxide solar cells. *Nature Photonics, 9*, 61. https://doi.org/10.1038/nphoton.2014.255

Ogale, S. B. (2005). *Thin films and heterostructures for oxide electronics*. Springer. https://doi.org/10.1007/b136780

Ohtomo, A., Muller, D. A., Grazul, Wang, J. L., & Wang, H. Y. (2002). Artificial charge-modulation in atomic-scale perovslite titanate superlattices. *Nature, 419*, 378–380. https://doi.org/10.1038/nature00977

Paillard, C., Bai, X., Infante, I. C., Guennou, M., Geneste, G., et al. (2016). Dkhil, photovoltaics with ferroelectrics: Current status and beyond. *Advanced Materials, 28*, 5153–5168. https://doi.org/10.1002/adma.201505215

Park, S. E., & Shrout, T. R. (1997). Ultrahigh strain and piezoelectric behavior in relaxor based ferroelectric single crystals. *Applied Physics, 82*, 1804–1811. https://doi.org/10.1063/1.365983

Park, T. J., Papaefthymiou, G. C., Viescas, A. J., Moodenbaugh, A. R., & Wong, S. S. (2007). Size-dependent magnetic properties of single-crystalline multiferroic $BiFeO_3$ nanoparticles. *Nano Letters, 7*, 766–772. https://doi.org/10.1021/nl063039w

Park, J. G., Le, M. D., Jeong, J., & Lee, S. (2014). Structure and spin dynamics of multiferroic $BiFeO_3$. *Journal of Physics: Condensed Matter, 26*, 433202. https://doi.org/10.1088/0953-8984/26/43/433202

Park, N. G., Grätzel, Miyasaka, M. T., Zhu, K., & Emery, K. (2016). Towards stable and commercially available perovskite solar cells. *Nature Energy, 1*, 16152. https://doi.org/10.1038/nenergy.2016.152

Pérez-Tomás, A., Mingorance, A., Tanenbaum, D., & Lira-Cantú, M. (2018). Chapter 8: Metal oxides in photovoltaics: All-oxide, ferroic, and perovskite solar cells. In M. Lira-Cantu (Ed.), *The future of semiconductor oxides in next-generation solar cells* (Metal oxides) (pp. 267–356). Elsevier.

Pettersson, L. A. A., Roman, L. S., & Inganäs, O. (1999). Modeling photocurrent action spectra of photovoltaic devices based on organic thin films. *Applied Physics, 86*, 487–496. https://doi.org/10.1063/1.370757

Qi, X., Dho, J., Tomov, R., Blamire, M. G., & MacManus-Driscoll, J. L. (2005). Greatly reduced leakage current and conduction mechanism in aliovalent-ion-doped $BiFeO_3$. *Applied Physics Letters, 86*, 062903. https://doi.org/10.1063/1.1862336

Raj, A., Kumar, M., & Anshul, A. (2021a). Recent advancement in inorganic-organic electron transport layers in perovskite solar cell: Current status and future outlook. *Materials Today Chemistry, 22*, 100595. https://doi.org/10.1016/j.mtchem.2021.100595

Raj, A., Kumar, M., Bherwani, H., Gupta, A., & Anshul, A. (2021b). Evidence of improved power conversion efficiency in lead-free $CsGeI_3$ based perovskite solar cell heterostructure via scaps simulation. *Journal of Vacuum Science & Technology B, 39*, 012401. https://doi.org/10.1116/6.0000718

Raj, A., Kumar, M., & Anshul, A. (2022). Recent progress in cesium-based lead-free halide double perovskite materials for photovoltaic applications. *Physica Status Solidi, 219*(22), 2200425. https://doi.org/10.1002/pssa.202200425

Ramirez, M. O., Krishnamurthi, M., Denev, S., Kumar, A., Yang, S. Y., et al. (2008). Two-phonon coupling to the antiferromagnetic phase transitionin multiferroic $BiFeO_3$. *Applied Physics Letters, 92*, 022511. https://doi.org/10.1063/1.2829681

Ramirez, M. O., Kumar, A., Denev, S. A., Chu, Y. H., et al. (2009a). Spin-charge-lattice coupling through resonant multi-magnon excitations in multiferroic $BiFeO_3$. *Applied Physics Letters, 94*, 161905. https://doi.org/10.1063/1.3118576

Ramirez, M. O., Kumar, A., Denev, S., Podraza, N., Xu, X. S., et al. (2009b). Magnon sidebands and spin-charge coupling in bismuth ferrite probed by nonlinear optical spectroscopy. *Physical Review B: Condensed Matter and Materials Physics, 79*, 224106.

Ramohlola, K. E., Monana, G. R., Hato, M. J., Modibane, K. D., Molapo, K. M., Masikini, M., Mduli, S. B., & Iwuoha, E. I. (2018). Polyaniline-metal organic framework nanocomposite as an efficient electrocatalyst for hydrogen evolution reaction. *Composites Part B: Engineering, 137*, 129–139. https://doi.org/10.1016/j.compositesb.2017.11.016

Rappe, A. M., Grinberg, I., & Spanier, J. E. (2017). Getting a charge out of hybrid perovskites. *Proceedings. National Academy of Sciences. United States of America, 114*, 7191–7193. https://doi.org/10.1073/pnas.1708154114

Rühle, S., Anderson, A. Y., Barad, H. N., Kupfer, B., et al. (2012). All-oxide photovoltaics. *The Journal of Physical Chemistry Letters, 3*, 3755–3764. https://doi.org/10.1021/jz3017039

Sando, D., Carrétéro, C., Grisolia, M. N., Barthélémy, A., Nagarajan, V., & Bibes, M. (2018). Revisiting the optical band gap in epitaxial BiFeO3 thin films. *Advanced Optical Materials, 6*, 1700836. https://doi.org/10.1002/adom.201700836

Sarkar, A., Singh, A. K., Sarkar, D., Khan, G. G., & Mandal, K. (2015). Three-dimensional nano-architecture of BiFeO3 anchored TiO2 nanotube arrays for electrochemical energy storage and solar energy conversion. *ACS Sustainable Chemistry & Engineering, 3*, 2254–2263. https://doi.org/10.1021/acssuschemeng.5b00519

Schmid, H. (1994). Multi-ferroic magnetoelectrics. *Ferroelectrics, 162*, 317. https://doi.org/10.1080/00150199408245120

Schmid, H. (2008). Some symmetry aspects of ferroics and single phase multiferroics. *Journal of Physics: Condensed Matter, 20*, 434201–434220. https://doi.org/10.1088/0953-8984/20/43/434201

Schmidt, L. J. (2011). Tracking down the truth of perovski. In *38th Rochester mineralogical symposium program notes* (pp. 31–32).

Scott, J. F. (2008). Ferroelectrics go bananas. *Journal of Physics: Condensed Matter, 20*(2), 021001. https://doi.org/10.1088/0953-8984/20/02/021001

Scullin, M. L., Yu, C., Huijben, M., Mukerjee, S., et al. (2008). Anomalously large measured thermoelectric power factor in $Sr_{1-x}La_xTiO_3$ thin films due to $SrTiO_3$ substrate reduction. *Applied Physics Letters, 92*, 202113. https://doi.org/10.1063/1.2916690

Scullin, M. L., Ravichandran, J., Yu, C., Huijben, M., et al. (2010). Pulsed laser deposition-induced reduction of $SrTiO_3$ crystals. *Acta Materialia, 58*, 457. https://doi.org/10.1016/j.actamat.2009.09.024

Seidel, J. (2012). Domain walls as nanoscale functional elements. *The Journal of Physical Chemistry Letters, 3*, 2905. https://doi.org/10.1021/jz3011223

Seidel, J., Fu, D., Yang, S. Y., Alarcón-Lladó, E., Wu, J., Ramesh, R., & Ager, J. W. (2011). Efficient photovoltaic current generation at ferroelectric domain walls. *Physical Review Letters, 107*, 126805. https://doi.org/10.1103/PhysRevLett.107.126805

Seidel, J., Yang, S. Y., Alarcón-Lladó, E., Ager, J. W., & Ramesh, R. (2012a). Nanoscale probing of high photovoltages at 109° domain walls. *Ferroelectrics, 433*, 123. https://doi.org/10.1080/00150193.2012.678156

Seidel, J., Luo, W., Suresha, S. J., Nguyen, P. K., Lee, A. S., Kim, S. Y., et al. (2012b). Prominent electrochromism through vacancy-order melting in a complex oxide. *Nature Communications, 3*, 799. https://doi.org/10.1038/ncomms1799

Seidel, J., Bhalla, G. S., He, Q., Yang, S. Y., Chu, Y. H., & Ramesh, R. (2013). Domain wall functionality in $BiFeO_3$. *Phase Transitions: A Multinational Journal, 86*(1), 53–66. https://doi.org/10.1080/01411594.2012.695371

Shang, M., Zhang, C., Zhang, T., Yuan, L., Ge, L., Yuan, H., & Feng, S. (2013). The multiferroic perovskite $YFeO_3$. *Applied Physics Letters, 102*, 062903. https://doi.org/10.1063/1.4791697

Smolenskiĭ, G. A., & Chupis, I. E. E. (1982). Ferroelectromagnets. *Soviet Physics Uspekhi, 25*(7), 475. https://doi.org/10.1070/PU1982v025n07ABEH004570

Spaldin, N. A., & Rammamoorthy, R. (2019). Advances in magnetoelectric multiferroics. *Nature Materials, 18*, 203–212. https://doi.org/10.1038/s41563-018-0275-2

Sun, B., Zhou, G., Sun, L., Zhao, H., Chen, Y., Yang, F., Zhaoab, Y., & Song, Q. (2021). ABO_3 multiferroic perovskite materials for memristive memory and neuromorphic computing. *Nanoscale Horizons, 6*, 939. https://doi.org/10.1039/d1nh00292a

Tokura, Y., & Nagaosa, N. (2000). Orbital physics in transition-metal oxides. *Science, 288*, 462–468. https://doi.org/10.1126/science.288.5465.462

Vaz, C. A. F., Hoffman, J., Ahn, C. H., & Ramesh, R. (2010). Magnetoelectric coupling effects in multiferroic complex oxide composite structures. *Advanced Materials, 22*, 2900–2918. https://doi.org/10.1002/adma.200904326

Wang, J., Neaton, J. B., Zheng, H., Nagarajan, V., et al. (2003). Epitaxial $BiFeO_3$ multiferroic thin film heterostructures. *Science, 299*, 1719–1722. https://doi.org/10.1126/science.1080615

Wang, F., Grinberg, I., & Rappe, A. M. (2014). Band gap engineering strategy via polarization rotation in perovskite ferroelectrics. *Applied Physics Letters, 104*, 152903. https://doi.org/10.1063/1.4871707

Wang, F., Grinberg, I., Jiang, L., Young, S. M., Davies, P. K., & Rappe, A. M. (2015). Materials design of visible-light ferroelectric photovoltaics from first principles. *Ferroelectrics, 483*, 1–12. https://doi.org/10.1080/00150193.2015.1058096

Weingart, C., Spaldin, N., & Bousquet, E. (2012). Non collinear magnetism and single-ion anisotropy in multiferroic perovskites. *Physical Review B, 86*(9), 094413. https://doi.org/10.1103/PhysRevB.86.094413

Xiao, Y., Shenoy, V. B., & Bhattacharya, K. (2005). Depletion layers and domain walls in semiconducting ferroelectric thin films. *Physical Review Letters, 95*, 247603. https://doi.org/10.1103/PhysRevLett.95.247603

Yamada, H., Ogawa, Y., Ishii, Y., Sato, H., Kawasaki, M., Akoh, H., & Tokura, Y. (2004). Engineered interface of magnetic oxides. *Science, 305*, 646–648. https://doi.org/10.1126/science.1098867

Yamasaki, Y., Sagayama, H., Goto, T., et al. (2007). Electric control of spin helicity in a magnetic ferroelectric. *Physical Review Letters, 98*(14), 147204. https://doi.org/10.1103/PhysRevLett.98.147204

Yang, C. H., Koo, T. Y., & Jeong, Y. H. (2005). How to obtain magnetocapacitance effects at room temperature: The case of Mn-doped $BiFeO_3$. *Solid State Communications, 134*(4), 299. https://doi.org/10.1016/j.ssc.2004.12.049

Yang, C. H., Koo, T. Y., Lee, S. H., Song, C., Lee, K. B., & Jeong, Y. H. (2006). Orbital ordering and enhanced magnetic frustration of strained $BiMnO_3$ thin films. *Europhysics Letters, 74*(2), 348.

Yang, C. H., Lee, S. H., Koo, T. Y., & Jeong, Y. H. (2007). Dynamically enhanced magnetodielectric effect and magnetic-field-controlled electric relaxations in La-doped $BiMnO_3$. *Physical Review B: Condensed Matter and Materials Physics, 75*, 140–104. https://doi.org/10.1103/PhysRevB.75.140104

Yang, C. H., Seidel, J., Kim, S. Y., Rossen, P. B., et al. (2009). Electric modulation of conduction in multiferroic Ca-doped $BiFeO_3$ films. *Nature Materials, 8*, 485–493. https://doi.org/10.1038/nmat2432

Yang, S. Y., Seidel, J., Byrnes, S. J., Shafer, P., Yang, C. H., et al. (2010). Above-bandgap voltages from ferroelectric photovoltaic devices. *Nature Nanotechnology, 5*, 143–147. https://doi.org/10.1038/nnano.2009.451

Yang, M., Bhatnagar, A., & Alexe, M. (2015). Electronic origin and tailoring of photovoltaic effect in $BiFeO_3$ single crystals. *Advanced Electronic Materials, 1*, 1500139. https://doi.org/10.1002/aelm.201500139

Yang, Y., Xu, W., Xu, X., Wang, Y., Yuan, G., Wang, Y., & Liu, Z. (2016). The enhanced photocurrent of epitaxial $BiFeO_3$ film at 130 °C. *Applied Physics, 119*, 044102. https://doi.org/10.1063/1.4940411

Yu, P., Lee, J. S., Okamoto, S., Rossell, M. D., et al. (2010). Interface ferromagnetism and orbital reconstruction in $BiFeO_3$–$La_{0.7}Sr_{0.3}MnO_3$ heterostructures. *Physical Review Letters, 105*, 027201. https://doi.org/10.1103/PhysRevLett.105.027201

Zener, C. (1951a). Interaction between the d shells in the transition metals. *Physics Review, 81*(3), 440–444. https://doi.org/10.1103/PhysRev.81.440

Zener, C. (1951b). Interaction between the d-shells in the transition metals. II. Ferromagnetic compounds of manganese with perovskite structure. *Physics Review, 82*(3), 403–405. https://doi.org/10.1103/PhysRev.82.403

Zenkevich, A., Matveyev, Y., Maksimova, K., Gaynutdinov, R., Tolstikhina, A., & Fridkin, V. (2014). Giant bulk photovoltaic effect in thin ferroelectric $BaTiO_3$ films. *Physical Review B: Condensed Matter and Materials Physics, 90*, 161409. https://doi.org/10.1103/PhysRevB.90.161409

Zhang, Y., Sahoo, M. P. K., & Wang, J. (2017). Tuning the band gap and polarization of BaSnO3/SrSnO3 superlattices for photovoltaic applications. *Physical Chemistry Chemical Physics, 19*, 7032–7039. https://doi.org/10.1039/C6CP06042K

Zheng, H., Wang, J., Lofland, S. E., Ma, Z., et al. (2004). Multiferroic $BaTiO_3$-$CoFe_2O_4$ nanostructures. *Science, 303*, 661. https://doi.org/10.1126/science.1094207

Zhu, L., Ran, R., Tade, M., Wang, W., & Shao, Z. (2016). Perovskite materials in energy storage and conversion. *Asia-Pacific Journal of Chemical Engineering, 11*(3), 338–369. https://doi.org/10.1002/apj.2000

Zubko, P., Gariglio, S., Gabay, M., Ghosez, P., & Triscone, J. M. (2011). Interface physics in complex oxide hetrostructures. *Annual Review of Condensed Matter Physics, 2*, 141–165. https://doi.org/10.1146/annurev-conmatphys-062910-140445

Index

A
Adsorption, v, 2–21, 102, 122, 132, 135, 140, 177, 186, 193, 200, 204, 205, 207, 208, 275
Annealing temperature, vi, 74, 251–264
Ayurveda, 216, 218

B
Bioactive components, 199–203, 207–210, 213, 215, 222–224
Biocompatibility, 12, 30, 94, 104, 106, 117, 122, 132–134, 140, 141, 156–157, 167, 169, 175–193, 229, 230, 240, 246
Biodegradability, 12
Biological pathogens, 32
Biomedicine, 105, 106, 133, 137, 156, 168, 169, 176

C
CdZnTe, vi, 251–264
Charge transfer, 128, 239, 277, 292, 335
Chemical bath deposition (CBD), 252, 254, 255, 264

D
Diagnosis, 93–111, 134, 168, 230, 240–242, 246
Dispersion, 9, 59, 61, 62, 64, 71, 122, 127, 128, 160, 176, 182, 190, 238, 246, 276, 283, 305, 306, 336
Dyes, 8, 9, 35

F
Fingerprint analysis, 30–35, 47
Forensics, 29–53

G
Graphene, 3, 4, 6, 19, 21, 36, 70, 160, 163, 267, 268, 275, 280, 281, 287–288, 290, 292
Green energy, 321–339
Green synthesis, 13, 231, 232, 234, 235, 239

H
Hair analysis, 32, 35
Heavy metals, 3, 12, 21, 62, 117, 135, 140
Herbal drugs, 200, 202, 204, 206–209, 223
High surface area, 20, 95, 135, 140, 168
Hydrogen production, 121, 142, 165, 303–316

I
Inorganic pollutants, 12–16

L
Laser irradiation, 303–316

M
MAX phase, 4, 115–121, 123, 124

S. Bansal et al. (eds.), *Emerging Applications of Novel Nanoparticles*, Lecture Notes in Nanoscale Science and Technology 37,
https://doi.org/10.1007/978-3-031-57843-4

Metal organic frameworks (MOFs), 3, 5, 8, 19–21
Metal oxide nanoparticles, 230
Metal oxides (MOs), vi, 8, 62, 137, 175–193, 230, 278, 303, 333
Molybdenum disulfide (MoS_2), vi, vii, 4, 9, 10, 13, 145–169
Multiferroics, 321–339
MXene, v, vi, 4, 5, 9, 12, 15, 19, 78, 82, 115–142

N
Nanocomposite, 19, 41, 71, 74–76, 81, 115, 117, 120, 122, 129, 132, 135, 184–186, 190, 281, 282, 289, 292, 303–316
Nanomaterials, v, vi, 2–21, 29–53, 94, 97, 107, 128, 159, 169, 175–179, 181, 189, 190, 193, 200–205, 207–209, 212, 214, 215, 218, 220, 222–225, 230, 232, 281, 304, 306
Nanomedicine, 109, 200, 211, 212, 214, 218, 230
Nanoparticles (NPs), 5, 29, 74, 93, 157, 176, 199, 229, 278, 304
Nanophytomedicine, 200, 204–206, 208, 216–223
Nanotherapy, 201–206, 212–215
Non-aqueous medium, 252, 264

O
Optoelectronic devices, vi, vii, 117, 131, 139, 141, 154, 251, 267–293
Organic pollutants, 2, 8–12, 18

P
Pathogen, 20, 81, 93, 94, 99, 100, 104, 105, 110, 132, 134, 202, 211
Pathogenic microbiology
Perovskites, vii, 290–291, 303–316, 321–339
Pharmacokinetic study, 203, 220, 223
Phenol, 2, 3, 10, 21, 235, 239
Photocatalysis, 315, 316
Precise, 33, 93, 95, 105, 107, 109, 110, 124, 125, 167, 220, 336, 338

R
Radioactive materials, 3, 15
Resistance, 4, 19, 41, 59, 70, 71, 76–78, 80–82, 94, 108, 109, 132, 135, 140, 141, 151, 157, 158, 184, 191, 222, 252, 279, 290

S
Surface modifications, 68, 79, 106, 109, 178, 179, 181, 183, 186, 246

T
Theranostics, 94, 110, 246
Titanium Carbide MXenes ($Ti_3C_2T_x$), vi, 9, 12, 15, 19, 115–142
Tool marks, 32, 50
Transition metal dichalcogenides (TMDs), 3, 4, 6, 7, 21, 147, 148, 268
Treatment, vi, 2, 4, 21, 30, 68, 93–111, 118, 120, 121, 125, 126, 135, 140, 157, 178, 179, 184, 199–225, 230, 236, 240–242, 245, 246, 257, 259, 305
Two-dimensional (2D), v, vi, 2–21, 115–119, 123–127, 131, 135, 139, 148, 159, 175–193, 267, 268, 272, 273, 275, 276, 278, 280–284, 287, 293
2D materials, v, vi, 2–16, 19–21, 115, 117, 120, 121, 128, 145–169, 268, 275, 282
2D nanosheets, 9, 192

W
Wastewater treatment, 2, 3, 17, 19, 21

The manufacturer's authorised representative in the EU is Springer Nature Customer Service Centre GmbH, Europaplatz 3, 69115 Heidelberg, Germany. If you have any concerns regarding our products, please contact ProductSafety@springernature.com

Printed and bound by CPI Group (UK) Ltd, Croydon, CR0 4YY
01/10/2026
02225906-0011